Concepts, Theories and Applications of Mathematical Analysis

Concepts, Theories and Applications of Mathematical Analysis

Edited by **Matt Ferrier**

Willford Press
New York

Published by Willford Press,
118-35 Queens Blvd., Suite 400,
Forest Hills, NY 11375, USA
www.willfordpress.com

Concepts, Theories and Applications of Mathematical Analysis
Edited by Matt Ferrier

International Standard Book Number: 978-1-68285-168-5 (Hardback)

Printed in China.

Contents

Preface

This book aims to highlight the current researches and provides a platform to further the scope of innovations in this area. This book is a product of the combined efforts of many researchers and scientists from different parts of the world. The objective of this book is to provide the readers with the latest information in the field.

Mathematical analyses have proved useful in many fields such as quantum mechanics, relativity, signal processing and various other fields. The topics covered in this extensive book deal with the core aspects of the subject such as primary branches of analysis (real, complex, functional, differential, etc.), metric spaces, current applications, etc. This systematic book includes contributions of experts and scientists which will provide innovative insights into this field. Coherent flow of topics, student-friendly language and extensive use of examples make this book an invaluable source of knowledge.

I would like to express my sincere thanks to the authors for their dedicated efforts in the completion of this book. I acknowledge the efforts of the publisher for providing constant support. Lastly, I would like to thank my family for their support in all academic endeavors.

Editor

1

A new PDE approach to the large time asymptotics of solutions of Hamilton–Jacobi equations

Guy Barles · Hitoshi Ishii · Hiroyoshi Mitake

Abstract We introduce a new PDE approach to establishing the large time asymptotic behavior of solutions of Hamilton–Jacobi equations, which modifies and simplifies the previous ones (Barles et al. in Arch Ration Mech Anal 204(2):515–558, 2012; Barles and Souganidis in SIAM J Math Anal 31(4):925–939, 2000), under a refined "strict convexity" assumption on the Hamiltonians. Not only such "strict convexity" conditions generalize the corresponding requirements on the Hamiltonians in

---

Dedicated to Professor Neil S. Trudinger on the occasion of his 70th birthday.

Communicated by S.K. Jain.

The work of HI was supported in part by KAKENHI #20340019, #21340032, #21224001, #23340028 and #23244015, JSPS. The work of HM was supported in part by KAKENHI #24840042, JSPS and Grant for Basic Science Research Projects from the Sumitomo Foundation.

G. Barles
Laboratoire de Mathématiques et Physique Théorique (UMR CNRS 7350), Fédération Denis Poisson, Université de Tours (FR CNRS 2964), Place de Grandmont, 37200 Tours, France
e-mail: barles@lmpt.univ-tours.fr
URL: http://www.lmpt.univ-tours.fr/~barles

H. Ishii (✉)
Faculty of Education and Integrated Arts and Sciences, Waseda University, Nishi-Waseda, Shinjuku, Tokyo 169-8050, Japan
e-mail: hitoshi.ishii@waseda.jp
URL: http://www.edu.waseda.ac.jp/~ishii/

H. Ishii
Faculty of Science, King Abdulaziz University, P.O. Box 80203, Jeddah 21589, Saudi Arabia

H. Mitake
Department of Applied Mathematics, Faculty of Science, Fukuoka University, Fukuoka 814-0180, Japan
e-mail: mitake@math.sci.fukuoka-u.ac.jp

Barles and Souganidis (SIAM J Math Anal 31(4):925–939, 2000), but also one of the most refined our conditions covers the situation studied in Namah and Roquejoffre (Commun Partial Differ Equ 24(5–6):883–893, 1999).

Keywords Asymptotic behavior · Hamilton–Jacobi equations · PDE approach

Mathematics Subject Classification (1991) Primary 35F21; Secondary 35B40 · 35D40 · 35F31 · 49L25

1 Introduction

In this article we introduce a new PDE approach to establishing the large time asymptotic behavior of solutions of Hamilton–Jacobi equations.

In the last two decades there have been major developments in the study of the large time asymptotics of solutions of Hamilton–Jacobi equations, initiated by the work by Namah and Roquejoffre [19] and by Fathi [9].

The approach by Fathi is based on the weak KAM theory and the representation of solutions of the Hopf-Lax-Oleinik type or, in other words, as the value functions of optimal control, and has a wide scope which is different from the one in Namah–Roquejoffre [19]. The optimal control/dynamical approach of Fathi has been subsequently developed for further applications and technical improvements by many authors (see, for instance, [8, 10, 12, 14, 15, 17, 18]).

At the beginning of the developments mentioned above, another approach has been introduced by the first author and Souganidis [5], which does not depend on the representation formulas of solutions and thus applies to a more general class of Hamilton–Jacobi equations including those with non-convex Hamiltonians. We refer for recent developments in this direction to [3, 4].

We also refer [3] for further comments and references related to the large time asymptotics of solutions of Hamilton–Jacobi equations and [6] for a new development on this study for the general degenerate viscous Hamilton-Jacobi equations.

Our aim here is to modify and slightly simplify the main ingredient in the PDE approach by the first author and Souganidis [5] as well as to refine the requirements on the Hamiltonians.

To clarify and simplify the presentation, we consider the asymptotic problem in the periodic setting. We are thus concerned with the Cauchy problem

$$\begin{cases} u_t(x,t) + H(x, D_x u(x,t)) = 0 \text{ in } Q, \\ u(x,0) = u_0(x) \qquad \text{for } x \in \mathbb{R}^n, \end{cases} \tag{CP}$$

where $Q := \mathbb{R}^n \times (0, \infty)$, u represents the unknown function on $\overline{Q}$, $u_t = u_t(x,t) = (\partial u/\partial t)(x,t)$, $D_x u(x,t) = ((\partial u/\partial x_1)(x,t), ..., (\partial u/\partial x_n)(x,t))$ and u_0 represents the initial data. The functions $u(x,t)$ and $u_0(x)$ are supposed to be periodic in x.

We make the following assumptions throughout this article:

(A1) The function u_0 is continuous in $\mathbb{R}^n$ and periodic with period $\mathbb{Z}^n$.
(A2) $H \in C(\mathbb{R}^n \times \mathbb{R}^n)$.

(A3) The Hamiltonian $H(x, p)$ is periodic in x with period $\mathbb{Z}^n$ for every $p \in \mathbb{R}^n$.
(A4) The Hamiltonian H is coercive. That is,

$$\lim_{r\to\infty} \inf\{H(x, p) : (x, p) \in \mathbb{R}^{2n}, |p| \geq r\} = \infty.$$

Our notational conventions are as follows. We may regard functions $f(x)$ on $\mathbb{R}^n$ (resp., $g(x, y)$ on $\mathbb{R}^n \times V$, where V is a subset of $\mathbb{R}^m$) periodic in $x \in \mathbb{R}^n$ with period $\mathbb{Z}^n$ as functions on the torus $\mathbb{T}^n$ (resp., $\mathbb{T}^n \times V$). In this viewpoint, we write $C(\mathbb{T}^n)$, $C(\mathbb{T}^n \times V)$, etc, for the subspaces of all functions $f(x)$ in $C(\mathbb{R}^n)$, of all functions $g(x, y)$ in $C(\mathbb{R}^n \times V)$, etc, periodic in x with period $\mathbb{Z}^n$. We denote the sup-norm (or the L^∞-norm) of a function f by $\|f\|_\infty$ and $\|f\|_{L^\infty}$ interchangeably. Regarding the notion of solution of Hamilton–Jacobi equations, in this article we will be only concerned with viscosity solutions, viscosity subsolutions and viscosity supersolutions, which we refer simply as solutions, subsolutions and supersolutions. For any $R > 0$, B_R denotes the open ball of $\mathbb{R}^n$ with center at the origin and radius R. For any $X \subset \mathbb{R}^n$, UC (X) and Lip(X) denote the spaces of all uniformly continuous functions and all Lipschitz continuous functions on X, respectively.

We now recall the following basic results.

Theorem 1 *Under the hypotheses (A1)–(A4), there exists a unique solution* $u \in$ UC $(\mathbb{T}^n \times [0, \infty))$ *of (CP). Furthermore, if* $u_0 \in$ Lip$(\mathbb{T}^n)$, *then* $u \in$ Lip$(\mathbb{T}^n \times [0, \infty))$.

Theorem 2 *Under the hypotheses (A2)–(A4), let* $u, v \in$ UC $(\mathbb{T}^n \times [0, \infty))$ *be solutions of*

$$u_t + H(x, D_x u) = 0 \quad in\ Q. \tag{HJ}$$

Then

$$\|u - v\|_{L^\infty(Q)} \leq \|u(\cdot, 0) - v(\cdot, 0)\|_{L^\infty(\mathbb{R}^n)}.$$

Theorem 3 *Under the hypotheses (A2)–(A4), there exists a unique constant* $c \in \mathbb{R}$ *such that the problem*

$$H(x, Dv(x)) = c \quad in\ \mathbb{R}^n \tag{EP}$$

has a solution $v \in$ Lip$(\mathbb{T}^n)$.

These theorems are classical results in viscosity solutions theory. For instance, the existence part of Theorem 1 is a consequence of Corollaire II.1 in [1]. Under assumptions (A2) and (A3), as is well known, the comparison principle holds between bounded semicontinuous sub and supersolutions of (CP) if one of them is Lipschitz continuous. This comparison result and the existence part of Theorem 1 assure that for each continuous solution u of (CP) there is a sequence $\{u_k\}_{k\in\mathbb{N}}$ of Lipschitz continuous solutions of (CP), with u_0 replaced by $u_k(\cdot, 0)$, which converges to u uniformly in $\overline{Q}$. The existence of such a sequence of Lipschitz continuous solutions of (CP) and the comparison principle for Lipschitz continuous solutions of (CP) guarantees the Theorem 2 holds. Theorem 3 and its proof can be found in [16].

The problem of finding a pair $(c, v) \in \mathbb{R} \times C(\mathbb{T}^n)$, where v satisfies (EP) in the viscosity sense, is called an additive eigenvalue problem or ergodic problem. Thus, for such a pair (c, v), the function v (resp., the constant c) is called an additive eigenfunction (resp., eigenvalue).

We note that the conditions (A2)–(A4) are invariant under addition of constants. Hence, by replacing H by $H - c$, with c being the additive eigenvalue of (EP), we may normalize so that the additive eigenvalue c is zero. Thus, in what follows, we always assume that

(A5) $c = 0$, where c denotes the additive eigenvalue.

Accordingly, problem (EP) becomes simply a stationary problem

$$H(x, Dv(x)) = 0 \quad \text{in } \mathbb{R}^n. \tag{1}$$

The crucial assumptions in this article are the following conditions.

(A6)$_+$ There exist constants $\eta_0 > 0$ and $\theta_0 > 1$ and for each $(\eta, \theta) \in (0, \eta_0) \times (1, \theta_0)$ a constant $\psi = \psi(\eta, \theta) > 0$ such that for all $x, p, q \in \mathbb{R}^n$, if $H(x, p) \le 0$ and $H(x, q) \ge \eta$, then

$$H(x, p + \theta(q - p)) \ge \eta\theta + \psi.$$

(A6)$_-$ There exist constants $\eta_0 > 0$ and $\theta_0 > 1$ and for each $(\eta, \theta) \in (0, \eta_0) \times (1, \theta_0)$ a constant $\psi = \psi(\eta, \theta) > 0$ such that for all $x, p, q \in \mathbb{R}^n$, if $H(x, p) \le 0$ and $H(x, q) \ge -\eta$, then

$$H(x, p + \theta(q - p)) \ge -\eta\theta + \psi.$$

We will furthermore modify and refine these conditions [see (A9)$_\pm$] in Sect. 4, one of which covers the situation studied by Namah–Roquejoffre [19]. An important consequence is that our PDE method gives a unified approach to most of the large time asymptotic convergence results for (CP) in the literature.

The assumptions above are some kind of strict convexity requirements and they are satisfied if H is strictly convex in p. Indeed in this case, since $q = \theta^{-1}(p+\theta(q-p)) + (1-\theta^{-1})p$,

$$\begin{aligned} H(x, q) &< \theta^{-1} H(x, p + \theta(q - p)) + (1 - \theta^{-1}) H(x, p) \\ &< \theta^{-1} H(x, p + \theta(q - p)), \end{aligned}$$

and ψ measures how strict is this inequality. We point out that, for (A6)$_-$, this argument is valid if $p \ne q$ and the inequality is obvious if $p = q$, while in the case of (A6)$_+$ clearly we have always $p \ne q$.

One may have another interpretation of these assumptions, namely that the function $H(x, r)$, as a function of r, grows more than linearly on the line segment connecting from q to $p + \theta_0(q - p)$ for some $\theta_0 > 1$ (notice that this growth rate is negative in the case of (A6)$_-$).

We conclude these remarks on (A6)$_\pm$ by pointing out that (A6)$_+$ is an assumption on the behavior of H on the set $\{H \geq 0\}$ while (A6)$_-$ is an assumption on the behavior of H on the set $\{H \leq 0\}$. We refer to Sect. 3 for more precise comments in this direction.

A condition similar to (A6)$_+$ has appeared first in Barles–Souganidis [5] (see (H4) in [5]). Our condition (A6)$_+$ is less stringent and has a wider application than (A6)$_+$ in [3]. For this comparison, see Sect. 3. Also, (A6)$_-$ is less stringent than (A6)$_-$ in [3]. A type of condition (A6)$_-$ has first introduced in Ichihara–Ishii [11] for convex Hamiltonians (see the condition (16) in [11]).

We establish the following theorem by a PDE approach which modifies and simplifies the previous ones in [3,5].

Theorem 4 *Assume that (A1)–(A5) hold and that either (A6)$_+$ or (A6)$_-$ holds. Then the unique solution $u(x, t)$ in* UC $(\mathbb{T}^n \times [0, \infty))$ *of (CP) converges uniformly in $\mathbb{R}^n$, as $t \to \infty$, to a function $u_\infty(x)$ in* Lip$(\mathbb{T}^n)$*, which is a solution of* (1).

A generalization of the theorem above is given in Sect. 4 (see Theorem 11), which covers the main result in [19] in the periodic setting.

In Sect. 2, we give an explanation of the new ingredient in our new PDE method, a (hopefully transparent) formal proof of Theorem 4 by the new PDE method and its exact version. In Sect. 3, we make comparisons between (A6)$_\pm$ and its classical versions, and discuss convexity-like properties of the Hamiltonians H implied by (A6)$_\pm$ as well as a couple of conditions equivalent to (A6)$_\pm$. In Sect. 4, we present a theorem, with (A6)$_\pm$ replaced by refined conditions, which includes the situation in [19] as a special case.

2 Proof of Theorem 4

Throughout this section, we assume that (A1)–(A5) hold. The first step consists in reducing to the case when $u_0 \in \mathrm{Lip}(\mathbb{T}^n)$ and therefore u is Lipschitz continuous on $\mathbb{T}^n \times [0, \infty)$.

Lemma 5 *If the result of Theorem 4 holds for any $u_0 \in \mathrm{Lip}(\mathbb{T}^n)$ then it holds for any $u_0 \in C(\mathbb{T}^n)$.*

Proof For a general $u_0 \in C(\mathbb{T}^n)$ we select a sequence $\{u_{0,j}\}_{j\in\mathbb{N}} \subset \mathrm{Lip}(\mathbb{T}^n)$ which converges to u_0 uniformly in $\mathbb{R}^n$. For each $j \in \mathbb{N}$ let $u_j \in \mathrm{Lip}(\mathbb{T}^n \times [0, \infty))$ be the unique solution of (CP), with $u_{0,j}$ in place of u_0. By Theorem 2, we have

$$\|u_j - u_k\|_{L^\infty(Q)} \leq \|u_{0,j} - u_{0,k}\|_{L^\infty(\mathbb{R}^n)} \quad \text{for all } j, k \in \mathbb{N}. \tag{2}$$

Since Theorem 4 holds for any initial data in Lip$(\mathbb{T}^n)$, we know that for each $j \in \mathbb{N}$ there exists a function $u_{\infty,j} \in C(\mathbb{T}^n)$ such that $\lim_{t\to\infty} u_j(x, t) = u_{\infty,j}(x)$ uniformly in $\mathbb{R}^n$. This implies

$$\|u_{\infty,j} - u_{\infty,k}\|_{L^\infty(\mathbb{R}^n)} \leq \|u_j - u_k\|_{L^\infty(Q)} \quad \text{for all } j, k \in \mathbb{N},$$

which together with (2) yields

$$\|u_{\infty,j} - u_{\infty,k}\|_\infty \le \|u_{0,j} - u_{0,k}\|_\infty \quad \text{for all } j, k \in \mathbb{N}.$$

Hence there is a function $u_\infty \in C(\mathbb{T}^n)$ such that $\lim_{j\to\infty} u_{\infty,j}(x) = u_\infty(x)$ uniformly in $\mathbb{R}^n$.

Observe by using Theorem 2 that for any $j \in \mathbb{N}$,

$$\begin{aligned}\|u(\cdot,t) - u_\infty\|_\infty &\le \|u(\cdot,t) - u_j(\cdot,t)\|_\infty + \|u_j(\cdot,t) - u_{\infty,j}\|_\infty + \|u_{\infty,j} - u_\infty\|_\infty \\ &\le \|u_0 - u_{0,j}\|_\infty + \|u_j(\cdot,t) - u_{\infty,j}\|_\infty + \|u_{\infty,j} - u_\infty\|_\infty,\end{aligned}$$

from which we conclude that $\lim_{t\to\infty} \|u(\cdot,t) - u_\infty\|_\infty = 0$. By the stability property of viscosity solutions, we see that u_∞ is a solution of (1) and, consequently, $u_\infty \in \mathrm{Lip}(\mathbb{R}^n)$ by Theorem 3. □

Now we turn to the proof of Theorem 4 when $u_0 \in \mathrm{Lip}(\mathbb{T}^n)$. By Theorem 1, there exists a unique solution $u \in \mathrm{Lip}(\mathbb{T}^n \times [0,\infty))$ of (CP) and we have to prove that $u(x,t)$ converges uniformly in $\mathbb{R}^n$ to a function $u_\infty(x)$ as $t \to \infty$.

Henceforth in this section we assume that $u_0 \in \mathrm{Lip}(\mathbb{T}^n)$ and hence the solution u of (CP) is in $\mathrm{Lip}(\mathbb{T}^n \times [0,\infty))$. Also, we fix a solution $v_0 \in \mathrm{Lip}(\mathbb{T}^n)$ of (1). Such a function v_0 exists thanks to Theorem 3. We set $L := \max\{\|D_x u\|_\infty, \|D_x v_0\|_\infty\}$.

If we set $z(x,t) = v_0(x)$ and invoke Theorem 2, then we get

$$\|u - z\|_{L^\infty(Q)} \le \|u_0 - v_0\|_{L^\infty(\mathbb{R}^n)},$$

which shows that u is bounded in $\overline{Q}$. We may assume by adding a constant to v_0 if needed that for some constant $C_0 > 0$,

$$0 \le u(x,t) - v_0(x) \le C_0 \quad \text{for all } (x,t) \in \overline{Q}.$$

2.1 Under assumption $(A6)_+$

Throughout this subsection we assume, in addition to (A1)–(A5), that $(A6)_+$ holds. Let $\eta_0 > 0$ and $\theta_0 > 1$ be the constants from $(A6)_+$.

For $(\eta,\theta) \in (0,\eta_0) \times (1,\theta_0)$, we define the function w on $\overline{Q}$ by

$$w(x,t) = \sup_{s \ge t}[u(x,t) - v_0(x) - \theta(u(x,s) - v_0(x) + \eta(s-t))]. \tag{3}$$

The following proposition is crucial in our proof of Theorem 4 under $(A6)_+$. To state the proposition, we introduce the functions $\omega_{H,R}$, with $R > 0$, as

$$\omega_{H,R}(r) = \sup\{|H(x,p) - H(x,q)| : x \in \mathbb{R}^n,\ p, q \in \overline{B}_R,\ |p-q| \le r\}.$$

Note that for each $R > 0$, the function $\omega_{H,R}$ is nonnegative and nondecreasing in $[0,\infty)$ and $\omega_{H,R}(0) = 0$.

Proposition 6 *Let $\psi = \psi(\eta, \theta) > 0$ be the constant from (A6)$_+$. Then the function w is a subsolution of*

$$\min\{w(x,t), w_t(x,t) - \omega_{H,R}(|D_x w(x,t)|) + \psi\} \leq 0 \quad in\ Q, \tag{4}$$

where $R := (2\theta_0 + 1)L$.

Our proof of Theorem 4 follows the outline of previous works like [3,5] where a key result is an asymptotic monotonicity property for u. This asymptotic monotonicity is a consequence of Proposition 6 which, roughly speaking, implies that $\min\{u_t, 0\} \to 0$ as $t \to \infty$. This is rigorously stated in Lemma 8 and its consequence in (27). With assumption (A6)$_-$, this is also the case but with a different monotonicity (i.e., $\max\{u_t, 0\} \to 0$ as $t \to \infty$).

For this reason, the function w defined by (3) is a kind of Lyapunov function in our asymptotic analysis in a broad sense. The main new aspect in this article, compared to [3,5], is indeed the simpler form of our w, which is defined by taking supremum in s of the function

$$u(x,t) - v_0(x) - \theta(u(x,s) - v_0(x) + \eta(s-t)),$$

whose functional dependence on u and v_0 is linear. In the previous works, the function

$$\sup_{s \geq t} \frac{u(x,s) - v_0(x) + \eta(s-t)}{u(x,t) - v_0(x)} \tag{5}$$

(one should assume here by adding a constant to v_0 if necessary that $\inf_{(x,t)\in Q}(u(x,t) - v_0(x)) > 0$), played the same role as our function w, and the value

$$\frac{u(x,s) - v_0(x) + \eta(s-t)}{u(x,t) - v_0(x)}$$

depends nonlinearly in u and v_0. One might see that the passage from the function given by (5) to w given by (3) bears a resemblance that from the Kruzkov transform to a linear change in [13] in the analysis of the comparison principle for stationary Hamilton–Jacobi equations.

From a technical point of view, they are a lot of variants for such results. For example, as it is the case in [5], one may look for a variational inequality for $m(t) := \max_{x\in\mathbb{R}^n} w(x,t)$ or for $m(t) := \max_{x\in\overline{\Omega}} w(x,t)$ where Ω is a suitable domain of $\mathbb{R}^n$. This last form can be typically useful when one wants to couple different assumptions on H on Ω and its complementary as in [5] where the coupling with Namah–Roquejoffre type assumptions was solved in that way, the point being to control the behavior of u on $\partial\Omega$.

For the connections between our assumptions and Namah–Roquejoffre type assumptions, we refer to Sect. 4.

2.1.1 A formal computation

Here we explain the algebra which bridges condition $(A6)_+$ to Proposition 6 under the strong regularity assumptions that $u,\ w \in C^1(\mathbb{T}^n \times [0, \infty))$ and $v_0 \in C^1(\mathbb{T}^n)$ and that for each $(x, t) \in Q$ there exists an $s > t$ such that

$$w(x, t) = u(x, t) - v_0(x) - \theta(u(x, s) - v_0(x) + \eta(s - t)). \tag{6}$$

Of course, these conditions do not hold in general.

Fix any $(x, t) \in Q$ and an $s > t$ so that (6) holds. If $w(x, t) \leq 0$, then (4) holds at (x, t). We thus suppose that $w(x, t) > 0$.

Setting

$$p = Dv_0(x), \quad q = D_x u(x, s), \quad r = D_x u(x, t), \quad a = u_s(x, s) \ \text{ and } \ b = u_t(x, t),$$

we have

$$H(x, p) \leq 0. \tag{7}$$

$$a + H(x, q) \geq 0, \tag{8}$$

$$b + H(x, r) \leq 0, \tag{9}$$

Also, by the choice of s, we get

$$D_x w(x, t) = r - p - \theta(q - p), \tag{10}$$

$$w_t(x, t) = b + \theta\eta, \tag{11}$$

$$0 = -\theta(a + \eta). \tag{12}$$

Combining (8) and (12) yields

$$H(x, q) \geq \eta. \tag{13}$$

Now, in view of inequalities (7) and (13), we may use assumption $(A6)_+$, to get

$$H(x, p + \theta(q - p)) \geq \theta\eta + \psi.$$

Using (10), we get

$$H(x, r) = H(x, D_x w(x, t) + p + \theta(q - p)).$$

Using the definition of $L > 0$, we clearly have

$$|r| = |D_x u(x, t)| \leq L \leq R, \quad |p + \theta(q - p)| \leq (1 + 2\theta)L \leq R$$

and therefore we get

$$\begin{aligned} H(x,r) &\ge H(x,p+\theta(q-p)) - \omega_{H,R}(|D_x w(x,t)|) \\ &\ge -\omega_{H,R}(|D_x w(x,t)|) + \theta\eta + \psi. \end{aligned}$$

This together with (9) and (11) yields

$$\begin{aligned} 0 \ge b + H(x,r) &= w_t(x,t) - \theta\eta + H(x,r) \\ &\ge w_t(x,t) - \theta\eta - \omega_{H,R}(|D_x w(x,t)|) + \theta\eta + \psi \\ &= w_t(x,t) - \omega_{H,R}(|D_x w(x,t)|) + \psi. \end{aligned}$$

This shows under our convenient regularity assumptions that (4) holds.

Remark 1 The actual requirement to v_0 is just the subsolution property in the above computation, which is true also in the following proof of Theorem 4. Some of subsolutions of (1) may have a better property, which solutions of (1) do not have. This is the technical insight in the generalization of Theorem 4 in Sect. 4.

2.1.2 Proof of Proposition 6

We begin with the following lemma.

Lemma 7 *We have*

$$-C_0(\theta-1) \le w(x,t) \le C_0 \quad \textit{for all } (x,t) \in \mathbb{R}^n \times [0,\infty).$$

Proof We just need to note that for all $(x,t) \in \mathbb{R}^n \times [0,\infty)$,

$$\begin{aligned} w(x,t) \ge u(x,t) - v_0(x) - \theta(u(x,t) - v_0(x)) &= (1-\theta)(u(x,t) - v_0(x)) \\ &\ge -C_0(\theta-1), \end{aligned}$$

and

$$w(x,t) \le \max_{s\ge t}(u(x,t) - v(x)) \le C_0.$$

□

Proof of Proposition 6 Noting that $u \in \mathrm{Lip}(\mathbb{T}^n \times [0,\infty))$ and $v_0 \in \mathrm{Lip}(\mathbb{T}^n)$ and rewriting w as

$$w(x,t) = \max_{r\ge 0}(u(x,t) - v_0(x) - \theta(u(x,r+t) - v_0(x) + \eta r)),$$

we deduce that $w \in \mathrm{Lip}(\mathbb{T}^n \times [0,\infty))$.

Fix any $\phi_0 \in C^1(Q)$ and $(\hat{x}, \hat{t}) \in Q$, and assume that

$$\max_Q (w - \phi_0) = (w - \phi_0)(\hat{x}, \hat{t}).$$

We intend to prove that for $R = (2\theta_0 + 1)L$,

$$\min\{w, \phi_{0,t} - \omega_{H,R}(|D\phi_0|) + \psi\} \leq 0 \quad \text{at } (\hat{x}, \hat{t}). \tag{14}$$

If $w(\hat{x}, \hat{t}) \leq 0$, then (14) clearly holds. We may thus suppose that $w(\hat{x}, \hat{t}) > 0$. We choose an $\hat{s} \geq \hat{t}$ so that

$$w(\hat{x}, \hat{t}) = u(\hat{x}, \hat{t}) - v_0(\hat{x}) - \theta(u(\hat{x}, \hat{s}) - v_0(\hat{x}) + \eta(\hat{s} - \hat{t})).$$

Observe that for any $s = \hat{t}$,

$$u(\hat{x}, \hat{t}) - v_0(\hat{x}) - \theta(u(\hat{x}, s) - v_0(\hat{x}) + \eta(s - \hat{t}) = (1 - \theta)(u(\hat{x}, \hat{t}) - v_0(\hat{x})) \leq 0,$$

which guarantees that $\hat{s} > \hat{t}$.

Define the function $\phi \in C^1(Q \times (0, \infty))$ by

$$\phi(x, t, s) = \phi_0(x, t) + |x - \hat{x}|^2 + (t - \hat{t})^2 + (s - \hat{s})^2.$$

Note that the function

$$u(x, t) - v_0(x) - \theta(u(x, s) - v_0(x) + \eta(s - t)) - \phi(x, t, s)$$

on $Q \times (0, \infty)$ attains a strict maximum at $(\hat{x}, \hat{t}, \hat{s})$, and that $D_x\phi(\hat{x}, \hat{t}, \hat{s}) = D_x\phi_0(\hat{x}, \hat{t})$, $\phi_t(\hat{x}, \hat{t}, \hat{s}) = \phi_{0,t}(\hat{x}, \hat{t})$ and $\phi_s(\hat{x}, \hat{t}, \hat{s}) = 0$.

Now, if B is an open ball of $\mathbb{R}^{3n+2}$ centered at $(\hat{x}, \hat{x}, \hat{x}, \hat{t}, \hat{s})$ with its closure $\overline{B}$ contained in $\mathbb{R}^{3n} \times (0, \infty)^2$, we use the technique of "tripling variables" and consider the function Φ on $\overline{B}$ given by

$$\begin{aligned}\Phi(x, y, z, t, s) = {} & u(x, t) - v_0(z) - \theta(u(y, s) - v_0(z) + \eta(s - t)) \\ & - \phi(x, t, s) - \alpha(|x - y|^2 + |x - z|^2),\end{aligned}$$

where $\alpha > 0$ is a (large) constant.

Let $(x_\alpha, y_\alpha, z_\alpha, t_\alpha, s_\alpha) \in \overline{B}$ be a maximum point of Φ. As usual in viscosity solutions theory, we observe that

$$\lim_{\alpha \to \infty} (x_\alpha, y_\alpha, z_\alpha, t_\alpha, s_\alpha) = (\hat{x}, \hat{x}, \hat{x}, \hat{t}, \hat{s}).$$

Consequently, if α is sufficiently large, then

$$(x_\alpha, y_\alpha, z_\alpha, t_\alpha, s_\alpha) \in B.$$

We assume henceforth that α is sufficiently large so that the above inclusion holds.

Next, setting

$$p_\alpha = 2(\theta-1)^{-1}\alpha(z_\alpha - x_\alpha) \quad \text{and} \quad q_\alpha = 2\theta^{-1}\alpha(x_\alpha - y_\alpha),$$

and noting that

$$\begin{aligned}\Phi(x,y,z,t,s) = u(x,t) - \theta u(y,s) + (\theta-1)v_0(z) - \theta\eta(s-t) \\ -\phi(x,t,s) - \alpha(|x-y|^2 + |x-z|^2),\end{aligned}$$

we observe that

$$p_\alpha \in D^+ v_0(z_\alpha), \tag{15}$$
$$\left(q_\alpha,\ -\theta^{-1}\phi_s(x_\alpha,t_\alpha,s_\alpha) - \eta\right) \in D^- u(y_\alpha, s_\alpha), \tag{16}$$
$$\left(D_x\phi(x_\alpha,t_\alpha,s_\alpha) + \theta q_\alpha - (\theta-1)p_\alpha,\ \phi_t(x_\alpha,t_\alpha,s_\alpha) - \theta\eta\right) \in D^+ u(x_\alpha,t_\alpha), \tag{17}$$

By the definition of L, we see as usual in viscosity solutions theory that $\max\{|p_\alpha|,\ |q_\alpha|\} \le L$. Sending $\alpha \to \infty$ in (15)–(17) along an appropriate sequence, we find points $\hat{p},\ \hat{q} \in \overline{B}_L$ such that

$$\hat{p} \in \overline{D}^+ v_0(\hat{x}), \tag{18}$$
$$\left(\hat{q},\ -\theta^{-1}\phi_s(\hat{x},\hat{t},\hat{s}) - \eta\right) \in \overline{D}^- u(\hat{x},\hat{s}), \tag{19}$$
$$\left(D_x\phi(\hat{x},\hat{t},\hat{s}) + \theta\hat{q} - (\theta-1)\hat{p},\ \phi_t(\hat{x},\hat{t},\hat{s}) - \theta\eta\right) \in \overline{D}^+ u(\hat{x},\hat{t}), \tag{20}$$

where $\overline{D}^\pm$ stand for the closures of $D^\pm$, for instance, $\overline{D}^+ u(\hat{x},\hat{s})$ denotes the set of points $(q,b) \in \mathbb{R}^n \times \mathbb{R}$ for which there are sequences $\{(q_j,b_j)\}_j \subset \mathbb{R}^n \times \mathbb{R}$ and $\{(x_j,s_j)\}_j \subset Q$ such that $\lim_j (q_j,b_j,x_j,s_j) = (q,b,\hat{x},\hat{s})$ and $(q_j,b_j) \in D^+ u(x_j,s_j)$ for all j. Here recall that $\phi_s(\hat{x},\hat{t},\hat{s}) = 0$, $\phi_t(\hat{x},\hat{t},\hat{s}) = \phi_{0,t}(\hat{x},\hat{t})$ and $D_x\phi(\hat{x},\hat{t},\hat{s}) = D_x\phi_0(\hat{x},\hat{t})$.

From (18) and (19), we get $H(\hat{x},\hat{p}) \le 0$ and

$$-\eta + H(\hat{x},\hat{q}) \ge 0.$$

By condition (A6)$_+$, we get

$$H(\hat{x},\hat{p} + \theta(\hat{q} - \hat{p})) \ge \theta\eta + \psi. \tag{21}$$

From (20), we get

$$0 \ge \phi_{0,t}(\hat{x},\hat{t}) - \theta\eta + H(\hat{x}, D_x\phi_0(\hat{x},\hat{t}) + \theta\hat{q} - (\theta-1)\hat{p}). \tag{22}$$

Noting that $|\hat{p} + \theta(\hat{q} - \hat{p})| \le (1+2\theta)L \le R$ and $|D_x\phi_0(\hat{x},\hat{t}) + \theta\hat{q} - (\theta-1)\hat{p}| \le L$ because of (20) and combining (22) and (21), we get

$$\begin{aligned} 0 &\geq \phi_{0,t}(\hat{x}, \hat{t}) - \theta\eta + H(\hat{x}, \hat{p} + \theta(\hat{q} - \hat{p})) - \omega_{H,R}(|D_x\phi_0(\hat{x}, \hat{t})|) \\ &\geq \phi_{0,t}(\hat{x}, \hat{t}) - \omega_{H,R}(|D_x\phi_0(\hat{x}, \hat{t})|) + \psi, \end{aligned}$$

which shows that (14) holds. □

2.1.3 Completion of the proof of Theorem 4 under (A6)$_+$

We set

$$w_\infty(x) = \limsup_{t\to\infty} w(x, t) \quad \text{for all } x \in \mathbb{R}^n.$$

Lemma 8 *We have*

$$w_\infty(x) \leq 0 \quad \textit{for all } x \in \mathbb{R}^n.$$

Moreover, the convergence

$$\lim_{t\to\infty} \max\{w(x, t), 0\} = 0 \tag{23}$$

is uniform in $x \in \mathbb{R}^n$.

Proof It is sufficient to prove that the convergence (23) holds uniformly in $x \in \mathbb{R}^n$. Contrary to this, we suppose that there is a sequence $(x_j, t_j) \in Q$ such that $\lim_{j\to\infty} t_j = \infty$ and $w(x_j, t_j) \geq \delta$ for all $j \in \mathbb{N}$ and some constant $\delta > 0$. In view of the periodicity of w, we may assume that $\lim_{j\to\infty} x_j = y$ for some $y \in \mathbb{R}^n$. Moreover, in view of the Ascoli–Arzela theorem, we may assume by passing to a subsequence of $\{(x_j, t_j)\}$ if needed that

$$\lim_{j\to\infty} w(x, t + t_j) = f(x, t) \quad \text{locally uniformly in } \mathbb{R}^n \times (-\infty, +\infty),$$

for some bounded function $f \in \mathrm{Lip}(\mathbb{T}^n \times \mathbb{R})$.

Now, note that $f(y, 0) \geq \delta$. By the stability of the subsolution property under uniform convergence, we see that f is a subsolution of

$$\min\{f(x, t),\ f_t(x, t) - \omega_{H,R}(|D_x f(x, t)|) + \psi\} \leq 0 \quad \text{in } \mathbb{R}^{n+1}. \tag{24}$$

Since $f \in C(\mathbb{T}^n \times \mathbb{R})$ and f is bounded on $\mathbb{R}^{n+1}$, for every $\varepsilon > 0$ the function $f(x, t) - \varepsilon t^2$ attains a maximum over $\mathbb{R}^{n+1}$ at a point $(x_\varepsilon, t_\varepsilon)$. Observe as usual in the viscosity solutions theory that

$$f(x_\varepsilon, t_\varepsilon) - \varepsilon t_\varepsilon^2 \geq f(y, 0) \geq \delta,$$

and therefore

$$f(x_\varepsilon, t_\varepsilon) \geq \delta \quad \text{and} \quad \varepsilon|t_\varepsilon| \leq (\varepsilon\|f\|_\infty)^{1/2}.$$

In particular, we have $\lim_{\varepsilon\to 0+} \varepsilon t_\varepsilon = 0$. In view of inequality (24), we get

$$2\varepsilon t_\varepsilon - \omega_{H,R}(0) + \psi \le 0,$$

which, in the limit as $\varepsilon \to 0+$, yields $\psi \le 0$, a contradiction. This shows that the uniform convergence (23) holds.

Proof of Theorem 4 under (A6)$_+$ Let w be the function defined by (3), with arbitrary $(\eta, \theta) \in (0, \eta_0) \times (0, \theta_0)$.

Fix any $\varepsilon > 0$. Thanks to (23), we may choose a constant $T_\varepsilon \equiv T_{\varepsilon,\eta,\theta} > 0$ so that for any $t \ge T_\varepsilon$,

$$w(x,t) \le \varepsilon \quad \text{for all } x \in \mathbb{R}^n.$$

Let $t \ge T_\varepsilon$ and $x \in \mathbb{R}^n$. From the above, for any $s \ge t$, we have

$$\begin{aligned} u(x,t) - v_0(x) &\le \varepsilon + \theta(u(x,s) - v_0(x)) + \theta\eta(s-t) \\ &= \varepsilon + u(x,s) - v_0(x) + (\theta - 1)(u(x,s) - v_0(x)) + \theta\eta(s-t) \\ &\le \varepsilon + u(x,s) - v_0(x) + (\theta - 1)C_0 + \theta\eta(s-t). \end{aligned}$$

Thus, for any $0 \le s \le 1$, we have

$$u(x,t) \le u(x,t+s) + (\theta - 1)C_0 + \theta\eta + \varepsilon. \tag{25}$$

Now, since u is bounded and Lipschitz continuous in $\overline{Q}$, in view of the Ascoli–Arzela theorem, we may choose a sequence $\tau_j \to \infty$ and a bounded function $z \in \mathrm{Lip}(\mathbb{T}^n \times (-\infty, +\infty))$ so that

$$\lim_{j\to\infty} u(x, t+\tau_j) = z(x,t) \quad \text{locally uniformly on } \mathbb{R}^{n+1}. \tag{26}$$

By (25) we get

$$z(x,t) \le z(x,t+s) + (\theta - 1)C_0 + \theta\eta + \varepsilon \quad \text{for all } (x,t,s) \in \mathbb{R}^{n+1} \times [0,\ 1].$$

This is valid for all $(\eta, \theta) \in (0, \eta_0) \times (1, \theta_0)$. Hence,

$$z(x,t) \le z(x,t+s) + \varepsilon \quad \text{for all } (x,t,s) \in \mathbb{R}^{n+1} \times [0,\ 1],$$

and moreover

$$z(x,t) \le z(x,t+s) \quad \text{for all } (x,t,s) \in \mathbb{R}^{n+1} \times [0,\ 1]. \tag{27}$$

Thus we find that the function $z(x,t)$ is nondecreasing in $t \in \mathbb{R}$ for all $x \in \mathbb{R}^n$. From this, we conclude that

$$\lim_{t\to\infty} z(x,t) = u_\infty(x) \quad \text{uniformly on } \mathbb{R}^n \tag{28}$$

for some function $u_\infty \in \mathrm{Lip}(\mathbb{T}^n)$.

Fix any $\delta > 0$. By (28) there is a constant $\tau > 0$ such that

$$\|z(\cdot,\tau) - u_\infty\|_{L^\infty(\mathbb{R}^n)} < \delta,$$

Then, by (26) there is a $j \in \mathbb{N}$ such that

$$\|z(\cdot,\tau) - u(\cdot,\tau+\tau_j)\|_\infty < \delta.$$

Hence,

$$\|u(\cdot,\tau+\tau_j) - u_\infty\|_\infty < 2\delta.$$

By the contraction property (Theorem 2), we see that for any $t \geq \tau + \tau_j$,

$$\|u(\cdot,t) - u_\infty\|_\infty \leq \|u(\cdot,\tau+\tau_j) - u_\infty\|_\infty < 2\delta,$$

which completes the proof.

2.2 Under assumption (A6)$_-$

In addition to (A1)–(A5), we assume throughout this subsection that (A6)$_-$ holds.

To accommodate the previous w to (A6)$_-$, we modify and replace it by the new function, which we denote by the same symbol, given by

$$w(x,t) = \max_{0\leq s\leq t}(u(x,t) - v_0(x) - \theta(u(x,s) - v_0(x) - \eta(s-t)),$$

where (η,θ) is chosen arbitrarily in $(0,\ \eta_0) \times (1,\ \theta_0)$ and the constants η_0 and θ_0 are those from (A6)$_-$.

Lemma 9 *We have*

$$-C_0(\theta - 1) \leq w(x,t) \leq C_0 \quad \textit{for all } (x,t) \in \overline{Q}.$$

Proof Recall that $0 \leq u(x,t) - v_0(x) \leq C_0$ for all $(x,t) \in \overline{Q}$, and note that for all $(x,t) \in \overline{Q}$,

$$\begin{aligned} w(x,t) \geq u(x,t) - v_0(x) - \theta(u(x,t) - v_0(x)) &= (1-\theta)(u(x,t) - v_0(x)) \\ &\geq -C_0(\theta - 1) \end{aligned}$$

and

$$w(x,t) \leq \max_{0\leq s\leq t}(u(x,t) - v_0(x)) \leq C_0.$$

□

We have the following proposition similar to Proposition 6.

Proposition 10 *The function w is a subsolution of*

$$\min\{w(x,t),\, w_t(x,t)-\omega_{H,R}(|D_x w(x,t)|)+\psi\} \le 0 \ \text{in}\ (x,t)\in\mathbb{R}^n \times (T,\ \infty), \quad (29)$$

where $\psi = \psi(\theta, \eta) > 0$ *is the constant from (A6)*$_-$, $T := C_0/\eta$ *and* $R := (2\theta_0 + 1)L$.

Since the proof of the above proposition is very similar to that of Proposition 6, we present just an outline of it.

Outline of proof Note that for any $(x, t) \in \mathbb{R}^n \times (T,\ \infty)$ and $s \in [0,\ t - T)$,

$$\begin{aligned} u(x,t)-v_0(x)-\theta(u(x,s)-v_0(x)-\eta(s-t)) &\le C_0-\theta\eta(t-s) \\ &< C_0-\theta\eta T=-(\theta-1)C_0. \quad (30) \end{aligned}$$

Hence, in view of Lemma 9, for any $(x, t) \in \mathbb{R}^n \times (T,\ \infty)$ we have

$$\begin{aligned} w(x,t) &= \max_{t-T\le s\le t} [u(x,t) - v_0(x) - \theta(u(x,s) - v_0(x) - \eta(s-t))] \\ &= \max_{-T\le s\le 0} [u(x,t) - v_0(x) - \theta(u(x,s+t) - v_0(x) - \eta s)]. \end{aligned}$$

From this latter expression of w, as the functions u and v_0 are Lipschitz continuous in $\overline{Q}$ and $\mathbb{R}^n$, respectively, we see that w is Lipschitz continuous in $\mathbb{R}^n \times [T,\ \infty)$. Also, from (30) we see that for any $(x, t) \in \mathbb{R}^n \times (T,\ \infty)$, if

$$w(x,t) = u(x,t) - v_0(x) - \theta(u(x,s) - v_0(x) - \eta(s-t))$$

for some $0 \le s \le t$, then $s \ge t - T > 0$.

To see that (29) holds, we fix any test function $\phi_0 \in C^1(\mathbb{R}^n \times (T,\ \infty))$ and assume that $w - \phi_0$ attains a strict maximum at a point $(\hat{x}, \hat{t})$.

Following the same arguments as in the proof under (A6)$_+$, we are led to the inclusions

$$\begin{cases} \hat{p} \in \overline{D}^+ v_0(\hat{x}), \\ (\hat{q}, \eta) \in \overline{D}^- u(\hat{x}, \hat{s}), \\ \big(D_x\phi_0(\hat{x},\hat{t}) + \theta\hat{q} - (\theta-1)\hat{p},\ \phi_{0,t}(\hat{x},\hat{t}) + \theta\eta\big) \in \overline{D}^+ u(\hat{x},\hat{t}) \end{cases} \quad (31)$$

for some $\hat{p}, \hat{q} \in \mathbb{R}^n$.

Using (31), we observe that $H(\hat{x}, \hat{p}) \le 0$ and $\eta + H(\hat{x}, \hat{q}) \ge 0$. Hence, by condition (A6)$_-$, we get

$$H(\hat{x}, \hat{p} + \theta(\hat{q} - \hat{p})) \ge -\theta\eta + \psi.$$

Moreover, we compute that

$$\begin{aligned} 0 &\ge \phi_t(\hat{x},\hat{t},\hat{s}) + \theta\eta + H(\hat{x}, D_x\phi(\hat{x},\hat{t},\hat{s}) + \theta\hat{q} - (\theta-1)\hat{p}) \\ &\ge \phi_{0,t}(\hat{x},\hat{t}) + \theta\eta - \omega_{H,R}(|D_x\phi_0(\hat{x},\hat{t})|) + H(\hat{x}, \theta\hat{q} - (\theta-1)\hat{p}) \\ &\ge \phi_{0,t}(\hat{x},\hat{t}) - \omega_{H,R}(|D_x\phi_0(\hat{x},\hat{t})|) + \psi. \end{aligned}$$

Note that, as above, $|\hat{p} + \theta(\hat{q} - \hat{p})| \leq R$ and $|D_x\phi_0(\hat{x}, \hat{t}) + \theta\hat{q} - (\theta - 1)\hat{p}| \leq L$. This completes the proof. □

Outline of proof of Theorem 4 under (A6)$_-$ Using Proposition 10 and arguing as the proof of Lemma 8, we deduce that

$$\lim_{t\to\infty} \max\{w(x,t), 0\} = 0 \quad \text{uniformly in } \mathbb{R}^n.$$

We fix any $\varepsilon > 0$ and choose a constant $T_\varepsilon \equiv T_{\varepsilon,\eta,\theta} > T$ so that for any $t \geq T_\varepsilon$,

$$w(x,t) \leq \varepsilon \quad \text{for all } x \in \mathbb{R}^n.$$

Let $t \geq T_\varepsilon$ and $x \in \mathbb{R}^n$. For any $0 \leq s \leq t$, we have

$$u(x,t) - v_0(x) \leq \varepsilon + u(x,s) - v_0(x) + (\theta - 1)C_0 + \theta\eta(t - s).$$

We may assume that $T_\varepsilon > 1$, and from the above, for any $0 \leq s \leq 1$, we have

$$u(x,t) \leq u(x,t-s) + (\theta - 1)C_0 + \theta\eta + \varepsilon. \tag{32}$$

Since $u \in \mathrm{Lip}(\mathbb{T}^n \times (0,\infty))$ and it is bounded in $\overline{Q}$, the Ascoli–Arzela theorem assures that there is a sequence $\{\tau_j\}_{j\in\mathrm{N}} \subset (0,\infty)$ diverging to infinity such that for some function $z \in \mathrm{Lip}(\mathbb{T}^n \times \mathbb{R})$,

$$\lim u(x, t+\tau_j) = z(x,t) \quad \text{locally uniformly in } \mathbb{R}^{n+1}.$$

We see immediately from (32) that the function $z(x,t)$ is nonincreasing in t for every x. Furthermore, we infer that for some function $u_\infty \in C(\mathbb{T}^n)$,

$$\lim_{t\to\infty} z(x,t) = u_\infty(x) \quad \text{uniformly in } \mathbb{R}^n.$$

As exactly under (A6)$_+$, we deduce from this that

$$\lim_{t\to\infty} u(x,t) = u_\infty(x) \quad \text{uniformly in } \mathbb{R}^n,$$

which completes the proof.

3 Conditions (A6)$_\pm$

First of all we restate the conditions (A6)$_\pm$ in [3] as (A)$_\pm$:
(A)$_+$ There exists $\eta_0 > 0$ such that, for any $\eta \in (0, \eta_0)$, there exists $\nu = \nu(\eta) > 0$ such that for all $x, p, q \in \mathbb{R}^n$ and $\theta > 1$, if $H(x,q) \geq \eta$ and $H(x,p) \leq 0$, then

$$H(x, p + \theta(q - p)) \geq \theta H(x,q) + \nu(\theta - 1).$$

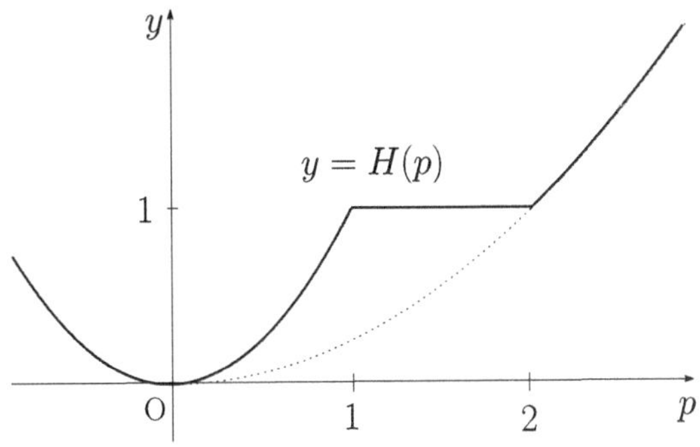

Fig. 1 Hamiltonian satisfying $(A6)_+$ and not $(A)_+$

$(A)_-$ There exists $\eta_0 > 0$ such that, for any $\eta \in (0, \eta_0)$, there exists $\nu = \nu(\eta) > 0$ such that for all $x, p, q \in \mathbb{R}^n$ and $\lambda \in [0, 1]$, if $H(x, q) \leq -\eta$ and $H(x, p) \leq 0$, then

$$H(x, (1-\lambda)p + \lambda q) \leq \lambda H(x, q) - \nu\lambda(1-\lambda).$$

Conditions $(A6)_\pm$ and $(A)_\pm$ can be considered as a sort of strict convexity requirements on the function $H(x, p)$ in p near the points where H vanishes ($(A6)_+$ and $(A)_+$ are the ones for those points (x, p) where $H(x, p) \geq 0$, while $(A6)_-$ and $(A)_-$ are for those points where $H \leq 0$).

The condition (H4) in [5] has a general feature more than $(A)_+$ above, and its additional generality is in the point that includes the key assumption in Namah–Roquejoffre [19]. If we push this point aside, then the condition (H4) in [5] is same as $(A)_+$ above.

Now, we give comparison between $(A6)_+$ and $(A)_+$. Let η_0, θ_0 and $\psi(\eta, \theta)$ be the positive constants from $(A6)_+$. Note that the key inequality in $(A6)_+$ holds with $\psi(\eta, \theta)$ replaced by $\min\{\psi(\eta, \theta), 1\}$. Thus, the behavior of the function H where the value of H is large (larger than $\eta_0\theta_0 + 1$), is irrelevant to condition $(A6)_+$, while $(A)_+$ requires a certain growth of the function H where its value is positive. The function H on $\mathbb{R}^n$ (see Fig. 1) given by

$$H(p) = \max\{\min\{|p|^2, 1\}, |p|^2/4\}$$

satisfies (A2)–(A5) and $(A6)_+$, as is easily checked. However, if $p = 0$, $|q| = 1$ and $1 < \theta < 2$, then we have

$$H(p + \theta(q - p)) = H(\theta q) = 1 < \theta = \theta H(q).$$

Therefore, $(A)_+$ does not hold with this Hamiltonian $H(x, p) = H(p)$.

The difference of two conditions observed above is concerned with the behavior of the Hamiltonian $H(x, p)$ where H is large.

The following example shows that $(A)_+$ is a stronger requirement on H than $(A6)_+$ even in a neighborhood of the points (x, p) where H vanishes. In this regard, the difference between two conditions is that the term $\psi(\eta, \theta)$ in $(A6)_+$ depends generally on η, θ while the term $\nu(\eta)(\theta - 1)$ in $(A)_+$ depends linearly in $\theta - 1$.

Fig. 2 Function H_0

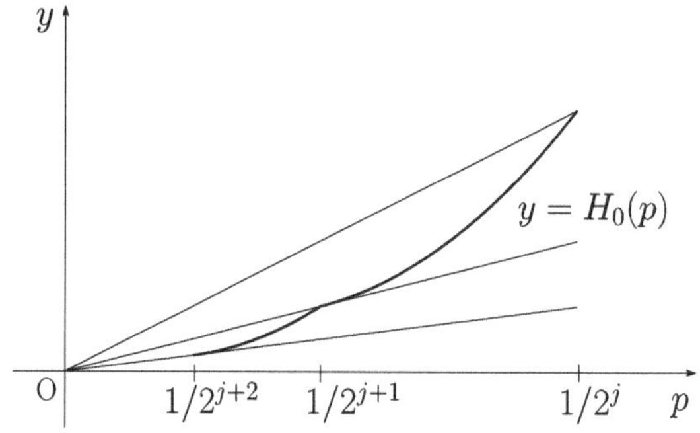

We define the function H_0 (see Fig. 2) and H in $C(\mathbb{R})$ by

$$H_0(p) = \begin{cases} 0 & \text{for } p \leq 0, \\ p + (p-1)^2 & \text{if } p \geq 1, \\ p/2 + 2(p - 1/2)^2 & \text{if } 1/2 \leq p \leq 1, \\ \vdots & \vdots \\ p/2^{j+1} + 2^{j+1}\left(p - 1/2^{j+1}\right)^2 & \text{if } 1/2^{j+1} \leq p \leq 1/2^j, \\ \vdots & \vdots \end{cases}$$

and

$$H(p) = |p+1| - 1 + H_0(p) + H_0(-p-1)$$

This Hamiltonian H satisfies (A2)–(A4), and the problem

$$H(u'(x)) = 0 \quad \text{in } \mathbb{R} \quad \text{and} \quad u \in C(\mathbb{T}),$$

where $u' = \mathrm{d}u/\mathrm{d}x$, has a solution $u(x) \equiv 0$. Thus, (A5) is satisfied with our function H. Moreover, it is easily seen that H satisfies $(A6)_+$. However, H does not satisfy condition $(A)_+$. To check this, fix any $j \in \mathbb{N}$ and choose $p = 0$ and $q = 1/2^{j+1}$. Note that

$$H(q) = \frac{1}{2^{j+1}} + \frac{1}{2^{2j+2}},$$

and that for any $\theta \in (1, 2)$, we have $1/2^{j+1} < \theta q < 1/2^j$ and

$$H(\theta q) = \frac{\theta}{2^{j+1}} + \frac{\theta}{2^{2j+2}} + 2^{j+1}\left(\frac{\theta}{2^{j+1}} - \frac{1}{2^{j+1}}\right)^2 = \theta H(q) + \frac{(\theta-1)^2}{2^{j+1}}.$$

Hence,

$$H(\theta q) - \theta H(q) = o(\theta - 1) \quad \text{as } \theta \to 1+,$$

which violates the inequality in $(A)_+$. Note finally that $q = 1/2^{j+1}$ can be taken as close to $p = 0$ as we wish.

Next, we show that if $H \in C(\mathbb{T}^n \times \mathbb{R}^n)$ satisfies $(A)_-$, then it satisfies $(A6)_-$.

For this, let $H \in C(\mathbb{T}^n \times \mathbb{R}^n)$ satisfy $(A6)_-$. Let $\eta_0 > 0$ be the constant and ν the function on $(0, \eta_0)$ given by $(A)_-$.

Fix any $\eta \in (0, \eta_0)$ and $\theta > 1$, and set $\lambda = \theta^{-1} \in (0, 1)$. Let $x, p, q \in \mathbb{R}^n$ and assume that $H(x, p) \le 0$ and $H(x, q) \ge -\eta$. Set

$$\psi = \psi(\eta, \theta) := (\theta - 1)\min\{\theta^{-1}\nu(\eta), \eta\} = \min\{(1-\lambda)\nu(\eta), (\theta - 1)\eta\}. \quad (33)$$

It is enough to show that

$$H(x, p + \theta(q - p)) \ge -\theta\eta + \psi. \quad (34)$$

To the contrary, we suppose that

$$H(x, p + \theta(q - p)) < -\theta\eta + \psi. \quad (35)$$

Set $r = p + \theta(q - p)$ and note that $q = \lambda r + (1 - \lambda)p$. Note by the choice of ψ that

$$H(x, r) < -\theta\eta + (\theta - 1)\eta = -\eta.$$

Hence, using $(A)_-$, (34) and (33), we deduce that

$$H(x, q) = H(x, \lambda r + (1-\lambda)p) \le \lambda H(x, r) - \nu(\eta)\lambda(1-\lambda) < \lambda(-\theta + \psi) - \psi\lambda = -\eta.$$

This is a contradiction, which shows that (34) holds.

Now, let $H \in C(\mathbb{T}^n \times \mathbb{R}^n)$ satisfy $(A6)_+$, and we show that for each $x \in \mathbb{R}^n$ the sublevel set $\{p \in \mathbb{R}^n : H(x, p) \le 0\}$ is convex.

To do this, we fix any $x \in \mathbb{R}^n$ and let $p_1, p_2 \in K := \{p \in \mathbb{R}^n : H(x, p) \le 0\}$. We need to show that

$$\lambda p_1 + (1 - \lambda)p_2 \in K \quad \text{for all } \lambda \in [0, 1]. \quad (36)$$

We suppose that this is not the case and will get a contradiction.

Let $\eta_0 > 0$ and $\theta_0 > 0$ be the constants from $(A6)_+$. Then, setting

$$\lambda_0 = \sup\{\lambda \in [0, 1] : \lambda p_1 + (1 - \lambda)p_2 \notin K\},$$

we have

$$\lambda_0 p_1 + (1 - \lambda_0)p_2 \in K \quad \text{by the continuity of } H.$$

By the definition of λ_0, we may select a $\lambda \in (0, \lambda_0)$ so that

$$\lambda p_1 + (1 - \lambda)p_2 \notin K \quad \text{and} \quad \lambda\theta_0 > \lambda_0.$$

Fig. 3 Hamiltonian satisfying (A6)$_-$ and not (A)$_-$

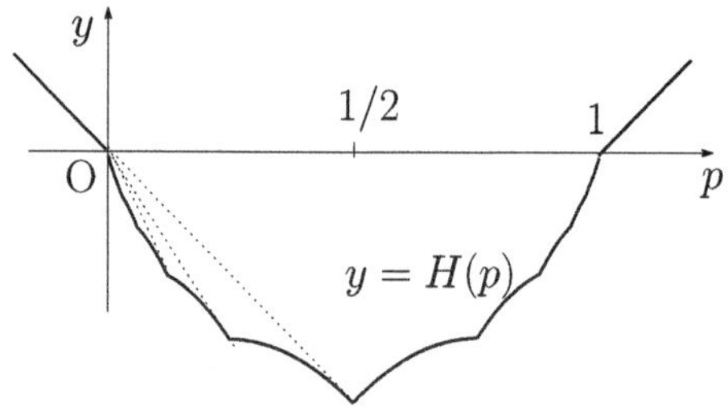

Set

$$q := \lambda p_1 + (1-\lambda)p_2 = p_2 + \lambda(p_1 - p_2),$$
$$\theta := \lambda_0/\lambda \in (1, \theta_0),$$

and note that $H(x, q) > 0$. Fix an $0 < \eta < \eta_0$ so that $H(x, q) \geq \eta$, and use condition (A6)$_+$, to get

$$H(x, p_2 + \theta(q - p_2)) > \theta\eta > 0,$$

and moreover,

$$0 < H(x, p_2 + \theta(q - p_2)) = H(x, \lambda_0 p_1 + (1-\lambda_0)p_2) \leq 0.$$

This is a contradiction.

An argument similar to the above guarantees that if $H \in C(\mathbb{T}^n \times \mathbb{R}^n)$ satisfies (A6)$_-$, then the sublevel set $\{p \in \mathbb{R}^n : H(x, p) < 0\}$ is convex for every $x \in \mathbb{R}^n$. We leave it for the interested reader to check this convexity property.

The following example of $H(x, p) = H(p)$ explicitly shows that condition (A)$_-$ is more stringent than (A6)$_-$. Define the functions $f, g \in C(\mathbb{R})$ by

$$f(p) = \begin{cases} 0 & \text{if } p \leq 0 \text{ or } p \geq 1, \\ -p/2 & \text{if } 0 \leq p \leq 1/2, \\ -(p-1)^2 & \text{if } 1/2 \leq p \leq 1, \end{cases}$$
$$g(p) = -p + \sum_{k=1}^{\infty} 2^{-k} f(2^k p),$$

and then $H \in C(\mathbb{R})$ (see Fig. 3) by

$$H(p) = \max\{g(p), g(1-p)\}.$$

We do not give the detail, but observing that in the py plane, for each slope $m < 0$, the half line $y = mp$, $p > 0$, meets the graph $y = H(p)$ at exactly one point, we can deduce that the function H satisfies (A6)$_-$. On the other hand, setting $p = 0$ and

$q = 1/2^k$, with $k \in \mathbb{N}$, observing that if $\frac{1}{2} \le \lambda \le 1$, then $1/2^{k+1} \le \lambda q \le 1/2^k$ and that for any $\frac{1}{2} \le \lambda \le 1$,

$$H(\lambda q) = -\frac{k+1}{2^{k+1}} - \frac{(\lambda - 1)^2}{2^k},$$

and hence,

$$H(q) = -\frac{k+1}{2^{k+1}},$$
$$H(\lambda q) = \lambda H(q) - \frac{(\lambda - 1)^2}{2^k},$$

we may deduce that $(A)_-$ does not hold with the current function H.

Next, we remark that under hypotheses (A2)–(A4), conditions $(A6)_+$ and $(A6)_-$ are equivalent to the following $(A7)_+$ and $(A7)_-$, respectively.

$(A7)_+$ There exist constants $\eta_0 > 0$ and $\theta_0 > 1$ such that for all $(\eta, \theta) \in (0, \eta_0) \times (1, \theta_0)$, $x, p, q \in \mathbb{R}^n$, if $H(x, p) \le 0$ and $H(x, q) \ge \eta$, then

$$H(x, p + \theta(q - p)) > \eta\theta.$$

$(A7)_-$ There exist constants $\eta_0 > 0$ and $\theta_0 > 1$ and for all $(\eta, \theta) \in (0, \eta_0) \times (1, \theta_0)$, $x, p, q \in \mathbb{R}^n$, if $H(x, p) \le 0$ and $H(x, q) \ge -\eta$, then

$$H(x, p + \theta(q - p)) > -\eta\theta.$$

Indeed, it is clear that $(A6)_\pm$ imply $(A7)_\pm$, respectively. On the other hand, assuming that $(A7)_+$ holds, choosing $R > 0$ so large that

$$H(x, p) > \eta_0\theta_0 \quad \text{if } |p| > R,$$

where $\eta_0 > 0$ and $\theta_0 > 1$ are the constants from $(A7)_+$, and setting

$$\psi(\eta, \theta) = \min\{H(x, p + \theta(q - p)) - \theta\eta \, : \, x \in \mathbb{T}^n, H(x, p) \le 0, H(x, q) \ge \eta\}$$

for any $(\eta, \theta) \in (0, \, \eta_0)(1, \theta_0)$ we observe that $\psi(\eta, \theta)$ is positive and satisfies

$$H(x, p + \theta(q - p)) \ge \eta\theta + \psi(\eta, \theta)$$

for all $(x, p, q) \in \mathbb{R}^{3n}$ such that $H(x, p) \le 0$ and $H(x, q) \ge \eta$, which shows that $(A6)_+$ holds. Similarly, we see that $(A7)_-$ implies $(A6)_-$.

Finally, we remark that under (A2)–(A4), conditions $(A6)_+$ and $(A6)_-$ are equivalent to the following $(A8)_+$ and $(A8)_-$, respectively.

$(A8)_+$ There exist constants $\eta_0 > 0$ and $\theta_0 > 1$ and for each $(\eta, \theta) \in (0, \eta_0) \times (1, \theta_0)$ a constant $\psi = \psi(\eta, \theta) > 0$ such that for all $x, p, q \in \mathbb{R}^n$, if $H(x, p) = 0$ and $H(x, q) = \eta$, then

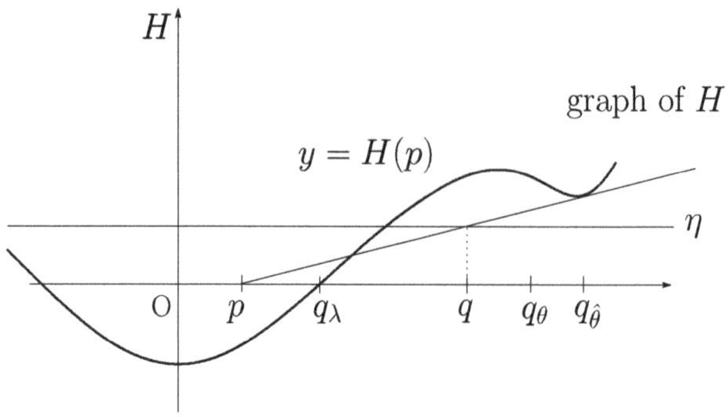

Fig. 4 Position of p, q, q_λ, etc

$$H(x, p + \theta(q - p)) \geq \eta\theta + \psi.$$

(A8)$_-$ There exist constants $\eta_0 > 0$ and $\theta_0 > 1$ and for each $(\eta, \theta) \in (0, \eta_0) \times (1, \theta_0)$ a constant $\psi = \psi(\eta, \theta) > 0$ such that for all $x, p, q \in \mathbb{R}^n$, if $H(x, p) = 0$ and $H(x, q) = -\eta$, then

$$H(x, p + \theta(q - p)) \geq -\eta\theta + \psi.$$

It is clear that (A6)$_\pm$ imply (A8)$_+$, respectively.

We next show that (A8)$_+$ implies (A7)$_+$, which is equivalent to (A6)$_+$. We leave it to the reader to check that (A8)$_-$ implies (A7)$_-$.

Let η_0 and θ_0 be the constants from (A8)$_+$. We may assume, by replacing θ_0 by a smaller one if needed, that $\theta_0 < 2$.

Fix any $0 < \eta < \eta_0/2$ and $(x, p, q) \in \mathbb{R}^{3n}$ such that $H(x, p) \leq 0$ and $H(x, q) \geq \eta$. It is enough to show that for all $1 < \theta < \theta_0$,

$$H(x, p + \theta(q - p) > \theta\eta. \tag{37}$$

We assume for contradiction that (37) does not hold. We set

$$\Theta = \{\theta \in (1, \theta_0) : H(x, p + \theta(q - p)) \leq \theta\eta\}.$$

Note by the above assumption that $\Theta \neq \emptyset$ and set $\hat{\theta} := \inf \Theta$. It is clear that $1 \leq \hat{\theta} < \theta_0$, $H(x, p + \hat{\theta}(q - p)) = \hat{\theta}\eta$ since $H(x, q) \geq \eta$ and $H(x, p + \theta(q - p)) > \theta\eta$ if $1 < \theta < \hat{\theta}$.

In what follows, we write $H(r) := H(x, r)$ and $q_\theta = p + \theta(q - p)$ for $0 \leq \theta < \theta_0$. We fix a $\lambda \in [0, 1)$ so that $H(p + \lambda(q - p)) = 0$. Note that $H(q_\lambda) = 0$.

Consider the case where $\hat{\theta} = 1$. In this case we have $q_{\hat{\theta}} = q$ and $H(q) = \eta$. By (A8)$_+$, we get

$$H(q_\lambda + \rho(q - q_\lambda)) > \rho\eta \quad \text{for all } \rho \in (1, \theta_0). \tag{38}$$

Noting that

$$\begin{aligned} q_\lambda + \rho(q - q_\lambda) &= p + \lambda(q-p) + \rho(q - p - \lambda(q-p)) \\ &= p + (\lambda + (1-\lambda)\rho)(q-p), \end{aligned}$$

from (38) we get

$$H(p + (\lambda + (1-\lambda)\rho)(q-p)) > \rho\eta \geq (\lambda + (1-\lambda)\rho)\eta \quad \text{for all } \rho \in (1, \theta_0),$$

which implies that $\Theta \cap (1,\ \lambda + (1-\lambda)\theta_0) = \emptyset$. This ensures that $\hat{\theta} \geq \lambda + (1-\lambda)\theta_0 > 1$, which contradicts that $\hat{\theta} = 1$.

Consider next the case where $\hat{\theta} > 1$ (see Fig. 4). Recall that $H(q_{\hat{\theta}}) = \hat{\theta}\eta < 2\eta < \eta_0$ and $H(q_\theta) > \theta\eta$ for all $\theta \in (1, \hat{\theta})$. Setting $\eta_\theta := H(q_\theta)$, we observe that if $1 < \theta < \hat{\theta}$ is close to $\hat{\theta}$, then $\theta\eta < \eta_\theta < \eta_0$. For any such θ, by $(A8)_+$, we get

$$H(q_\lambda + \rho(q_\theta - q_\lambda)) > \rho\eta_\theta \quad \text{for all } \rho \in (1, \theta_0). \tag{39}$$

Note that $q_\lambda + \rho(q_\theta - q_\lambda) = p + (\lambda + \rho(\theta - \lambda))(q - p)$. We select $\hat{\rho}$ so that $\hat{\theta} = \lambda + \hat{\rho}(\theta - \lambda)$ or, equivalently, $\hat{\rho} = (\hat{\theta} - \lambda)/(\theta - \lambda)$. Since θ is assumed to be close enough to $\hat{\theta}$, we may assume that $\hat{\rho} \in (1,\ \theta_0)$. Thanks to (39), we get

$$\hat{\theta}\eta = H(q_{\hat{\theta}}) = H(q_\lambda + \hat{\rho}(q_\theta - q_\lambda)) > \hat{\rho}\eta_\theta > \frac{\hat{\theta} - \lambda}{\theta - \lambda}\eta\theta.$$

Thus, we get $\hat{\theta}(\theta - \lambda) > \theta(\hat{\theta} - \lambda)$ or, equivalently, $\lambda(\hat{\theta} - \theta) < 0$. This is a contradiction. We thus see that $(A8)_+$ implies $(A7)_+$.

4 A generalization of $(A6)_\pm$

We recall that the following conditions on the Hamiltonian $H \in C(\mathbb{T}^n \times \mathbb{R}^n)$ has been introduced by Namah–Roquejoffre [19] in their study of the large time asymptotic behavior of solutions of (CP).

(NR1) The function $H(x, p)$ is convex in $p \in \mathbb{R}^n$ for every $x \in \mathbb{R}^n$.
(NR2) $\min_{p \in \mathbb{R}^n} H(x, p) = H(x, 0)$ for all $x \in \mathbb{R}^n$.
(NR3) $\max_{x \in \mathbb{R}^n} H(x, 0) = 0$.
(NR4) $\lim_{r \to \infty} \inf\{H(x, p) : (x, p) \in \mathbb{T}^n \times \mathbb{R}^n,\ |p| \geq r\} = \infty$.

Assume for the moment that $H \in C(\mathbb{T}^n \times \mathbb{R}^n)$ satisfies (NR3). Then the function $v(x) \equiv 0$ solves in the classical sense

$$H(x, Dv(x)) = H(x, 0) \quad \text{in } \mathbb{R}^n.$$

Here, if $H(x, 0) < 0$ for some points x, then v is a "strict" subsolution of $H(x, Du) = 0$ in the set $\{x \in \mathbb{R}^n\ :\ H(x, 0) < 0\}$.

We take this observation into account and modify conditions $(A6)_\pm$ as follows. The new conditions depend on our choice of a subsolution v_0 of (1), which plays the same role as the function v_0 in the proof of Theorem 4. As we have already noted in Remark 1, the function v_0 in the proof of Theorem 4 is needed to be just a subsolution of (1) and the outcome may depend on our choice of v_0. Now we fix a subsolution $v_0 \in C(\mathbb{T}^n)$ of (1) and choose a nonnegative function $f \in C(\mathbb{T}^n)$ so that v_0 is a subsolution of

$$H(x, Dv_0(x)) \leq -f(x) \quad \text{in } \mathbb{R}^n.$$

$(A9)_+$ There exist constants $\eta_0 > 0$ and $\theta_0 > 1$ and for each $(\eta, \theta) \in (0, \eta_0) \times (1, \theta_0)$ a constant $\psi = \psi(\eta, \theta) > 0$ such that for all $x, p, q \in \mathbb{R}^n$, if $H(x, p) \leq -f(x)$ and $H(x, q) \geq \eta$, then

$$H(x, p + \theta(q - p)) \geq \eta\theta + \psi.$$

$(A9)_-$ There exist constants $\eta_0 > 0$ and $\theta_0 > 1$ and for each $(\eta, \theta) \in (0, \eta_0) \times (1, \theta_0)$ a constant $\psi = \psi(\eta, \theta) > 0$ such that for all $x, p, q \in \mathbb{R}^n$, if $H(x, p) \leq -f(x)$ and $H(x, q) \geq -\eta$, then

$$H(x, p + \theta(q - p)) \geq -\eta\theta + \psi.$$

The same proof as that of Theorem 4 yields the following proposition. We do not repeat its proof here, and leave it to the reader to check the detail.

Theorem 11 *The assertion of Theorem 4, with $(A9)_\pm$ in place of $(A6)_\pm$, holds.*

In the following, we show that if $H \in C(\mathbb{T}^n \times \mathbb{R}^n)$ satisfies (NR1)–(NR3), then $(A9)_-$ holds.

We choose v_0 to be the function $v_0(x) \equiv 0$. This function v_0 satisfies

$$H(x, Dv_0(x)) = H(x, 0) = -f(x) \quad \text{for all } x \in \mathbb{R}^n,$$

where $f(x) := -H(x, 0)$.

Fix any $x, p, q \in \mathbb{R}^{3n}$, $\eta > 0$ such that $H(x, p) \leq -f(x)$ and $H(x, q) \geq -\eta$. To prove that $(A9)_-$ holds with $f(x) = -H(x, 0)$, it is enough to show that there is a constant $\psi(\eta, \theta) > 0$ such that

$$H(x, p + \theta(q - p)) \geq -\theta\eta + \psi(\eta, \theta).$$

Since

$$H(x, p) \leq -f(x) = H(x, 0) = \min_{r \in \mathbb{R}^n} H(x, r),$$

we have $H(x, p) = -f(x) = H(x, 0)$. Fix any $\theta > 1$. By the convexity of H, we have

$$\begin{aligned} H(x, p + \theta(q - p)) &\geq H(x, p) + \theta(H(x, q) - H(x, p)) \\ &= -f(x) + \theta(-\eta + f(x)) = -\theta\eta + (\theta - 1)f(x), \end{aligned}$$

while we have

$$H(x, p + \theta(q - p)) \geq H(x, 0) = -f(x) = -\theta\eta + (\theta\eta - f(x)).$$

Setting

$$\psi(\eta, \theta) = \min_{x \in \mathbb{T}^n} \max\{(\theta - 1)f(x), \theta\eta - f(x)\},$$

we observe that $\psi(\eta, \theta) > 0$ and

$$H(x, p + \theta(q - p)) \geq -\theta\eta + \psi(\eta, \theta).$$

Thus, H satisfies (A9)$_-$.

References

1. Barles, G.: Remarques sur des résultats d'existence pour les équations de Hamilton–Jacobi du premier ordre. Ann. Inst. H. Poincaré Anal. Non Linéaire **2**(1), 21–32 (1985)
2. Barles, G.: Solutions de viscosité des équations de Hamilton–Jacobi, Math. Appl. (Berlin), 17. Springer, Paris (1994)
3. Barles, G., Ishii, H., Mitake, H.: On the large time behavior of solutions of Hamilton–Jacobi equations associated with nonlinear boundary conditions. Arch. Ration. Mech. Anal. **204**(2), 515–558 (2012)
4. Barles, G., Mitake, H.: A PDE approach to large-time asymptotics for boundary-value problems for nonconvex Hamilton–Jacobi equations. Commun. Partial Differ. Equ. **37**(1), 136–168 (2012)
5. Barles, G., Souganidis, P.E.: On the large time behavior of solutions of Hamilton–Jacobi equations. SIAM J. Math. Anal. **31**(4), 925–939 (2000)
6. Cagnetti, F., Gomes, D.A., Mitake, H., Tran, H.V.: A new method for large time behavior of convex Hamilton–Jacobi equations I: degenerate equations and weakly coupled systems, submitted (arXiv: 1212.4694)
7. Crandall, M.G., Ishii, H., Lions, P.-L.: User's guide to viscosity solutions of second order partial differential equations. Bull. Am. Math. Soc. (N.S.) **27**(1), 1–67 (1992)
8. Davini, A., Siconolfi, A.: A generalized dynamical approach to the large time behavior of solutions of Hamilton–Jacobi equations. SIAM J. Math. Anal. **38**(2), 478–502 (2006)
9. Fathi, A.: Sur la convergence du semi-groupe de Lax-Oleinik. C. R. Acad. Sci. Paris Sér. I Math. **327**(3), 267–270 (1998)
10. Fujita, Y., Ishii, H., Loreti, P.: Asymptotic solutions of Hamilton–Jacobi equations in Euclidean n space. Indiana Univ. Math. J. **55**(5), 1671–1700 (2006)
11. Ichihara, N., Ishii, H.: The large-time behavior of solutions of Hamilton–Jacobi equations on the real line. Methods Appl. Anal. **15**(2), 223–242 (2008)

12. Ichihara, N., Ishii, H.: Long-time behavior of solutions of Hamilton–Jacobi equations with convex and coercive Hamiltonians. Arch. Ration. Mech. Anal. **194**(2), 383–419 (2009)
13. Ishii, H.: A simple, direct proof of uniqueness for solutions of the Hamilton–Jacobi equations of eikonal type. Proc. Am. Math. Soc. **100**(2), 247–251 (1987)
14. Ishii, H.: Asymptotic solutions for large time of Hamilton–Jacobi equations in Euclidean n space. Ann. Inst. H. Poincaré Anal. Non Linéaire **25**(2), 231–266 (2008)
15. Ishii, H.: Long-time asymptotic solutions of convex Hamilton–Jacobi equations with Neumann type boundary conditions. Calc. Var. Partial Differ. Equ. **42**(1–2), 189–209 (2011)
16. Lions, P.-L., Papanicolaou, G., Varadhan, S.R.S.: Homogenization of Hamilton–Jacobi equations. Unpublished work
17. Mitake, H.: Asymptotic solutions of Hamilton–Jacobi equations with state constraints. Appl. Math. Optim. **58**(3), 393–410 (2008)
18. Mitake, H.: The large-time behavior of solutions of the Cauchy–Dirichlet problem for Hamilton–Jacobi equations. NoDEA Nonlinear Differ. Equ. Appl. **15**(3), 347–362 (2008)
19. Namah, G., Roquejoffre, J.-M.: Remarks on the long time behaviour of the solutions of Hamilton–Jacobi equations. Commun. Partial Differ. Equ. **24**(5–6), 883–893 (1999)

2

On s-semipermutable or ss-quasinormal subgroups of finite groups

Ping Kang

Abstract Suppose that G is a finite group and H is a subgroup of G. H is said to be s-semipermutable in G if $HG_p = G_pH$ for any Sylow p-subgroup G_p of G with $(p, |H|) = 1$; H is said to be an ss-quasinormal subgroup of G if there is a subgroup B of G such that $G = HB$ and H permutes with every Sylow subgroup of B. We will study finite groups G saisfying the following: for each noncyclic Sylow subgroup P of G, there exists a subgroup D of P such that $1 < |D| < |P|$ and every subgroup H of P with order $|D|$ is s-semipermutable or ss-quasinormal in G. Some recent results are generalized and unified.

Keywords s-Semipermutable subgroup · ss-Quasinormal subgroup · Saturated formation

Mathematics Subject Classification 20D10 · 20D20

1 Introduction

All groups considered in this paper are finite. G always means a group, $|G|$ is the order of G, $\pi(G)$ denotes the set of all primes dividing $|G|$ and G_p is a Sylow p-subgroup of G for some $p \in \pi(G)$.

Let $\mathscr{F}$ be a class of groups. We call $\mathscr{F}$ a formation, provided that (1) if $G \in \mathscr{F}$ and $H \trianglelefteq G$, then $G/H \in \mathscr{F}$, and (2) if G/M and G/N are in $\mathscr{F}$, then $G/(M \cap N)$

Communicated by S. K. Jain.

P. Kang (✉)
Department of Mathematics, Tianjin Polytechnic University, Tianjin 300387, People's Republic of China
e-mail: kangping2929@163.com

is in $\mathscr{F}$ for any normal subgroups M, N of G. A formation $\mathscr{F}$ is said to be saturated if $G/\Phi(G) \in \mathscr{F}$ implies that $G \in \mathscr{F}$. $\mathscr{U}$ will denote the class of all supersolvable groups. Clearly, $\mathscr{U}$ is a saturated formation.

A subgroup H of G is called s-quasinormal (or s-permutable, π-quasinormal) in G provided H permutes with all Sylow subgroups of G, i.e, $HP=PH$ for any Sylow subgroup P of G. This concept was introduced by Kegel in [5] and has been studied extensively by Deskins [1] and Schmidt [11]. More recently, Zhang and Wang [14] generalized s-quasinormal subgroups to s-semipermutable subgroups. A subgroup H is said to be s-semipermutable in G if $HG_p = G_pH$ for any Sylow p-subgroup G_p of G with $(p, |H|) = 1$. Clearly, every s-quasinormal subgroup of G is an s-semipermutable subgroup of G, but the converse does not hold. Many authors consider minimal or maximal subgroups of a Sylow subgroup of a group when investigating the structure of G, such as in [1] and [4–14], etc. For example, Han in [4] proves the following result.

Theorem 1.1 *Let p be a prime dividing the order of a group G satisfying $(|G|, p-1) = 1$ and P a Sylow p-subgroup of G. Suppose there exists a nontrivial subgroup D such that $1 < |D| < |P|$ and every subgroup H of P with order $|H| = |D|$ or with order $2|D|$ (if P is a nonabelian 2-group and $|P : D| > 2$) is s-semipermutable in G, then G is p-nilpotent.*

As another generalization of the s-quasinormality, Li et al. [7] introduce the following concept: A subgroup H of G is called an ss-quasinormal subgroup of G if there is a subgroup B of G such that $G = HB$ and H permutes with every Sylow subgroup of B. Many authors consider minimal or maximal subgroups of a Sylow subgroup of a group when investigating the structure of G, such as in [6–9], [12], etc. In [13], Wei and Guo provide a result as follows.

Theorem 1.2 *Let $\mathscr{F}$ be a saturated formation containing $\mathscr{U}$ and E be a normal subgroup of a group G such that $G/E \in \mathscr{F}$. Then $G \in \mathscr{F}$ if only if for every non-cyclic Sylow subgroup P of $F^*(E)$, there is a subgroup D of P with $1 < |D| < |P|$ such that every subgroup H of P with order $|D|$ or $2|D|$ whenever $|D| = 2$ is ss-quasinormal in G.*

The aim of this article is to unify and improve above Theorems using s-semipermutable and ss-quasinormal subgroups. Our main theorems are as follows:

Theorem 1.3 (i.e., Theorem 3.5) *Let $\mathscr{F}$ be a saturated formation containing $\mathscr{U}$, the class of all supersolvable groups and G a group with E as a normal subgroup of G such that $G/E \in \mathscr{F}$. Suppose that every non-cyclic Sylow subgroup P of $F^*(E)$ has a subgroup D such that $1 < |D| < |P|$ and every subgroup H of P with order $|H| = |D|$ or with order $2|D|$ (if P is a nonabelian 2-group and $|P : D| > 2$) is either s-semipermutable or ss-quasinormal in G, where $F^*(E)$ is the generalized Fitting subgroup of E. Then $G \in \mathscr{F}$.*

2 Basic definitions and preliminary results

In this section, we collect some known results that are useful later.

Lemma 2.1 *Suppose that H is an s-semipermutable subgroup of G. Then the following assertions hold.*

(i) *If $H \leq K \leq G$, then H is s-semipermutable in K;*
(ii) *Let N be a normal subgroup of G. If H is a p-group for some prime $p \in \pi(G)$, then HN/N is s-semipermutable in G/N;*
(iii) *If $H \leq O_p(G)$, then H is s-permutable in G*

Proof (i) is [14, Property 1], (ii) is [14, Property 2], and (iii) is [14, Lemma 3]. □

Lemma 2.2 *Let H be an ss-quasinormal subgroup of a group G.*

(i) *If $H \leq L \leq G$, then H is ss-quasinormal in L;*
(ii) *If N is normal in G, then HN/N is ss-quasinormal in G/N;*
(iii) *If $H \leq F(G)$, then H is s-quasinormal in G;*
(iv) *If H is a p-subgroup(p a prime), then H permutes with every Sylow q-subgroup of G with $q \neq p$.*

Proof (i) and (ii) are [7, Lemma 2.1], (iii) is [7, Lemma 2.2], and (iv) is [7, Lemma 2.5]. □

Lemma 2.3 ([12]) *Let G be a group and P a Sylow p-subgroup of G, where p is the smallest prime dividing $|G|$. If every maximal subgroup of P is s-semipermutable in G, then G is p-nilpotent.*

Lemma 2.4 ([2] III, 5.2 and IV, 5.4). *Suppose that p is a prime and G is a minimal non-p-nilpotent group, i.e., G is not a p-nilpotent group but whose proper subgroups are all p-nilpotent.*

(i) *G has a normal Sylow p-subgroup P for some prime p and $G = PQ$, where Q is a non-normal cyclic q-subgroup for some prime $q \neq p$.*
(ii) *$P/\Phi(P)$ is a minimal normal subgroup of $G/\Phi(P)$.*
(iii) *The exponent of P is p or 4.*

Lemma 2.5 ([4]) *Let N be an elementary abelian normal p-subgroup of a group G. If there exists a subgroup D in N such that $1 < |D| < |N|$ and every subgroup H of N with $|H| = |D|$ is s-semipermutable in G, then there exists a maximal subgroup M of N such that M is normal in G.*

Lemma 2.6 ([2], VI, 4.10) *Assume that A and B are two subgroups of a group G and $G \neq AB$. If $AB^g = B^gA$ holds for any $g \in G$, then either A or B is contained in a nontrivial normal subgroup of G.*

The generalized Fitting subgroup $F^*(G)$ of G is the unique maximal normal quasinilpotent subgroup of G. Its definition and important properties can be found in [3, X, 13]. We would like to give the following basic facts we will use in our proof.

Lemma 2.7 ([3], X, [13]) *Let G be a group and M a subgroup of G.*

(i) *If M is normal in G, then $F^*(M) \leq F^*(G)$;*

(ii) $F^*(G) \neq 1$ *if* $G \neq 1$*; in fact,* $F^*(G)/F(G) = Soc(F(G)C_G(F(G))/F(G))$*;*
(iii) $F^*(F^*(G)) = F^*(G) \geq F(G)$*; if* $F^*(G)$ *is solvable, then* $F^*(G) = F(G)$.

Lemma 2.8 ([10]) *Let* $\mathscr{F}$ *be a saturated formation containing* $\mathscr{U}$*, the class of all supersolvable groups and* G *a group with* E *as a normal subgroup of* G *such that* $G/E \in \mathscr{F}$*. Suppose that every non-cyclic Sylow subgroup* P *of* $F^*(E)$ *has a subgroup* D *such that* $1 < |D| < |P|$ *and every subgroup* H *of* P *with order* $|H| = |D|$ *or with order* $2|D|$ *(if* P *is a nonabelian 2-group and* $|P : D| > 2$*) is weakly s-permutable in* G*, where* $F^*(E)$ *is the generalized Fitting subgroup of* E*. Then* $G \in \mathscr{F}$.

3 Main results

In this section, we will prove our main results.

Theorem 3.1 *Let* p *be the smallest prime dividing the order of a group* G *and* P *be a Sylow* p*-subgroup of* G*. If every maximal subgroup of* P *is either* s*-semipermutable or* ss*-quasinormal in* G*. Then* G *is* p*-nilpotent.*

Proof Assume that the theorem is not true and let G be a counterexample of minimal order. We derive a contradiction in several steps. □

By Lemmas 2.1 and 2.2, the following two steps are obvious.

Step 1. $O_{p'}(G) = 1$.

Step 2. G has a unique minimal normal subgroup N and G/N is p-nilpotent. Moreover, $\Phi(G) = 1$.

Step 3. $O_p(G) = 1$.

If $O_p(G) \neq 1$, then step 2 yields $N \leq O_p(G)$ and $\Phi(O_p(G)) \leq \Phi(G) = 1$. Therefore, G has a maximal subgroup M such that $G = MN$ and $G/N \cong M$ is p-nilpotent. Since $O_p(G) \cap M$ is normalized by N and M, we conclude that $O_p(G) \cap M$ is normal in G. The uniqueness of N yields $N = O_p(G)$. Clearly, $P = N(P \cap M)$. Furthermore, $P \cap M < P$, and, thus there exists a maximal subgroup P_1 of P such that $P \cap M \leq P_1$. Hence, $P = NP_1$. By hypothesis, P_1 is s-semipermutable or ss-quasinormal in G. In either case, we have that P_1M_q is a group for $q \neq p$. Hence

$$P_1\langle M_p, M_q | q \in \pi(M), q \neq p\rangle = P_1M$$

is a group. Then $P_1M = M$ or G by maximality of M. If $P_1M = G$, then

$$P = P \cap P_1M = P_1(P \cap M) = P_1,$$

a contradiction. If $P_1M = M$, then $P_1 \leq M$. Therefore, $P_1 \cap N = 1$ and N is of prime order. Then the p-nilpotency of G/N implies the p-nilpotency of G, a contradiction. Therefore, $O_p(G) = 1$.

Step 4. The final contradiction.

Assume that P has a maximal subgroup P_1 which is ss-quasinormal in G, then there exists a subgroup B of G such that $G = P_1B$ and P_1 permutes with every Sylow subgroup of B. Noticing that $H \cap B$ is s-quasinormal in B, we have $P_1 \cap B \leq O_p(B)$. Since p is the smallest prime dividing the order of G and $H \cap B$ is a maximal subgroup of a Sylow p-subgroup of B, we see that $B/O_p(B)$ is p-nilpotent and so B is p-solvable. By [2, Chapter VI, Hauptsatz 1.7], we may assume that K is a Hall p'-subgroup of B, $\pi(K) = \{p_2, \ldots, p_s\}$ and $P_i \in \mathrm{Syl}_{p_i}(K)$. Since P_1 permutes with every P_i, $P_1K = KP_1$ and therefore P_1K is normal in G. Now let $\{H_j : j = 1, \ldots, t\}$ be the set of all maximal subgroup of P. Then, by the above, there is K_j such that H_jK_j is normal in G for $j = 1, \ldots, t$, respectively. Let $N = \bigcap\{H_jK_j : j = 1, \ldots, t\}$. Then $N \trianglelefteq G$ and G/N is a p-group. Thus we have $P \cap N = \bigcap_{j=1}^{t}((H_jK_j) \cap P) = \bigcap_{j=1}^{t}((P \cap K_j)H_j) = \bigcap_{j=1}^{t} H_j \leq \Phi(P)$. By Tate's theorem [2, Chapter IV, Satz 4.7], N is p-nilpotent and hence G is p-nilpotent, a contradiction. Now we may assume that all maximal subgroups of P are s-semipermutable in G. Then G is p-nilpotent by Lemma 2.3, a contradiction.

Theorem 3.2 *Let p be the smallest prime dividing the order of a group G and P be a Sylow p-subgroup of G. If P has a subgroup D such that $1 < |D| < |P|$ and every subgroup H of P with order $|H| = |D|$ or with order $2|D|$ (if P is a nonabelian 2-group and $|P : D| > 2$) is either s-semipermutable or ss-quasinormal in G. Then G is p-nilpotent.*

Proof Suppose that the theorem is false and let G be a counterexample of minimal order. We will derive a contradiction in several steps. □

Step 1. $O_{p'}(G) = 1$.

If $O_{p'}(G) \neq 1$, Lemma 2.1 (ii) and Lemma 2.2 (iii) guarantee that $G/O_{p'}(G)$ satisfies the hypotheses of the theorem. Thus $G/O_{p'}(G)$ is p-nilpotent by the choice of G. Then G is p-nilpotent, a contradiction.

Step 2. $|D| > p$.

Suppose that $|D| = p$. Since G is not p-nilpotent, G has a minimal non-p-nilpotent subgroup G_1. By Lemma 2.4 (i), $G_1 = [P_1]Q$, where $P_1 \in \mathrm{Syl}_p(G_1)$ and $Q \in \mathrm{Syl}_q(G_1)$, $p \neq q$. Let $X/\Phi(P_1)$ be a subgroup of $P_1/\Phi(P_1)$ of order p, $x \in X \backslash \Phi(P_1)$ and $L = \langle x \rangle$. Then L is of order p or 4 by Lemma 2.4 (iii). By the hypotheses, L is either s-semipermutable or ss-quasinormal in G, thus in G_1 by Lemmas 2.1 (i) and 2.2 (i). First, suppose that L is ss-quasinormal in G_1, then L is s-quasinormal in G_1 by Lemma 2.2 (iii). Thus $LQ \leq G_1$. Therefore, $LQ = L \times Q$. Then $G_1 = P_1 \times Q$. Hence G_1 is p-nilpotent, contrary to the choice of G_1. Therefore, $L = \langle x \rangle$ is s-semipermutable in G_1 for every element $x \in P_1$. Thus $LQ \leq G_1$. Therefore, $LQ = L \times Q$. Then $G_1 = P_1 \times Q$, a contradiction.

Step 3. $|P : D| > p$.

By Theorem 3.1.

Step 4. P has a subgroup D such that $1 < |D| < |P|$ and every subgroup H of P with order $|H| = |D|$ or with order $2|D|$ (if P is a nonabelian 2-group and $|P : D| > 2$) is s-semipermutable in G. Assume that $H \leq P$ such that $|H| = |D|$ and H is ss-quasinormal in G. Then we may assume G has a normal subgroup M such that $|G : M| = p$ and $G = HM$. Since $|P : D| > p$ by Step 3, M satisfies the hypotheses of the theorem. The choice of G yields that M is p-nilpotent. It is easy to see that G is p-nilpotent, contrary to the choice of G.

Step 5. If $N \leq P$ and N is minimal normal in G, then $|N| \leq |D|$.

Suppose that $|N| > |D|$. Since $N \leq O_p(G)$, N is elementary abelian. By Lemma 2.5, N has a maximal subgroup which is normal in G, contrary to the minimality of N.

Step 6. Suppose that $N \leq P$ and N is minimal normal in G. Then G/N is p-nilpotent.

If $|N| < |D|$, G/N satisfies the hypotheses of the theorem by Lemma 2.1 (ii). Thus G/N is p-nilpotent by the minimal choice of G. So we may suppose that $|N| = |D|$ by Step 5. We will show that every cyclic subgroup of P/N of order p or order 4 (when P/N is a non-abelian 2-group) is s-semipermutable in G/N. Let $K \leq P$ and $|K/N| = p$. By Step 2, N is non-cyclic, so are all subgroups containing N. Hence there is a maximal subgroup $L \neq N$ of K such that $K = NL$. Of course, $|N| = |D| = |L|$. Since L is s-semipermutable in G by the hypotheses, $K/N = LN/N$ is s-semipermutable in G/N by Lemma 2.1 (ii). If $p = 2$ and P/N is non-abelian, take a cyclic subgroup X/N of P/N of order 4. Let K/N be maximal in X/N. Then K is maximal in X and $|K/N|$=2. Since X is non-cyclic and X/N is cyclic, there is a maximal subgroup L of X such that N is not contained in L. Thus $X = LN$ and $|L| = |K| = 2|D|$. By the hypotheses, L is s-semipermutable in G. By Lemma 2.1 (ii), $X/N = LN/N$ is s-semipermutable in G/N. Hence G/N satisfies the hypotheses. By the minimal choice of G, G/N is p-nilpotent.

Step 7. $O_p(G) = 1$.

Suppose that $O_p(G) \neq 1$. Take a minimal normal subgroup N of G contained in $O_p(G)$. By Step 6, G/N is p-nilpotent. It is easy to see that N is the unique minimal normal subgroup of G contained in $O_p(G)$. Furthermore, $O_p(G) \cap \Phi(G) = 1$. Hence $O_p(G)$ is an elementary abelian p-group. On the other hand, G has a maximal subgroup M such that $G = MN$ and $M \cap N = 1$. It is easy to deduce that $O_p(G) \cap M = 1$, $N = O_p(G)$ and $M \cong G/N$ is p-nilpotent. Then G can be written as $G = N(M \cap P)M_{p'}$, where $M_{p'}$ is the normal p-complement of M. Pick a maximal subgroup S of $M_p = P \cap M$. Then $NSM_{p'}$ is a subgroup of G with index p. Since p is the minimal prime in $\pi(G)$, we know that $NSM_{p'}$ is normal in G. Now by Step 3 and the induction, we have $NSM_{p'}$ is p-nilpotent. Therefore, G is p-nilpotent, a contradiction.

Step 8. The minimal normal subgroup L of G is not p-nilpotent.

If L is p-nilpotent, then it follows from the fact that $L_{p'}$ char $L \lhd G$ that $L_{p'} \leq O_{p'}(G) = 1$. Thus L is a p-group. Whence $L \leq O_p(G) = 1$ by Step 7, a contradiction.

Step 9. G is a non-abelian simple group.

Suppose that G is not a simple group. Take a minimal normal subgroup L of G. Then $L < G$. If $|L|_p > |D|$, then L is p-nilpotent by the minimal choice of G, contrary to Step 8. If $|L|_p \leq |D|$. Take $P_* \geq L \cap P$ such that $|P_*| = p|D|$. Hence P_* is a Sylow p-subgroup of P_*L. Since every maximal subgroup of P_* is of order $|D|$, every maximal subgroup of P_* is s-semipermutable in G by hypotheses, thus in P_*L by Lemma 2.1 (i). Now applying Theorem 3.1, we get P_*L is p-nilpotent. Therefore, L is p-nilpotent, contrary to Step 8.

Step 10. The final contradiction.

Suppose that H is a subgroup of P with $|H| = |D|$ and Q is a Sylow q-subgroup with $q \neq p$. Then $HQ^g = Q^gH$ for any $g \in G$ by the hypotheses that H is s-

semipermutable in G. Since G is simple by Step 9, $G = HQ$ from Lemma 2.6, the final contradiction.

The following corollary is immediate from Theorem 3.2.

Corollary 3.3 *Suppose that G is a group. If every non-cyclic Sylow subgroup of G has a subgroup D such that $1 < |D| < |P|$ and every subgroup H of P with order $|H| = |D|$ or with order $2|D|$ (if P is a nonabelian 2-group and $|P : D| > 2$) is either s-semipermutable or ss-quasinormal in G, then G has a Sylow tower of supersolvable type.*

Theorem 3.4 *Let $\mathscr{F}$ be a saturated formation containing $\mathscr{U}$, the class of all supersolvable groups and G a group with E as a normal subgroup of G such that $G/E \in \mathscr{F}$. Suppose that every non-cyclic Sylow subgroup of E has a subgroup D such that $1 < |D| < |P|$ and every subgroup H of P with order $|H| = |D|$ or with order $2|D|$ (if P is a nonabelian 2-group and $|P : D| > 2$) is either s-semipermutable or ss-quasinormal in G. Then $G \in \mathscr{F}$.*

Proof Suppose that P is a non-cyclic Sylow p-subgroup of E, $\forall p \in \pi(E)$. Since P has a subgroup D such that $1 < |D| < |P|$ and every subgroup H of P with order $|H| = |D|$ or with order $2|D|$ (if P is a nonabelian 2-group and $|P : D| > 2$) is either s-semipermutable or ss-quasinormal in G by hypotheses, thus in E by Lemma 2.1 (i) and Lemma 2.2 (i). Applying Corollary 3.3, we conclude that E has a Sylow tower of supersolvable type. Let q be the maximal prime divisor of $|E|$ and $Q \in \mathrm{Syl}_q(E)$. Then $Q \trianglelefteq G$. Since $(G/Q, E/Q)$ satisfies the hypotheses of the theorem, by induction, $G/Q \in \mathscr{F}$. For any subgroup H of Q with $|H| = |D|$, since $Q \leq O_q(G)$, H is s-permutable in G by Lemmas 2.1 (iii) and 2.2 (iii). Since s-permutable implies weakly s-permutable and $F^*(Q) = Q$ by Lemma 2.7, we get $G \in \mathscr{F}$ by applying Lemma 2.8. □

Theorem 3.5 *Let $\mathscr{F}$ be a saturated formation containing $\mathscr{U}$, the class of all supersolvable groups and G a group with E as a normal subgroup of G such that $G/E \in \mathscr{F}$. Suppose that every non-cyclic Sylow subgroup of $F^*(E)$ has a subgroup D such that $1 < |D| < |P|$ and every subgroup H of P with order $|H| = |D|$ or with order $2|D|$ (if P is a nonabelian 2-group and $|P : D| > 2$) is either s-semipermutable or ss-quasinormal in G. Then $G \in \mathscr{F}$.*

Proof We distinguish two cases:

Case 1. $\mathscr{F} = \mathscr{U}$.

Let G be a minimal counter-example.

Step 1. Every proper normal subgroup N of G containing $F^*(E)$ (if it exists) is supersolvable. □

If N is a proper normal subgroup of G containing $F^*(E)$, then $N/N \cap E \cong NE/E$ is supersolvable. By Lemma 2.7 (iii), $F^*(E) = F^*(F^*(E)) \leq F^*(E \cap N) \leq F^*(E)$, so $F^*(E \cap N) = F^*(E)$. For any Sylow subgroup P of $F^*(E \cap N) = F^*(E)$, P has a subgroup D such that $1 < |D| < |P|$ and every subgroup H of P with order $|H| = |D|$ or with order $2|D|$ (if P is a nonabelian 2-group and $|P : D| > 2$) is either s-semipermutable or ss-quasinormal in G by hypotheses, thus in N by Lemma 2.1

(i) and Lemma 2.2 (i). So N and $N \cap H$ satisfy the hypotheses of the theorem, the minimal choice of G implies that N is supersolvable.

Step 2. $E = G$.

If $E < G$, then $E \in \mathscr{U}$ by Step 1. Hence $F^*(E) = F(E)$ by Lemma 2.7. It follows that every Sylow subgroup of $F^*(E)$ is normal in G. By Lemmas 2.1 (iii) and 2.2 (iii), every non-cyclic Sylow subgroup of $F^*(E)$ has a subgroup D such that $1 < |D| < |P|$ and every subgroup H of P with order $|H| = |D|$ or with order $2|D|$ (if P is a nonabelian 2-group and $|P : D| > 2$) is s-permutable in G. Applying Lemma 2.8 for the special case $\mathscr{F} = \mathscr{U}$, $G \in \mathscr{U}$, a contradiction.

Step 3. $F^*(G) = F(G) < G$.

If $F^*(G) = G$, then $G \in \mathscr{F}$ by Theorem 3.4, contrary to the choice of G. So $F^*(G) < G$. By Step 1, $F^*(G) \in \mathscr{U}$ and $F^*(G) = F(G)$ by Lemma 2.7.

Step 4. The final contradiction.

Since $F^*(G) = F(G)$, each non-cyclic Sylow subgroup of $F^*(G)$ has a subgroup D such that $1 < |D| < |P|$ and every subgroup H of P with order $|H| = |D|$ or with order $2|D|$ (if P is a nonabelian 2-group and $|P : D| > 2$) is s-permutable in G by Lemmas 2.1 (iii) and 2.2 (iii). Applying Lemma 2.8, $G \in \mathscr{U}$, a contradiction.

Case 2. $\mathscr{F} \neq \mathscr{U}$.

By hypotheses, every non-cyclic Sylow subgroup of $F^*(E)$ has a subgroup D such that $1 < |D| < |P|$ and every subgroup H of P with order $|H| = |D|$ or with order $2|D|$ (if P is a nonabelian 2-group and $|P : D| > 2$) is either s-semipermutable or ss-quasinormal in G, thus in E Lemma 2.1 (i) and Lemma 2.2 (i). Applying Case 1, $E \in \mathscr{U}$. Then $F^*(E) = F(E)$ by Lemma 2.7. It follows that each Sylow subgroup of $F^*(E)$ is normal in G. By Lemmas 2.1 (iii) and 2.2 (iii), each non-cyclic Sylow subgroup of $F^*(E)$ has a subgroup D such that $1 < |D| < |P|$ and every subgroup H of P with order $|H| = |D|$ or with order $2|D|$ (if P is a nonabelian 2-group and $|P : D| > 2$) is s-permutable in G. Applying Lemma 2.8, $G \in \mathscr{F}$. These complete the proof of the theorem.

The following corollaries are immediate from Theorem 3.5.

Corollary 3.6 *Let $\mathscr{F}$ be a saturated formation containing $\mathscr{U}$. Suppose that G is a group with a normal subgroup E such that $G/E \in \mathscr{F}$. Then $G \in \mathscr{F}$ if and only if every maximal subgroup of any Sylow subgroup of $F^*(E)$ is s-semipermutable in G.*

Corollary 3.7 *Let $\mathscr{F}$ be a saturated formation containing $\mathscr{U}$. Suppose that G is a group with a normal subgroup E such that $G/E \in \mathscr{F}$. Then $G \in \mathscr{F}$ if and only if every cyclic subgroup of any Sylow subgroup of $F^*(E)$ of prime order or order 4 is s-semipermutable in G.*

Corollary 3.8 ([6], Theorem 3.3) *Let $\mathscr{F}$ be a saturated formation containing $\mathscr{U}$. Suppose that G is a group with a normal subgroup H such that $G/H \in \mathscr{F}$. Then $G \in \mathscr{F}$ if and only if every maximal subgroup of any Sylow subgroup of $F^*(H)$ is ss-quasinormal in G.*

Corollary 3.9 ([6], Theorem 3.7) Let $\mathscr{F}$ be a saturated formation containing $\mathscr{U}$. Suppose that G is a group with a normal subgroup H such that $G/H \in \mathscr{F}$. Then $G \in \mathscr{F}$ if and only if every cyclic subgroup of any Sylow subgroup of $F^*(H)$ of prime order or order 4 is ss-quasinormal in G.

Corollary 3.10 ([8], Theorem 3.4) Let $\mathscr{F}$ be a saturated formation containing $\mathscr{U}$. Suppose that G is a group with a normal subgroup E such that $G/E \in \mathscr{F}$. Then $G \in \mathscr{F}$ if and only if every maximal subgroup of any Sylow subgroup of $F^*(E)$ is s-quasinormal in G.

Corollary 3.11 ([9], Theorem 3.3) Let $\mathscr{F}$ be a saturated formation containing $\mathscr{U}$. Suppose that G is a group with a normal subgroup E such that $G/E \in \mathscr{F}$. Then $G \in \mathscr{F}$ if and only if every cyclic subgroup of any Sylow subgroup of $F^*(E)$ of prime order or order 4 is s-quasinormal in G.

Acknowledgments The author is very grateful to the referee who provides a lot of valuable suggestions and useful comments. The paper is dedicated to Professor Geoffrey Robinson for his 60th birthday.

References

1. Deskins, W.E.: On quasinormal subgroups of finite groups. Math. Z. **82**, 125–132 (1963)
2. Huppert, B.: Endliche Gruppen I. Springer, Berlin, Heidelberg, New York (1967)
3. Huppert, B., Blackburn, N.: Finite Groups III. Springer, Berlin, New York (1982)
4. Han, Z.: On s-semipermutable subgroups of finite groups and p-nilpotency. Proc. Indian Acad. Sci. (Math. Sci.) **120**, 141–148 (2010)
5. Kegel, O.H.: Sylow Gruppen und subnormalteiler endlicher Gruppen. Math. Z. **78**, 205–221 (1962)
6. Li, S., Shen, Z., Kong, X.: On ss-quasinormal subgroups of finite subgroups. Commun. Algebra **36**, 4436–4447 (2008)
7. Li, S., Shen, Z., Liu, J.: The influence of ss-quasinormality of some subgroups on the structure of finite groups. J. Algebra **319**, 4275–4287 (2008)
8. Li, Y., Wei, H., Wang, Y.: The influence of π-quasinormality of some subgroups of a finite group. Arch. Math. **81**, 245–252 (2003)
9. Li, Y., Wang, Y.: The influence of minimal subgroups on the structure of a finite group. Proc. Am. Math. Soc. **131**, 337–341 (2002)
10. Skiba, A.N.: On weakly s-permutable subgroups of finite groups. J. Algebra **315**, 192–209 (2007)
11. Schmid, P.: Subgroups permutable with all Sylow subgroups. J. Algebra **207**, 285–293 (1998)
12. Wang, L., Wang, Y.: On s-semipermutable maximal and minimal subgroups of Sylow p-groups of finite groups. Commun. Algebra **34**, 143–149 (2006)
13. Wei, X., Guo, X.: On ss-quasinormal subgroups and the structure of finite groups. Sci. China Math. **54**(3), 449–456 (2011)
14. Zhang, Q., Wang, L.: The influence of s-semipermutable subgroups on the structure of a finite group. Acta Math. Sinica **48**, 81–88 (2005)

3

The Auslander bijections: how morphisms are determined by modules

Claus Michael Ringel

Abstract Let Λ be an artin algebra. In his seminal Philadelphia Notes published in 1978, Auslander introduced the concept of morphisms being determined by modules. Auslander was very passionate about these investigations (they also form part of the final chapter of the Auslander–Reiten–Smalø book and could and should be seen as its culmination). The theory presented by Auslander has to be considered as an exciting frame for working with the category of Λ-modules, incorporating all what is known about irreducible maps (the usual Auslander–Reiten theory), but the frame is much wider and allows for example to take into account families of modules—an important feature of module categories. What Auslander has achieved is a clear description of the poset structure of the category of Λ-modules as well as a blueprint for interrelating individual modules and families of modules. Auslander has subsumed his considerations under the heading of "morphisms being determined by modules". Unfortunately, the wording in itself seems to be somewhat misleading, and the basic definition may look quite technical and unattractive, at least at first sight. This could be the reason that for over 30 years, Auslander's powerful results did not gain the attention they deserve. The aim of this survey is to outline the general setting for Auslander's ideas and to show the wealth of these ideas by exhibiting many examples.

Communicated by S.K. Jain.

C. M. Ringel (✉)
Department of Mathematics, Shanghai Jiao Tong University,
Shanghai 200240, People's Republic of China
e-mail: ringel@math.uni-bielefeld.de

C. M. Ringel
King Abdulaziz University, P O Box 80200, Jeddah, Saudi Arabia

Keywords Auslander bijections · Auslander–Reiten theory · Right factorization lattice · Morphisms determined by modules · Finite length categories: global directedness · Local symmetries · Representation type · Brauer–Thrall conjectures · Riedtmann–Zwara degenerations · Hammocks · Kronecker quiver · Quiver Grassmannians · Auslander varieties · Modular lattices · Meet semi-lattices

Mathematics Subject Classification (2010) Primary 16G70 · 18E10 · 18A25 · 18A32 · 16G60; Secondary 16G20 · 18A20 · 06C05 · 14M15 · 19A49 · 03C60

1 Introduction

There are two basic mathematical structures: groups and lattices, or, more generally, semigroups and posets. A first glance at any category should focus the attention on these two structures: to symmetry groups (for example the automorphism groups of the individual objects), as well as to the posets given by suitable sets of morphisms, for example by looking at inclusion maps (thus dealing with the poset of all subobjects of an object), or at the possible factorizations of morphisms. In this way, one distinguishes between local symmetries and global directedness.

The present survey deals with the category mod Λ of finite length modules over an artin algebra Λ. Its aim is to report on the work of Auslander in his seminal Philadelphia Notes published in 1978. Auslander was very passionate about these investigations and they also form part of the final chapter of the Auslander–Reiten–Smalø book: there, they could (and should) be seen as a kind of culmination. It seems to be surprising that the feedback until now is quite meager. After all, the theory presented by Auslander has to be considered as an exciting frame for working with the category mod Λ, incorporating what is called the Auslander–Reiten theory (to deal with the irreducible maps), but this frame is much wider and allows for example to take into account families of modules—an important feature of a module category. Indeed, many of the concepts which are relevant when considering the categories mod Λ fit into the frame! What Auslander has achieved (but he himself may not have realized it) was a clear description of the poset structure of mod Λ and of the interplay between families of modules.

Auslander's considerations are subsumed under the heading of morphisms being determined by modules, but the wording in itself seems to be somewhat misleading, and the basic definition looks quite technical and unattractive, at least at first sight. This could be the reason that for over 30 years, Auslander's powerful results did not gain the attention they deserve.

Here is a short summary: Let Λ be an artin algebra. The modules which we consider will be left Λ-modules of finite length, and maps (or morphisms) will be Λ-module homomorphisms, unless otherwise specified. Auslander asks for a description of the class of maps ending in a fixed module Y. Two maps $f : X \to Y$ and $f' : X' \to Y$ are said to be *right equivalent* provided there are maps $h : X \to X'$ and $h' : X' \to X$ such that $f = f'h$ and $f' = fh'$. The right equivalence class of f will be denoted by $[f\rangle$. The object studied by Auslander is the set of right equivalence classes of maps ending in Y, we denote this set by

$$[\to Y\rangle.$$

It is a poset via the relation $\leq$ which is defined as follows:

$$[f : X \to Y\rangle \;\leq\; [f' : X' \to Y\rangle$$

provided there is a homomorphism $h : X \to X'$ with $f = f'h$, thus provided the following diagram commutes:

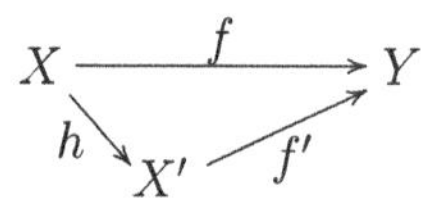

It is easy to see that the poset $[\to Y\rangle$ is a lattice, thus we call it the *right factorization lattice* for Y.

Looking at maps $f : X \to Y$, we may (and often will) assume that f is *right minimal*, thus that there is no non-zero direct summand X' of X with $f(X') = 0$. Note that any right equivalence class contains a right minimal map, and if $f : X \to Y$ and $f' : X' \to Y$ are right minimal maps, then any $h : X \to X'$ with $f = f'h$ has to be an isomorphism.

Of course, to analyze the poset $[\to Y\rangle$ is strongly related to a study of the contravariant Hom-functor $\mathrm{Hom}(-, Y)$, however the different nature of these two mathematical structures should be stressed: $\mathrm{Hom}(-, Y)$ is an additive functor whereas $[\to Y\rangle$ is a poset, and it is the collection of these posets $[\to Y\rangle$ which demonstrates the global directedness.

In general, the right factorization lattice $[\to Y\rangle$ is very large and does not satisfy any chain condition. The main idea of Auslander is to write $[\to Y\rangle$ as the filtered union of the subsets ${}^C[\to Y\rangle$, where ${}^C[\to Y\rangle$ is given by those maps f which are "right C-determined". These posets are again lattices and they are of finite height, we call ${}^C[\to Y\rangle$ the *right C-factorization lattice* for Y. Since the concept of "right determination" looks (at least at first sight) technical and unattractive, let us first describe the set ${}^C[\to Y\rangle$ only in the important case when C is a generator: in this case, ${}^C[\to Y\rangle$ consists of the (right equivalence classes of the right minimal) maps f ending in Y with kernel in $\mathrm{add}\,\tau C$ (we denote by $\tau = D\,\mathrm{Tr}$ and $\tau^- = \mathrm{Tr}\,D$ the Auslander–Reiten translations). Here is Auslander's first main assertion:

$$[\to Y\rangle = \bigcup\nolimits_C {}^C[\to Y\rangle, \tag{1}$$

where C runs through all isomorphism classes of Λ-modules; or, in the formulation of Auslander: any map in mod Λ is right determined by some module C. Note that the inclusion of the lattice ${}^C[\to Y\rangle$ into the lattice $[\to Y\rangle$ preserves meets, but usually not joins.

Auslander's second main assertion describes the right C-factorization lattice ${}^C[\to Y\rangle$ as follows: There is a lattice isomorphism

$$\eta_{CY} : {}^C[\to Y\rangle \longrightarrow \mathcal{S}\,\mathrm{Hom}(C, Y), \tag{2}$$

where $\mathrm{Hom}(C, Y)$ is considered as an $\mathrm{End}(C)^{\mathrm{op}}$-module, and where $\mathcal{S}M$ denotes the submodule lattice of a module M. Actually, the map η_{CY} is easy to describe, namely $\eta_{CY}(f) = \mathrm{Im}\,\mathrm{Hom}(C, f)$ for f a morphism ending in Y. The essential assertion is the surjectivity of η_{CY}, thus to say that any submodule of $\mathrm{Hom}(C, Y)$ is of the form $\mathrm{Im}\,\mathrm{Hom}(C, f)$ for some f.

What is the relevance? As we have mentioned, usually the lattice $[\to Y\rangle$ itself will not satisfy any chain conditions, but all the lattices ${}^C[\to Y\rangle$ are of finite height and often can be displayed very nicely: according to (2) we deal with the submodule lattice $\mathcal{S}M$ of some finite length module M over an artin algebra (namely over $\Gamma(C) = \mathrm{End}(C)^{\mathrm{op}}$) and it is easy to see that any submodule lattice arises in this way. Using the Auslander bijections η_{CY}, one may transfer properties of submodule lattices to the right C-factorization lattices ${}^C[\to Y\rangle$, this will be one of the aims of this paper. Given a submodule U of $\mathrm{Hom}(C, Y)$, let f be a right C-determined map ending in Y such that $\eta_{CY}(f) = U$. The composition series of the factor module $\mathrm{Hom}(C, Y)/U$ correspond to certain factorizations of f (to the "maximal C-factorizations"), and we may define the C-type of f so that it is equal to the dimension vector of the module $\mathrm{Hom}(C, Y)/U$ (recall that the dimension vector of a module M has as coefficients the Jordan–Hölder multiplicities of the various simple modules occurring in M).

Submodule lattices have interesting combinatorial features, and it seems to interesting that Auslander himself looked mainly at combinatorial properties (for example at waists in submodule lattices). But we should stress that we really are in the realm of algebraic geometry. Thus, let us assume for a moment that Λ is a k-algebra where k is an algebraically closed field. If M is a finite-dimensional Λ-module, the set $\mathcal{S}M$ of all submodules of M is the disjoint union of the sets $\mathbb{G}_{\mathbf{e}}(M)$ consisting of all submodules of M with fixed dimension vector $\mathbf{e}$. It is well-known that $\mathbb{G}_{\mathbf{e}}(M)$ is in a natural way a projective variety, called nowadays a *quiver Grassmannian*. Given Λ-modules C and Y, the Auslander bijections draw the attention on the $\mathrm{End}(C)^{\mathrm{op}}$-module $M = \mathrm{Hom}(C, Y)$, let $\mathbf{d}$ be its dimension vector and let $\mathbf{e}, \mathbf{e}'$ be dimension vectors with $\mathbf{e} + \mathbf{e}' = \mathbf{d}$. The quiver Grassmannians $\mathbb{G}_{\mathbf{e}'}(\mathrm{Hom}(C, Y))$ corresponds under the Auslander bijection η_{CY} to the set ${}^C[\to Y\rangle^{\mathbf{e}}$ of all right equivalence classes of right C-determined maps which end in Y and have type $\mathbf{e}$. We call ${}^C[\to Y\rangle^{\mathbf{e}}$ an *Auslander variety.* These Auslander varieties have to be considered as an important tool for studying the right equivalence classes of maps ending in a given module.

We end this summary by an outline in which way the Auslander bijections (2) incorporate the existence of minimal right almost split maps: we have to look at the special case where Y is indecomposable and $C = Y$ and to deal with the submodule $\mathrm{rad}(Y, Y)$ of $\mathrm{Hom}(Y, Y)$. The bijection (2) yields an element $f : X \to Y$ in ${}^Y[\to Y\rangle$ such that $\eta_{YY}(f) = \mathrm{rad}(Y, Y)$; to say that f is right Y-determined means that f is right almost split.

The survey is divided into three parts. Part I presents the general setting, it comprises the Sects. 2 to 10. The Sects. 11 to 15 form Part II, here we show in which way the

Auslander bijections deal with families of modules. Finally, in Parts III, we discuss some special cases; these are the Sects. 16 to 18.

I. The setting

2 The right factorization lattice $[\to Y\rangle$

Let Y be a Λ-module. Let $\bigsqcup_X \mathrm{Hom}(X, Y)$ be the class of all homomorphisms $f : X \to Y$ with arbitrary modules X (such homomorphisms will be said to be the homomorphisms *ending in* Y). We define a preorder $\preceq$ on this class as follows: Given $f : X \to Y$ and $f' : X' \to Y$, we write $f \preceq f'$ provided there is a homomorphism $h : X \to X'$ such that $f = f'h$ (clearly, this relation is reflexive and transitive). As usual, such a preorder defines an equivalence relation (in our setting, we call it *right equivalence*) by saying that f, f' are *right equivalent* provided we have both $f \preceq f'$ and $f' \preceq f$, and it induces a poset relation $\leq$ on the set $[\to Y\rangle$ of right equivalence classes of homomorphisms ending in Y. Given a morphism $f : X \to Y$, we denote its right equivalence class by $[f\rangle$ and by definition $[f\rangle \leq [f'\rangle$ if and only if $f \preceq f'$. As we will see in Proposition 2.2, the poset $[\to Y\rangle$ is a lattice, thus we will call it the *right factorization lattice* for Y.

It should be stressed that $[\to Y\rangle$ is a set, not only a class: namely, the isomorphism classes of Λ-modules form a set and for every module X, the homomorphisms $X \to Y$ form a set; we may choose a representative from each isomorphism class of Λ-modules and given a homomorphism $f : X \to Y$, then there is an isomorphism $h : X' \to X$ where X' is such a representative, and f is right equivalent to fh.

Recall that a map $f : X \to Y$ is said to be *right minimal* provided any direct summand X' of X with $f(X') = 0$ is equal to zero. If $f : X \to Y$ is a morphism and $X = X' \oplus X''$ such that $f(X'') = 0$ and $f|X' : X' \to Y$ is right minimal, then $f|X'$ is called a *right minimalisation of f.* The kernel of a right minimalisation of f will be called the *intrinsic kernel off*, it is unique up to isomorphism.

Proposition 2.1 *Every right equivalence class $[f\rangle$ in $[\to Y\rangle$ contains a right minimal morphism, namely $[f'\rangle$, where f' is a right minimalisation of f. Given right minimal morphisms $f : X \to Y$ and $f' : X' \to Y$, then f, f' are right equivalent if and only if there is an isomorphism $h : X \to X'$ such that $f = f'h$.*

Proof Let $f : X \to Y$ be a homomorphism ending in Y. Write $X = X_1 \oplus X_2$ such that $f(X_2) = 0$ and $f|X_1 : X_1 \to Y$ is right minimal. Let $u : X_1 \to X_1 \oplus X_2$ be the canonical inclusion, $p : X_1 \oplus X_2 \to X_1$ the canonical projection. Then $pu = 1_{X_1}$ and $f = fup$ (since $f(X_2) = 0$). We see that $fu \preceq f$ and $f = fup \preceq fu$, thus f and fu are right equivalent and $fu = f|X_1$ is right minimal. If the right minimal morphisms $f : X \to Y$ and $f' : X' \to Y$ are right equivalent, then there are morphisms $h : X \to X'$ and $h' : X' \to Y$ such that $f = f'h$ and $f' = fh'$. But $f = fh'h$ implies that $h'h$ is an automorphism, and $f' = f'hh'$ implies that hh' is an automorphism, thus h, h' have to be isomorphisms (see [5] I.2). □

Remark Monomorphisms $X \to Y$ are always right minimal, and the right equivalence classes of monomorphisms ending in Y may be identified with the submodules of Y

(here, we identify the right equivalence class of the monomorphism $f : X \to Y$ with the image of X).

Proposition 2.2 *The poset $[\to Y\rangle$ is a lattice with zero and one. Given $f_1 : X_1 \to Y$ and $f_2 : X_2 \to Y$, say with pullback $g_1 : X \to X_1, g_2 : X \to X_2$, the meet of $[f_1\rangle$ and $[f_2\rangle$ is given by the map $f_1 g_1 : X \to Y$, the join of $[f_1\rangle$ and $[f_2\rangle$ is given by $[f_1, f_2] : X_1 \oplus X_2 \to Y$.*

Proof (a trivial verification) Write $f = f_1 g_1 = f_2 g_2$. We have $f = f_1 g_1 \preceq f_1$ and $f = f_2 g_2 \preceq f_2$, thus $[f\rangle \leq [f_1\rangle$ and $[f\rangle \leq [f_2\rangle$. If $f' : X' \to Y$ is a morphism with $[f'\rangle \leq [f_1\rangle$ and $[f'\rangle \leq [f_2\rangle$, then $f' \preceq f_1$ and $f' \preceq f_2$, thus there are morphisms ϕ_i with $f' = f_i \phi_i$, for $i = 1, 2$ Since $f_1 \phi_1 = f_2 \phi_2$, the pullback property yields a morphism $\phi : X' \to X$ such that $\phi_i = g_i \phi$ for $i = 1, 2$. Thus $f' = f_1 \phi_1 = f_1 g_1 \phi = f \phi$ shows that $f' \preceq f$, thus $[f'\rangle \leq [f\rangle$. This shows that $[f\rangle$ is the meet of $[f_1\rangle$ and $[f_2\rangle$.

Second, denote the canonical inclusion maps $X_i \to X_1 \oplus X_2$ by u_i, for $i = 1, 2$, thus $[f_1, f_2]u_i = f_i$ and therefore $[f_i\rangle \leq [[f_1, f_2]\rangle$ for $i = 1, 2$. Assume that there is given a morphism $g : X'' \to Y$ with $[f_i\rangle \leq [g\rangle$ for $i = 1, 2$. This means that there are morphisms $\psi_i : X_i \to X''$ such that $f_i = g\psi_i$ for $i = 1, 2$. Let $\psi = [\psi_1, \psi_2] : X_1 \oplus X_2 \to X''$ (with $\psi u_i = \psi_i$). Then $[f_1, f_2] = g[\psi_1, \psi_2] = g\psi$ shows that $[f_1, f_2] \preceq g$, thus $[[f_1, f_2]\rangle \leq [g\rangle$. This shows that $[[f_1, f_2]\rangle$ is the join of $[f_1\rangle$ and $[f_2\rangle$.

It is easy to check that the map $0 \to Y$ is the zero element of $[\to Y\rangle$ and that the identity map $Y \to Y$ is its unit element. □

It should be stressed that *if $f_1 : X_1 \to Y$ and $f_2 : X_2 \to Y$ are right minimal, say with pullback $g_1 : X \to X_1, g_2 : X \to X_2$, then neither the map $f_1 g_1$ nor the direct sum map $[f_1, f_2] : X_1 \oplus X_2 \to Y$ will be right minimal, in general.* Thus if one wants to work with right minimal maps, one has to right minimalise the maps in question. Here are corresponding examples:

Example 1 Let Λ be the path algebra of the quiver

$$a \longleftarrow b$$

of type $\mathbb{A}_2$.

All path algebras of quivers considered in the paper will have coefficients in an arbitrary field k, unless we specify some further conditions. When dealing with the path algebra of a quiver Δ, and x is a vertex of Δ, we denote by $S(x)$ (or also just by x) the simple module corresponding to x, by $P(x)$ and $Q(x)$ the projective cover or injective envelope of $S(x)$, respectively.

Take as maps f_1, f_2 the canonical projection $f_1 = f_2 : P(b) \to S(b)$, this is a right minimal map. The pullback U of f_1 and f_2 is a submodule of $P(b) \oplus P(b)$ which is isomorphic to $S(a) \oplus P(b)$. Since any map $S(a) \to S(b)$ is zero, there is no right minimal map $U \to S(b)$.

Also, the map $[f_1, f_2] : P(b) \oplus P(b) \to S(b)$ is not right minimal, since we have $\dim \operatorname{Hom}(P(b), S(b)) = 1$.

As we have seen, the poset $[\to Y\rangle$ is a lattice. What will be important in the following discussion is the fact that we deal with a meet-semilattice (these are the posets such that any pair of elements has a meet). Note that all the semilattices which we deal with turn out to be lattices, however the poset maps to be considered will preserve meets, but usually not joins, thus we really work in the category of meet-semilattices.

Proposition 2.3 *The lattice $[\to Y\rangle$ is modular.*

Proof Let $f_i : X_i \to Y$ be maps with target Y, where $1 \le i \le 3$, such that $f_1 \preceq f_3$. We want to show that

$$([f_1\rangle \vee [f_2\rangle) \wedge [f_3\rangle \;\le\; [f_1\rangle \vee ([f_2\rangle) \wedge [f_3\rangle),$$

the reverse inequality being trivial. Since $f_1 \preceq f_3$, there is $h : X_1 \to X_3$ such that $f_1 = f_3 h$.

First, let us construct an element f in $[f_1\rangle \vee ([f_2\rangle) \wedge [f_3\rangle)$. A pullback diagram

$$\begin{array}{ccc} P & \xrightarrow{p_2} & X_2 \\ {\scriptstyle p_3}\downarrow & & \downarrow{\scriptstyle f_2} \\ X_3 & \xrightarrow{f_3} & Y \end{array}$$

yields a map $f_3 p_3$ such that $[f_3 p_3\rangle = [f_2\rangle) \wedge [f_3\rangle$, thus the map $f = [f_3 h, f_3 p_3] : X_1 \oplus P \to Y$ belongs to $[f_3 h\rangle \vee ([f_2\rangle) \wedge [f_3\rangle)$.

Next, we construct an element f' in $([f_1\rangle \vee [f_2\rangle) \wedge [f_3\rangle$. The map $[f_3 h, f_2] : X_1 \oplus X_2 \to Y$ belongs to $[f_1\rangle \vee [f_2\rangle$, now we form the pullback

$$\begin{array}{ccc} P' & \xrightarrow{\begin{bmatrix} g_1 \\ g_2 \end{bmatrix}} & X_1 \oplus X_2 \\ {\scriptstyle g_3}\downarrow & & \downarrow{\scriptstyle \begin{bmatrix} f_3 h & f_2 \end{bmatrix}} \\ X_3 & \xrightarrow{f_3} & Y \end{array}$$

the map $f' = f_3 g_3 = [f_3 h, f_2]\begin{bmatrix} g_1 \\ g_2 \end{bmatrix}$ is an element of $([f_3 h\rangle \vee [f_2\rangle) \wedge [f_3\rangle$.

It follows from $f_3 g_3 = [f_3 h, f_2]\begin{bmatrix} g_1 \\ g_2 \end{bmatrix} = f_3 h g_1 + f_2 g_2$ that $f_2 g_2 = f_3(g_3 - h g_1)$, thus the pair of maps $g_2, g_3 - h g_1$ factors through the pullback P of f_2, f_3. Thus there is $g : P' \to P$ such that $p_2 g = g_2$ and $p_3 g = g_3 - h g_1$. It follows that $g_3 = h g_1 + p_3 g$ and therefore

$$f = f_3 g_3 = f_3 h g_1 + f_3 p_3 g = [f_3 h, f_3 p_3]\begin{bmatrix} g_1 \\ g \end{bmatrix} = f'\begin{bmatrix} g_1 \\ g \end{bmatrix}.$$

This shows that $f \preceq f'$. □

Example 2 **Failure of the chain conditions.** Here are examples which show that in general $[\to Y\rangle$ *neither satisfies the ascending nor the descending chain condition. Let*

Λ be the Kronecker algebra, this is the path algebra of the quiver

$$a \leftleftarrows b$$

The Λ-modules are also called *Kronecker modules* (basic facts concerning the Kronecker modules will be recalled in Sect. 14). Let $Y = S(b)$, the simple injective module.

We denote by Q_n the indecomposable preinjective module of length $2n+1$ (thus $Q_0 = S(b) = Q(b),\ Q_1 = Q(a)$). There is a chain of epimorphisms

$$\cdots \longrightarrow Q_2 \xrightarrow{f_2} Q_1 \xrightarrow{f_1} Q_0 = Y,$$

thus we have the descending chain

$$\cdots [f_1 f_2 f_3\rangle < [f_1 f_2\rangle < [f_1\rangle < [1_Y\rangle$$

in $[\to Y\rangle$.

Here, we can assume that all the kernels $f_n : Q_n \to Q_{n-1}$ are equal to R, where R is a fixed indecomposable module of length 2. Also, if the ground field k is infinite, then there is such a chain of epimorphisms such that all the kernels are pairwise different and of length 2. In the first case, the kernels of the maps $f_1 f_2 \cdots f_n$ are all indecomposable (namely of the form $R[n]$ for $n \in \mathbb{N}$), in the second, they are direct sums of pairwise non-isomorphic modules of length 2.

In order to look at the ascending chain condition, let P_n be indecomposable preprojective of length $2n+1$. For $i \geq 1$, there are epimorphisms $f_i : P_i \to Y$ and monomorphisms $u_i : P_i \to P_{i+1}$ such that $f_{i+1} u_i = f_i$. Thus, there is a commutative diagram of maps

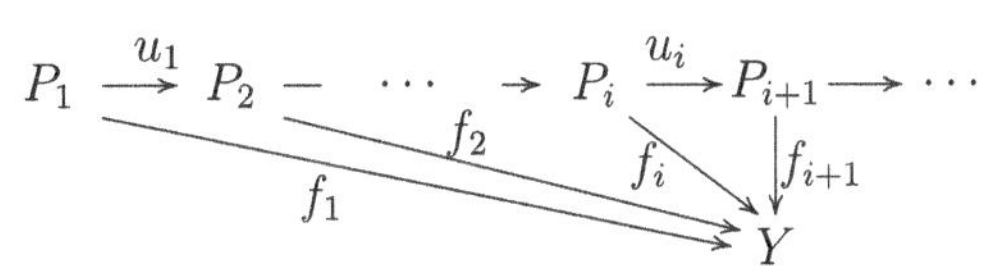

and we obtain an ascending chain in $[\to Y\rangle$.

$$[f_1\rangle < [f_2\rangle < [f_3\rangle < \cdots$$

Thus, looking at the sequence of maps

$$P_1 \xrightarrow{u_1} P_2 \xrightarrow{u_2} \cdots \longrightarrow Y$$

we obtain an ascending chain in $[\to Y\rangle$.

$$[f_1\rangle < [f_2\rangle < [f_3\rangle < \cdots.$$

There are similar chains in $[\to Y\rangle$ consisting of elements $[f : X \to Y\rangle$ with X regular Kronecker modules.

Remark If we fix Y (as we usually do), we may consider the class of right minimal morphisms $f : X \to Y$ as the objects of a category, with maps from $(f : X \to Y)$ to $(f' : X' \to Y)$ being given by the maps $h : X \to X'$ such that $f = f'h$. According to Proposition 2.1, this category is a groupoid (this means that all morphisms are isomorphisms), and any connected component of the category is just the class of all the right minimal maps which belong to a right equivalence class. If we work with a skeleton of the category mod Λ, then we only have to consider the sets

$$\text{r-Aut}(f) = \{h \in \text{End}(X) \mid fh = h\},$$

this is a subgroup of $\text{Aut}(X)$; we may call it the *right automorphism group* of f. The classification problem for the right minimal maps ending in Y is divided in this way into two problems: to determine, on the one hand, the structure of the right factorization lattice $[\to Y\rangle$ and, on the other hand, to determine r-Aut(f) for every right minimal map f ending in Y. This provides a nice separation of the local symmetries and the global directedness, as mentioned at the beginning of the paper.

3 Morphisms determined by modules: Auslander's First Theorem

Here is the decisive definition. Let $f : X \to Y$ be a morphism and C a module. Then f is said to be *right C-determined* (or right determined by C) provided the following condition is satisfied: given any morphism $f' : X' \to Y$ such that $f'\phi$ factors through f for all $\phi : C \to X'$, then f' itself factors through f. Thus one deals with the following diagrams:

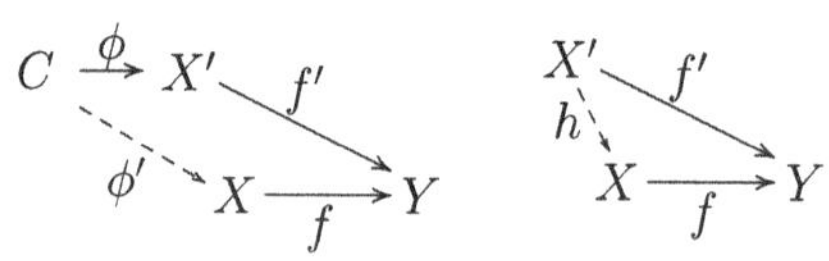

The existence of the dashed arrow ϕ' on the left for all possible maps $\phi : C \to X'$ shall imply the existence of the dashed arrow h on the right (of course, the converse implication always holds true: if $f' = fh$ for some morphism h, then $f'\phi = f(h\phi)$ for all morphisms $\phi : C \to X'$).

Proposition 3.1 *Let* : $X \to Y$ *be a morphism, let* C, C' *be modules.*

(a) *Assume that* $\text{add}\, C = \text{add}\, C'$. *Then* f *is right* C*-determined if and only if* f *is right* C'*-determined.*
(b) *If* f *is right* C*-determined, then* f *is also right* $(C \oplus C')$*-determined.*

Proof Trivial verification. □

We denote by ${}^{C}[\to Y\rangle$ the set of the right equivalence classes of the morphisms ending in Y which are right C-determined. We will see below that also ${}^{C}[\to Y\rangle$ is a lattice, thus we call it the *right C-factorization lattice* for Y.

Note that ${}^{C}[\to Y\rangle$ is usually not closed under predecessors or successors inside $[\to Y\rangle$. But there is the following important property:

Proposition 3.2 *The subset* ${}^C[\to Y\rangle$ *of* $[\to Y\rangle$ *is closed under meets.*

Proof Let $f_1 : X_1 \to Y$ and $f_2 : X_2 \to Y$ be right C-determined. As we know, the meet of $[f_1\rangle$ and $[f_2\rangle$ is given by forming the pullback of f_1 and f_2. Thus assume that X is the pullback with maps $g_1 : X \to X_1$ and $g_2 : X \to X_2$ and let $f = f_1g_1 = f_2g_2$. We want to show that f is right C-determined. Thus, assume that there is given $f' : X' \to Y$ such that for any $\phi : C \to X'$, there exists $\phi' : C \to X$ such that $f'\phi = f\phi'$. Then we see that for any $\phi : C \to X'$, we have $f'\phi = f\phi = f_1(g_1\phi)$, thus $f'\phi$ factors through f_1. Since f_1 is right C-determined, it follows that f' factors through f_1, say $f' = f_1h_1$ for some $h_1 : X' \to X_1$. Similarly, for any $\phi : C \to X'$, the morphism $f'\phi$ factors through f_2 and therefore $f' = f_2h_2$ for some $h_2 : X' \to X_2$. Now $f_1h_1 = f' = f_2h_2$ implies that there is $h : X' \to X$ such that $g_1h = h_1$, and $g_2h = h_2$. Thus $f' = f_1h_1 = f_1g_1h = fh$ shows that f' factors through f. □

We should stress that ${}^C[\to Y\rangle$ usually is not closed under joins, see the examples at the end of the section. One of these examples is chosen in order to convince the reader that this is not at all a drawback, but an important feature if we want to work with lattices of finite height.

Theorem 3.3 (Auslander's First Theorem) *For any* Λ*-module* Y*, one has*

$$\boxed{[\to Y\rangle = \bigcup_C {}^C[\to Y\rangle,}$$

where C *runs through all the* Λ*-modules (or just through representatives of all multiplicity-free* Λ*-modules) and this is a filtered union of meet-semilattices.*

By definition, the sets ${}^C[\to Y\rangle$ are subsets of $[\to Y\rangle$. By Proposition 3.1(a), we know that ${}^C[\to Y\rangle$ only depends on add C, thus we may restrict to look at representatives of multiplicity-free Λ-modules C. Proposition 3.1(b) asserts that both ${}^C[\to Y\rangle$ and ${}^{C'}[\to Y\rangle$ are contained in ${}^{C\oplus C'}[\to Y\rangle$, thus we deal with a filtered union. According to Proposition 3.2, we deal with embeddings of meet-semilattices. The essential assertion of Theorem 3.3 is that any morphism is right determined by some module, the usual formulation of Auslander's First Theorem. A discussion of this assertion and its proof follows.

There is a precise formula which yields for f the smallest possible module $C(f)$ which right determines f. We will call it the minimal right determiner of f, any other right determiner of f will have $C(f)$ as a direct summand.

We need another definition. An indecomposable projective module P is said to *almost factor through* f, provided there is a commutative diagram of the following form

$$\begin{array}{ccc} \operatorname{rad} P & \xrightarrow{\iota} & P \\ \downarrow & & \downarrow \eta \\ X & \xrightarrow{f} & Y \end{array}$$

where ι is the inclusion map, such that the image of η is not contained in the image of f. Let us mention the following: *If the indecomposable projective module* P *almost factors through f, then* $P/\operatorname{rad} P$ *embeds into the cokernel* $\operatorname{Cok}(f)$. Namely, given a map $\eta : P \to Y$ such that the image of η is not contained in the image of f, as well as the commutative diagram above, we may complete the diagram by adding the cokernels of the horizontal maps:

$$\begin{array}{ccccccc} \operatorname{rad} P & \xrightarrow{\iota} & P & \longrightarrow & P/\operatorname{rad} P & \longrightarrow & 0 \\ \downarrow & & \downarrow \eta & & \downarrow \eta' & & \\ X & \xrightarrow{f} & Y & \longrightarrow & \operatorname{Cok}(f) & \longrightarrow & 0 \end{array}$$

Since the image of η is not contained in the image of f, we see that η' is non-zero, thus $P/\operatorname{rad} P$ is a submodule of $\operatorname{Cok}(f)$.

Theorem 3.4 (Determiner formula of Auslander–Reiten–Smalø) *Let f be a morphism ending in Y. Let $C(f)$ be the direct sum of the indecomposable modules of the form $\tau^- K$, where K is an indecomposable direct summand of the intrinsic kernel of f and of the indecomposable projective modules which almost factor through f, one from each isomorphism class. Then f is right C-determined if and only if $C(f) \in \operatorname{add} C$.*

The theorem suggests to call $C(f)$ the *minimal right determiner of f*. For the proof of Theorem 3.4, see [5] and also [38].

Corollary 3.5 *Any morphism f is right C-determined by some C, for example by the module*

$$\boxed{\tau^- \operatorname{Ker}(f) \oplus P(\operatorname{soc} \operatorname{Cok}(f)).}$$

Proof We have to show that $C(f)$ is a direct summand of $\tau^- \operatorname{Ker}(f) \oplus P(\operatorname{soc} \operatorname{Cok}(f))$. The intrinsic kernel of f is a direct summand of $\operatorname{Ker}(f)$, thus if K is an indecomposable direct summand of the intrinsic kernel of f, then $\tau^- K$ is a direct summand of $\tau^- \operatorname{Ker}(f)$. Now assume that S is a simple module such that $P(S)$ almost factors through f. Then S is a submodule of $\operatorname{Cok}(f)$, thus $P(S)$ is a direct summand of $P(\operatorname{soc} \operatorname{Cok}(f))$. □

Corollary 3.6 (Auslander) *The module $\tau^- \operatorname{Ker}(f) \oplus \Lambda$ right determines f.*

Corollary 3.7 *Let P be a projective module and $f : X \to Y$ a right minimal morphism. Then f is right P-determined if and only if f is a monomorphism and the socle of the cokernel of f is generated by P.*

Proof This is an immediate consequence of the determiner formula: First, assume that f is right P-determined. Then the intrinsic kernel of f has to be zero. Since we assume that f is right minimal, f must be a monomorphism. If S is a simple submodule of the cokernel of f, then $P(S)$ almost factors through f, thus $P(S)$ is a direct summand of P. This shows that the socle of the cokernel of f is generated by P. Conversely, assume that f is a monomorphism and the socle of the cokernel of f is generated by P. Since f is a monomorphism, $C(f)$ is the direct sum of all indecomposable projective modules P' which almost factor through f. Such a module P' is the projective cover of a simple submodule of $\mathrm{Cok}(f)$. Since P generates the socle of the cokernel of f, it follows that P' is a direct summand of P. Thus $C(f)$ is in add P, therefore f is right P-determined. □

Corollary 3.8 *A right minimal morphism $f : X \to Y$ is a monomorphism if and only if it is right Λ-determined.*

Example 3 The subset ${}^{C}[\to Y\rangle$ of $[\to Y\rangle$ is usually not closed under joins, as the following example shows: Let Λ be the path algebra of the quiver of type $\mathbb{A}_3$ with two sources, namely of

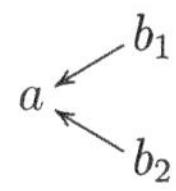

Let $f_i : P(b_i) \to Q(a)$ be non-zero maps for $i = 1, 2$, these are monomorphisms, thus they are right Λ-determined. The join of $[f_1\rangle$ and $[f_2\rangle$ in $[\to Y\rangle$ is given by the map $[f_1, f_2] : P(b_1) \oplus P(b_2) \to Q(a)$. Clearly, this map is right minimal, but it is not injective. Thus $[f_1, f_2]$ is not right Λ-determined.

Example 4 This example indicates that in general, it may not be advisable to ask for closure under joins. Consider again the Kronecker algebra Λ as exhibited in example 2. Let $Y = Q(a)$ and $C = \Lambda$, thus all the maps in ${}^{C}[\to Y\rangle$ are given by inclusion maps $f : X \to Y$, where X is a submodule of Y. In fact, we may identify ${}^{C}[\to Y\rangle$ with the submodule lattice of Y, it is a lattice of height 3 which looks as follows:

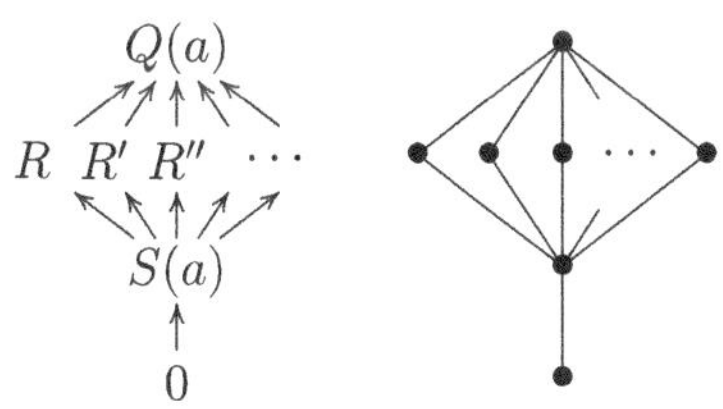

Here, the modules $R, R', R'', \dots$ are the indecomposable representations of length 2, one from each isomorphism class and all the arrows are inclusion maps.

The join in ${}^{C}[\to Y\rangle$ of two different maps f_1, f_2 in the height 2 layer is just the identity map $Y \to Y$, whereas the join of f_1, f_2 in $[\to Y\rangle$ is the direct sum map $[f_1, f_2] : R_1 \oplus R_2 \to Y$. More generally, if there are given n pairwise different regular modules $R_1, \dots, R_n$ of length 2 with inclusion maps $f_i : R_i \to Y$, then the join in $[\to Y\rangle$ is the direct sum map $[f_1, \dots, f_n] : R_1 \oplus \cdots \oplus R_n \to Y$. Let us stress that all these direct sum maps are right minimal (thus here we deal with a cofork as defined in Sect. 13). Thus, if the base field k is infinite, the smallest subposet of $[\to Y\rangle$ closed under meets and joins and containing the inclusion maps $R \to Y$ with R regular of length 2 will have infinite height.

Proposition 3.9 *Let $f : X \to Y$ be a morphism. If C' is an indecomposable direct summand of $C(f)$, then* $\mathrm{Hom}(C', Y) \neq 0$.

Proof By definition, there are two kinds of indecomposable direct summands of $C(f)$, the non-projective ones are of the form $\tau^- K'$, where K' is an indecomposable direct summand of the intrinsic kernel of f, the remaining ones are the indecomposable projective modules which almost factor through f. Of course, if P is an indecomposable projective module which almost factors through f, then $\mathrm{Hom}(P, Y) \neq 0$.

Thus, we have to consider a module of the form $C' = \tau^- K'$ with K' an indecomposable direct summand of the intrinsic kernel K of f and note that K' cannot be injective. Let us denote by $u : K \to X$ and $u' : K' \to K$ the inclusion maps. Since u' is split mono, there is $r : K \to K'$ with $ru' = 1_{K'}$. Let

$$0 \longrightarrow K' \xrightarrow{\mu} M \xrightarrow{\epsilon} C' \longrightarrow 0$$

be the Auslander–Reiten sequence starting with K'. Since the composition uu' is not split mono, there is a map $\phi : M \to X$ with $\phi\mu = uu'$. Thus, there is the following commutative diagram with exact rows

$$\begin{array}{ccccccccc} 0 & \longrightarrow & K' & \xrightarrow{\mu} & M & \xrightarrow{\epsilon} & C' & \longrightarrow & 0 \\ & & {\scriptstyle u'}\downarrow & & \downarrow{\scriptstyle \phi} & & \downarrow{\scriptstyle \phi'} & & \\ 0 & \longrightarrow & K & \xrightarrow{u} & X & \xrightarrow{f} & Y & & \end{array}$$

If we assume that $\phi' = 0$, then $f\phi = 0$, thus ϕ factors through the kernel of f, say $\phi = u\phi''$. Consequently, $uu' = \phi\mu = u\phi''\mu$. But u is injective, thus $u' = \phi''\mu$ and therefore $1_{K'} = ru' = r\phi''\mu$. But this means that μ is split mono, a contradiction. It follows that $\phi' \neq 0$, thus $\mathrm{Hom}(C', Y) \neq 0$. □

Remark Proposition 3.9 asserts that all the indecomposable direct summands C' of the minimal right determiner $C(f)$ of a map $f : X \to Y$ satisfy $\mathrm{Hom}(C', Y) \neq 0$. Actually, according to [5], Proposition XI.2.4 (see also [38]), such a module C' is equipped with a distinguished non-zero map $C' \to Y$ which is said to "almost factor through" f. At the beginning of this section we gave a corresponding definition in the special case when C' is projective. See also the Remark 3 at the end of Sect. 4.

4 The Auslander bijection. Auslander's Second Theorem

Let C, Y be objects. Let $\Gamma(C) = \mathrm{End}(C)^{\mathrm{op}}$. We always will consider $\mathrm{Hom}(C, Y)$ as a $\Gamma(C)$-module. For any module M, we denote by $\mathcal{S}M$ the set of all submodules (it is a lattice with respect to intersection and sum of submodules).

Define

$$\eta_{CY} : \bigcup_X \mathrm{Hom}(X, Y) \longrightarrow \mathcal{S}(\mathrm{Hom}(C, Y))$$

by $\eta_{CY}(f) = \mathrm{Im}\,\mathrm{Hom}(C, f) = f \cdot \mathrm{Hom}(C, X)$ for $f : X \to Y$ (note that $f \cdot \mathrm{Hom}(C, X)$ clearly is a $\Gamma(C)$-submodule).

Here is a reformulation of the definition of η_{CY}.

Proposition 4.1 *Let $f : X \to Y$. Then $\eta_{CY}(f)$ is the set of all $h \in \mathrm{Hom}(C, Y)$ which factor through f. This subset of $\mathrm{Hom}(C, Y)$ is a $\Gamma(C)$-submodule.*

Proof We have mentioned already, that $\eta_{CY}(f)$ is a $\Gamma(C)$-submodule of $\mathrm{Hom}(C, Y)$. Also, if $h \in \eta_{CY}(f) = f\,\mathrm{Hom}(C, X)$, then h factors through f. And conversely, if h factors through f, then h belongs to $f\,\mathrm{Hom}(C, X) = \eta_{CY}(f)$. □

Lemma 4.2 *If $X = X_0 \oplus X_1$ and $f(X_0) = 0$, then $\eta_{CY}(f) = \eta_{CY}(f|X_1)$.*

Proof Let $X = X_0 \oplus X_1$ and write $f = \begin{bmatrix} f_0 & f_1 \end{bmatrix} = \begin{bmatrix} 0 & f_1 \end{bmatrix}$ with $f_i : X_i \to Y$. Then, for $\phi_i : C \to X_i$, we have $\begin{bmatrix} 0 & f_1 \end{bmatrix} \begin{bmatrix} \phi_0 \\ \phi_1 \end{bmatrix} = f_1\phi_1$. Thus

$$\eta_{CY}(f) = f\,\mathrm{Hom}(C, X_0 \oplus X_1) = f_1\,\mathrm{Hom}(C, X_1) = \eta_{CY}(f_1).$$

□

In particular: *If f_1 is a right minimal version of f, then* $\eta_{CY}(f) = \eta_{CY}(f_1)$. Thus, η_{CY} is constant on right equivalence classes and we can define $\eta_{CY}([f\rangle) = \eta_{CY}(f)$. We obtain in this way a map

$$\eta_{CY} : [\to Y\rangle \to \mathcal{S}(\mathrm{Hom}(C, Y)).$$

Of special interest is the restriction of η_{CY} to ${}^{C}[\to Y\rangle$.

Proposition 4.3 *Let C, Y be modules. The map*

$$\eta_{CY} : {}^{C}[\to Y\rangle \to \mathcal{S}\,\mathrm{Hom}(C, Y).$$

is injective and preserves meets. As a consequence, it preserves and reflects the ordering.

Proof A trivial verification: First, let us show that η_{CZ} is injective. Consider maps $f : X \to Y$ and $f' : X' \to Y$ such that $f\,\mathrm{Hom}(C, X) = f'\,\mathrm{Hom}(C, X')$. Since f is right C-determined and $f'\,\mathrm{Hom}(C, X') \subseteq f\,\mathrm{Hom}(C, X)$, we see that $f' \in f\,\mathrm{Hom}(X', X)$. Since f' is right C-determined and $f\,\mathrm{Hom}(C, X) \subseteq f'\,\mathrm{Hom}(C, X')$, we see that $f \in f\,\mathrm{Hom}(X, X')$. But this means that $f' \preceq f \preceq f'$, thus $[f\rangle = [f'\rangle$. □

Next, consider the following pullback diagram

$$\begin{array}{ccc} X & \xrightarrow{g_1} & X_1 \\ {\scriptstyle g_2}\downarrow & & \downarrow{\scriptstyle f_1} \\ X_2 & \xrightarrow{f_2} & Y \end{array}$$

such that f_1, f_2 both are right C-determined. Let $f = f_1 g_1 = f_2 g_2$, thus the meet of $[f_1\rangle$ and $[f_2\rangle$ is $[f\rangle$. Now $f = f_1 g_1$ shows that $f\,\mathrm{Hom}(C, X) \subseteq f_1\,\mathrm{Hom}(C, X_1)$.

Similarly, $f = f_2 g_2$ shows that $f\,\mathrm{Hom}(C, X) \subseteq f_2\,\mathrm{Hom}(C, X_2)$. Both assertions together yield

$$f\,\mathrm{Hom}(C, X) \subseteq f_1\,\mathrm{Hom}(C, X_1) \cap f_2\,\mathrm{Hom}(C, X_2).$$

Conversely, take an element in $f_1\,\mathrm{Hom}(C, X_1) \cap f_2\,\mathrm{Hom}(C, X_2)$, say $f_1\phi_1 = f_2\phi_2$ with $\phi_i : C \to X_i$, for $i = 1, 2$. The pullback property yields a morphism $\phi : C \to X$ such that $g_i\phi = \phi_i$ for $i = 1, 2$. Therefore $f_1\phi_1 = f_1 g_1 \phi = f\phi$ belongs to $f\,\mathrm{Hom}(C, X)$. Thus,

$$f_1\,\mathrm{Hom}(C, X_1) \cap f_2\,\mathrm{Hom}(C, X_2) \subseteq f\,\mathrm{Hom}(C, X),$$

and therefore

$$f_1\,\mathrm{Hom}(C, X_1) \cap f_2\,\mathrm{Hom}(C, X_2) = f\,\mathrm{Hom}(C, X).$$

In general, assume that L, L' are posets with meets and $\eta : L \to L'$ is a set-theoretical map which preserves meets. Then $a \le b$ in L implies $\eta(a) \le \eta(b)$ in L'. Namely, $a \le b$ gives $a \wedge b = a$, thus $\eta(a) \wedge \eta(b) = \eta(a)$ and therefore $\eta(a) \le \eta(b)$. Conversely, if a, b are arbitrary elements in L with $\eta(a) \le \eta(b)$, let $c = a \wedge b$. Then $\eta(c) = \eta(a) \wedge \eta(b) = \eta(a)$. Thus, if η is injective, then $c = a$, and $a \wedge b = a$ implies $a \le b$. Altogether, we see that η_{CY} preserves and reflects the ordering.

Auslander's Second Theorem (as established in [2]) asserts:

Theorem 4.4 (Auslander's Second Theorem) *The map*

$$\eta_{CY} : [\to Y\rangle \to \mathcal{S}(\mathrm{Hom}(C, Y))$$

is surjective.

Altogether we see: *The map η_{CY} defined by $\eta_{CY}(f) = \mathrm{Im}\,\mathrm{Hom}(C, f)$ yields a lattice isomorphism*

$$\boxed{\eta_{CY} : {}^{C}[\to Y\rangle \longrightarrow \mathcal{S}\,\mathrm{Hom}(C, Y).}$$

Better: *The composition*

$$\boxed{{}^{C}[\to Y\rangle \;\subseteq\; [\to Y\rangle \;\xrightarrow{\eta_{CY}}\; \mathcal{S}_{\Gamma(C)}\,\mathrm{Hom}(C, Y)}$$

of the inclusion map and the map η_{CY} defined by $\eta_{CY}(f) = \mathrm{Im}\,\mathrm{Hom}(C, f)$ *is a lattice isomorphism.*

Convention. In the following, several examples of Auslander bijections will be presented. When looking at the submodule lattice $\mathcal{S}M$ of a module M, we usually will mark (some of) the elements of $\mathcal{S}M$ by bullets • and connect comparable elements by a solid lines. Here, going upwards corresponds to the inclusion relation.

For the corresponding lattices ${}^{C}[\rightarrow Y\rangle$, we often will mark an element $[f : X \rightarrow Y\rangle$ (with f a right minimal map) by just writing X and we will connect neighboring pairs $[f : X \rightarrow Y\rangle \leq [f' : X' \rightarrow Y\rangle$ by drawing an (upwards) arrow $X \rightarrow Y$. On the other hand, sometimes it seems to be more appropriate to refer to the right minimal map $f : X \rightarrow Y$ with kernel K' and image Y' by using the short exact sequence notation $K' \rightarrow X \rightarrow Y'$.

Note that the lattice $\mathcal{S}\,\mathrm{Hom}(C, Y)$ has two distinguished elements, namely $\mathrm{Hom}(C, Y)$ itself as well as its zero submodule. Under the bijection η_{CY} the total submodule $\mathrm{Hom}(C, Y)$ corresponds to the identity map 1_Y of Y, this is not at all exciting. But of interest seem to be the maps in $\eta_{CY}^{-1}(0)$, we will discuss them in this will be discussed in Proposition 5.5

The special case $C = {}_{\Lambda}\Lambda$. It is worthwhile to draw the attention on the special case when $C = {}_{\Lambda}\Lambda$.

Proposition 4.5 *The special case of the Auslander bijection $\eta_{\Lambda Y}$ is the obvious identification of both ${}^{\Lambda}[\rightarrow Y\rangle$ and $\mathcal{S}\,\mathrm{Hom}(\Lambda, Y)$ with $\mathcal{S}Y$.*

Proof First, consider ${}^{\Lambda}[\rightarrow Y\rangle$: The determiner formula asserts: a right minimal morphism is right Λ-determined if and only if it is a monomorphism. Thus ${}^{\Lambda}[\rightarrow Y\rangle$ is just the set of right equivalence classes of monomorphisms ending in Y, and the map $f \mapsto \mathrm{Im}(f)$ yields an identification between the set of right equivalence classes of monomorphisms ending in Y and the submodules of Λ.

Next, we deal with $\mathcal{S}\,\mathrm{Hom}(\Lambda, Y)$. Note that $\Gamma({}_{\Lambda}\Lambda) = \mathrm{End}({}_{\Lambda}\Lambda)^{\mathrm{op}} = \Lambda$ and there is a canonical identification $\epsilon : \mathrm{Hom}(\Lambda, Y) \simeq Y$ (given by $\epsilon(h) = h(1)$ for $h \in \mathrm{Hom}(\Lambda, Y)$), thus $\mathcal{S}\epsilon : \mathcal{S}\,\mathrm{Hom}(\Lambda, Y) \simeq \mathcal{S}Y$ (with $\mathcal{S}\epsilon(U) = \{h(1) \mid h \in U\}$ for U a submodule of $\mathrm{Hom}(\Lambda, Y)$).

The Auslander bijection $\eta_{\Lambda,Y}$ attaches to $f : X \rightarrow Y$ the submodule $f\,\mathrm{Hom}(\Lambda, X)$ and there is the following commutative diagram:

$$\begin{array}{ccccc} {}^{\Lambda}[\rightarrow Y\rangle & \xrightarrow{\quad\eta_{\Lambda Y}\quad} & \mathcal{S}\,\mathrm{Hom}(\Lambda, Y) & \qquad & \mathrm{Hom}(\Lambda, Y) \\ & {\scriptstyle \mathrm{Im}}\searrow \quad \swarrow{\scriptstyle \mathcal{S}\epsilon} & & & \downarrow{\scriptstyle \epsilon} \\ & \mathcal{S}Y & & & Y \end{array}$$

Namely, for $f : X \rightarrow Y$ we have

$$\begin{aligned}(\mathcal{S}\epsilon)\eta_{\Lambda Y}(f) &= (\mathcal{S}\epsilon)(f\,\mathrm{Hom}(\Lambda, X)) \\ &= \{fh(1) \mid h \in \mathrm{Hom}(\Lambda, X)\} = \{f(x) \mid x \in X\} = \mathrm{Im}(f).\end{aligned}$$ □

As a consequence, we see that *all possible submodule lattices $\mathcal{S}Y$ occur as images under the Auslander bijections.* This assertion can be strengthened considerably, as we want to show now.

By definition, an artin algebra Λ is an artin k-algebra for some commutative artinian ring k (this means that Λ is a k-algebra and that it is finitely generated as a k-module). Such an algebra is said to be *strictly wild* (or better *strictly k-wild*), provided for any artin k-algebra Γ, there is a full exact embedding $\operatorname{mod}\Gamma \to \operatorname{mod}\Lambda$. If M is a Λ-module and M' is a Λ'-module, a *semilinear isomorphism* from M to M' is a pair (α, f), where $\alpha : \Lambda \to \Lambda'$ is an algebra isomorphism, and $f : M \to M'$ is an isomorphism of abelian groups such that $f(\lambda m) = \alpha(\lambda) f(m)$ for all $\lambda \in \Lambda$ and $m \in M$. It is clear that any semilinear isomorphism from M to M' induces a lattice isomorphism $\mathcal{S}M \to \mathcal{S}M'$.

Proposition 4.6 *Let Λ be an artin k-algebra which is strictly k-wild. Let Γ be an artin k-algebra and M a Γ-module. Then there are Λ-modules C, Y such that the $\Gamma(C)$-module* $\operatorname{Hom}(C, Y)$ *is semilinearly isomorphic to M. Thus there is a lattice isomorphism ${}^{C}[\to Y\rangle \to \mathcal{S}M$.*

Proof Let $F : \operatorname{mod}\Gamma \to \operatorname{mod}\Lambda$ be a full embedding (we do not need that it is exact). Let $C = F({}_{\Gamma}\Gamma)$ and $Y = F(M)$. Let $\alpha : \Gamma = \operatorname{End}({}_{\Gamma}\Gamma)^{\mathrm{op}} \to \operatorname{End}(C)^{\mathrm{op}} = \Gamma(C)$ as well as $f : M = \operatorname{Hom}({}_{\Gamma}\Gamma, M) \to \operatorname{Hom}(C, M)$ both be given by applying the functor F. Since F is a full embedding, α is an algebra isomorphism and f is an isomorphism of abelian groups. The functoriality of F asserts that we also have $f(\gamma m) = \alpha(\gamma) f(m)$ for all $\gamma \in \Gamma$ and $m \in M$. This shows that the pair (α, f) is a semilinear isomorphism. □

Remark If $F : \operatorname{mod}\Gamma \to \operatorname{mod}\Lambda$ is a full embedding functor, and C, Y are Γ-modules, then we obtain a bijection

$$ {}^{C}[\to y\rangle \xrightarrow{\eta_{CY}} \mathcal{S}\operatorname{Hom}(C,Y) \xrightarrow{F} \mathcal{S}\operatorname{Hom}(F(C),F(Y)) \xrightarrow{\eta^{-1}_{F(C),F(Y)}} {}^{F(C)}[\to F(Y), $$

but even if F is exact, such a bijection will not be given by applying directly F. Namely, if $f : X \to Y$ is right minimal and right C-determined, then the kernel of f belongs to add τC, thus the kernel of $F(f)$ belongs to add $F(\tau C)$, whereas the intrinsic kernel of any right $F(C)$-determined map has to belong to add $\tau F(C)$ and the Λ-modules $F(\tau C)$ and $\tau F(C)$ may be very different, as the obvious embeddings of the category of n-Kronecker modules into the category of $(n + 1)$-Kronecker modules (using for one arrow the zero map) show.

Note that under a full exact embedding functor $F : \operatorname{mod}\Gamma \to \operatorname{mod}\Lambda$, submodule lattices are usually not preserved: given a Γ-module M, the functor F yields an embedding of $\mathcal{S}({}_{\Gamma}M)$ into $\mathcal{S}({}_{\Lambda}F(M))$, but usually this is a proper embedding. Actually, for any finite-dimensiona algebra Λ, there are submodule lattices $\mathcal{S}({}_{\Gamma}M)$ which cannot be realized as the submodule lattice of any Λ-module. Namely, assume that the length of the indecomposable projective Λ-modules is bounded by t and take a finite-dimensional algebra Γ with a local Γ-module M of length $t + 1$. Then $\mathcal{S}M$ is a modular lattice of height $t + 1$ with a unique element of height t (the radical of the module M). If $\mathcal{S}({}_{\Lambda}Y)$ is of the form $\mathcal{S}M$, then Y has to be a local Λ-module of length $t + 1$, thus a factor module of an indecomposable projective Λ-module. But by assumption, the indecomposable projective Λ-modules have length at most t.

Remark 1 Let $Y = \bigoplus Y_i$, then the subsets $\operatorname{Hom}(C, Y_i)$ of $\operatorname{Hom}(C, Y)$ are actually $\Gamma(C)$-submodules and there is an isomorphism of $\Gamma(C)$-modules $\operatorname{Hom}(C, Y) \simeq \bigoplus_i \operatorname{Hom}(C, Y_i)$. Thus η_{CY} maps the lattice ${}^{C}[\to Y\rangle$ bijectively onto the submodule lattice $\mathcal{S}(\bigoplus_i \operatorname{Hom}(C, Y_i)$. The lattice $\mathcal{S}(\bigoplus_i \operatorname{Hom}(C, Y_i)$ contains $\prod_i \mathcal{S}\operatorname{Hom}(C, Y_i)$ as a sublattice and both have the same height. However, $\prod_i \mathcal{S}\operatorname{Hom}(C, Y_i)$ may be a proper sublattice of $\mathcal{S}(\bigoplus_i \operatorname{Hom}(C, Y_i)$, since isomorphisms of subfactors of the various modules $\operatorname{Hom}(C, Y_i)$ yield diagonals in $\mathcal{S}(\bigoplus_i \operatorname{Hom}(C, Y_i)$.

Remark 2 When dealing with the Auslander bijections $\eta_{CY} : {}^C[\to Y\rangle \to \mathcal{S}\operatorname{Hom}(C, Y)$, we always can assume that C is multiplicity-free and supporting, here *supporting* means that $\operatorname{Hom}(C_i, Y) \neq 0$ for any indecomposable direct summand C_i of C. Namely, let C' be the direct sum of all indecomposable direct summands C_i of C with $\operatorname{Hom}(C_i, Y) \neq 0$, one from each isomorphism class. Then, on the one hand, ${}^C[\to Y\rangle = {}^{C'}[\to Y\rangle$ (since a map f ending in Y is right C-determined if and only if it is right C'-determined. On the other hand, there is an idempotent $e \in \Gamma(C)$ such that $e\Gamma(C)e = \Gamma(C')$ and $e\operatorname{Hom}(C, Y)e = \operatorname{Hom}(C', Y)$, and there is a lattice isomorphism $\mathcal{S}\operatorname{Hom}(C, Y) \to \mathcal{S}\operatorname{Hom}(C', Y)$, given by $U \mapsto eU$, where U is a submodule of $\operatorname{Hom}(C, Y)$.

Remark 3 Both objects ${}^C[\to Y\rangle$ and $\mathcal{S}\operatorname{Hom}(C, Y)$ related by the Auslander bijection η_{CY} concern morphisms ending in Y. Of course, in Proposition 3.9 we have seen already that all the indecomposable direct summands C' of the minimal right determiner $C(f)$ of a map $f : X \to Y$ satisfy $\operatorname{Hom}(C', Y) \neq 0$.
Looking at ${}^C[\to Y\rangle$, we deal with morphisms ending in Y and which are right C-determined. Looking at $\operatorname{Hom}(C, Y)$, we deal with maps ending in Y and starting in C. One should be aware that a right minimal map ending in Y and right C-determined usually will **not** start at C, thus the relationship between the elements of ${}^C[\to Y\rangle$ and the submodules of $\operatorname{Hom}(C, Y)$ is really of interest! Note however that in case we deal with a map $f : C \to Y$ which is right C-determined (and starts in C), then

$$\eta_{CY}(f) = f\operatorname{Hom}(C, C)$$

is just the $\Gamma(C)$-submodule of $\operatorname{Hom}(C, Y)$ generated by f.
We use the next two sections in order to transfer well-known properties of the lattice of submodules of a finite length module to the right C-factorization lattices, in particular the Jordan–Hölder theorem. In Sect. 5, we introduce the right C-length of a right C-determined map f ending in Y, it corresponds to the the length of the factor module $\operatorname{Hom}(C, Y)/\eta_{CY}(f)$. In Sect. 6 we will define the C-type of f as the dimension vector of $\operatorname{Hom}(C, Y)/\eta_{CY}(f)$.

5 Right C-factorizations and right C-length

The Auslander bijection asserts that the lattice ${}^C[\to Y\rangle$ is a modular lattice of finite height, thus there is a Jordan–Hölder Theorem for ${}^C[\to Y\rangle$; it can be obtained from the corresponding Jordan–Hölder Theorem for the submodule lattice $\mathcal{S}\operatorname{Hom}(C, Y)$. In Sects. 5 and 6, we are going to formulate the assertions for ${}^C[\to Y\rangle$ explicitly. Here we consider composition series of submodules and factor modules of ${}^C[\to Y\rangle$.

Let $h_i : X_i \to X_{i-1}$ be maps, where $1 \leq i \leq t$, with composition $f = h_1 \dots h_t$. The sequence $(h_1, h_2, \dots, h_t)$ is called a *right C-factorization of f of length t* provided the maps h_i are non-invertible and the compositions $f_i = h_1 \cdots h_i$ are right minimal and right C-determined, for $1 \le i \le t$. It sometimes may be helpful to deal also with right C-factorizations of length 0; by definition these are just the identity maps (or, if you prefer, the isomorphisms).

If $(h_1, \ldots, h_t)$ is a right C-factorization of a map f, then any integer sequence $0 = i(0) < i(1) < \cdots < i(s) = t$ defines a sequence of maps $(h'_1, h'_2, \ldots, h'_s)$ with $h'_j = h_{i(j-1)+1} \cdots h_{i(j)}$ for $1 \le j \le s$. We use the following lemma inductively, in order to show that $(h'_1, h'_2, \ldots, h'_s)$ is again a right C-factorization of f and we say that $(h_1, h_2, \ldots, h_t)$ is a *refinement* of $(h'_1, h'_2, \ldots, h'_s)$. In particular, any right C-factorization $(h_1, \ldots, h_t)$ of f is a refinement of f.

Lemma 5.1 *If $(h_1, \ldots, h_t)$ is a right C-factorization of length $t \ge 2$,then*

$$(h_1, \ldots, h_{i-1}, h_i h_{i+1}, h_{i+2}, \ldots, h_t)$$

is a right C-factorization (of length $t - 1$).

Proof We only have to check that $h_i h_{i+1}$ cannot be invertible. Assume $h_i h_{i+1}$ is invertible. Then h_i is a split epimorphism. Since $f_i = h_1 \cdots h_i$ is right minimal, it follows that h_i is invertible, a contradiction. □

We say that a right C-factorization $(h_1, h_2, \ldots, h_t)$ is *maximal* provided it does not have a refinement of length $t + 1$.

Proposition 5.2 *If $(h_1, \ldots, h_t)$ is a right C-factorization of a map f, and $f_i = h_1 \cdots h_i$ for $0 \le i \le t$, then*

$$\eta_{CY}(f_t) \subset \cdots \subset \eta_{CY}(f_1) \subset \eta_{CY}(f_0) = \mathrm{Hom}(C, Y)$$

is a chain of proper inclusions of submodules and any such chain is obtained in this way. The refinement of right C-factorizations corresponds to the refinement of submodule chains.

Proof This is a direct consequence of Auslander's Second Theorem. □

Corollary 5.3 *Any right C-factorization $(h_1, \ldots, h_t)$ has a refinement which is a maximal right C-factorization and all maximal right C-factorizations of $(h_1, \ldots, h_t)$ have the same length.*

Proof This follows from Proposition 5.2 and the Jordan–Hölder theorem. □

In particular, any right minimal right C-determined map f has a refinement which is a maximal right C-factorization, say $(h_1, \ldots, h_t)$ and its length t will be called the *right C -length* of f, we write $|f|_C$ for the right C-length of f. There is the following formula:

Proposition 5.4 *Let $f : X \to Y$ be right minimal and right C-determined. Then*

$$\boxed{|f|_C = |\operatorname{Hom}(C,Y)| - |\eta_{CY}(f)|,}$$

where $|\operatorname{Hom}(C, Y)|$ *denotes the length of the* $\Gamma(C)$*-module* $\operatorname{Hom}(C, Y)$ *and* $|\eta_{CY}(f)|$ *the length of its* $\Gamma(C)$*-submodule* $\eta_{CY}(f)$.

The right equivalence class $\eta_{CY}^{-1}(0)$. As we have mentioned in Sect. 4, it is of interest to determine the maps in the right equivalence class $\eta_{CY}^{-1}(0)$.

Proposition 5.5 *Let* C, Y *be modules. Up to right equivalence, there is a unique right* C*-determined map* f *ending in* Y *with* $|f|_C$ *maximal. The submodule* $\eta_{CY}(f)$ *of* $\operatorname{Hom}(C, Y)$ *is the zero module. If* f' *is any right* C*-determined map ending in* Y*, then* $f = f'h$ *for some* h.

Proof The lattice ${}^C[\to Y\rangle$ has a unique zero element, namely $\eta_{CY}^{-1}(0)$. Let $\eta_{CY}^{-1}(0) = [f\rangle$ for some right minimal map f. Then, the right C-length of f has to be maximal and $[f\rangle \le [f'\rangle$ for any right C-determined map f' ending in Y. □

In general it seems to be quite difficult to describe the maps f such that $[f\rangle = \eta_{CY}^{-1}(0)$. But one should be aware that such a map f always does exist: any pair C, Y of Λ-modules determines uniquely up to right equivalence a map f ending in Y, namely the right minimal, right C-determined map f with $\eta_{CY}(f) = 0$.

Proposition 5.6 *Let* C, Y *be modules. The set* $\eta_{CY}^{-1}(0)$ *is the right equivalence class of the zero map* $0 \to Y$ *if and only if* $P(\operatorname{soc} Y)$ *belongs to* $\operatorname{add} C$.

Proof This is a direct consequence of Corollary 3.7. □

The special case of C being projective. For an arbitrary projective module C, there is the following description of the right C-length of a right minimal, right C-determined morphism f. Here, we denote by $[M : S]$ the Jordan–Hölder multiplicity of the simple module S in the module M, this is the number of factors in a composition series of M which are isomorphic to S.

Proposition 5.7 *Let* C *be projective. The right minimal, right* C*-determined maps* $f : X \to Y$ *are up to right equivalence just the inclusion maps of submodules* X *of* Y *such that the socle of* Y/X *is generated by* C.

If $f : X \to Y$ *is right minimal and right* C*-determined, then* f *is injective and*

$$\boxed{|f|_C = \sum_{P(S)|C} [\operatorname{Cok}(f) : S].}$$

The minimal element $\eta_{CY}^{-1}(0)$ *of* ${}^C[\to Y\rangle$ *is the inclusion map* $X \to Y$*, where* X *is the intersection of the kernels of all maps* $Y \to Q(S)$*, where* S *is a simple module with* $P(S)$ *a direct summand of* C.

Proof Let $\mathcal{Q}$ be the set of modules $Q(S)$, where S is a simple module with $P(S)$ a direct summand of C. Let X be the intersection of the kernels of all maps $Y \to Q$ with $Q \in \mathcal{Q}$. Since Y is of finite length, there are finitely many maps $g_i : Y \to Q(S_i)$ with $Q(S_i) \in \mathcal{Q}$, say $1 \leq i \leq m$, such that $X = \bigcap_{i=1}^{m} \mathrm{Ker}(g_i)$. Then Y/X embeds into $\bigoplus_{i=1}^{m} Q(S_i)$, thus its socle is generated by C. It follows that the inclusion map $X \to Y$ is right C-determined. On the other hand, if $X' \to Y$ is right minimal and right C-determined, then it is a monomorphism, thus we can assume that it is an inclusion map. In addition, we know that the socle of Y/X' is generated by C, thus Y/X' embeds into a finite direct sum of modules in $\mathcal{Q}$. It follows that X' is the intersection of some maps $Y \to Q$, where $Q \in \mathcal{Q}$, thus $X \subseteq X'$. □

There is the following consequence: *The* ${}_\Lambda\Lambda$*-length of any inclusion map* $X \to Y$ (such a map is obviously right minimal and right Λ-determined) *is precisely the length of* Y/X.

For further results concerning the right C-length of maps, see Sect. 9.

6 The right C-type of a right C-determined map

Recall that we consider $\mathrm{Hom}(C, Y)$ as a $\Gamma(C)$-module, where $\Gamma(C) = \mathrm{End}(C)^{\mathrm{op}}$. The indecomposable projective $\Gamma(C)$-modules are of the form $\mathrm{Hom}(C, C_i)$, where C_i is an indecomposable direct summand of C, thus the simple $\Gamma(C)$-modules are of the form $S(C_0) = \mathrm{top}\,\mathrm{Hom}(C, C_0)$.

Given an artin algebra Γ, we denote by $K_0(\Gamma)$ its Grothendieck group (of all Γ-modules modulo all exact sequences), it is the free abelian group with basis the set of isomorphism classes $[S]$ of the simple Γ-modules S. Given a Γ-module M, we denote by $\mathbf{dim}M$ the corresponding element in $K_0(\Gamma)$, called the *dimension vector* of M. Of course, $\mathbf{dim}M$ can be written as an integral linear combination $\mathbf{dim}M = \sum_{[S]}[M : S][S]$, where the coefficient of $[S]$ is just the Jordan–Hölder multiplicity $[M : S]$ of S in M. The elements of $K_0(\Gamma)$ with non-negative coefficients will be said to be the Γ-dimension vectors. If $\mathbf{e}$ is a Γ-dimension vector and M is a Γ-module, we denote by $\mathcal{S}_{\mathbf{e}}M$ the subset of $\mathcal{S}M$ consisting of all submodules of M with dimension vector $\mathbf{e}$.

Let us return to the artin algebra $\Gamma(C)$, where C is a Λ-module. Thhe Grothendieck group $K_0(\Gamma(C))$ is the free abelian group with basis the set of modules $S(C_i)$, where C_i runs through a set of representatives of the isomorphism classes of the indecomposable direct summands C_i of C. We are interested here in the dimension vectors of $\mathrm{Hom}(C, Y)$ and of its factor modules. Actually, we want to attach to each right C-determined map ending in Y its right C-type $\mathbf{type}_C(f)$ so that $\mathbf{type}_C(f) = \mathbf{dim}\,\mathrm{Hom}(C, Y)/\eta_{CY}(f)$. We start with pairs of neighbors in the right C-factorization lattice ${}^C[\to Y\rangle$, since they correspond under η_{CY} to the composition factors of $\mathrm{Hom}(C, Y)$.

Let $f = hf'$ and f' be right minimal, right C-determined maps. We say that the pair (f, f') is a pair of *C-neighbors* provided $|f|_C = |f'|_C + 1$. Note that the pair (f, f') in ${}^C[\to Y\rangle$ is a pair of C-neighbors provided $[f\rangle < [f'\rangle$ and there is no f'' with $[f\rangle < [f''\rangle < [f'\rangle$ (of course, it is the condition $[f\rangle < [f'\rangle$ which implies that there is a map h with $f = f'h$).

Remark Let us consider a composition $f = f'h$, where f' both are right minimal and right C-determined. It can happen that h is also right minimal and right C-determined, but $f = f'h$ is not right C-determined. Also it can happen that both maps f' and $f = f'h$ are right minimal and right C-determined, whereas h is not right C-determined. Here are corresponding examples.

Example 5 We consider the path algebra Λ of the linearly directed quiver Δ of type $\mathbb{A}_3$

$$a \xrightarrow{\alpha} b \xrightarrow{\beta} c.$$

Let $X = Q(a)$, $X' = Q(b)$, $Y = Q(c) = S(c)$. There are non-zero maps

$$Q(a) \xrightarrow{h} Q(b) \xrightarrow{f'} S(c),$$

and we let $f = f'h$. All three maps f', h, f are surjective and right minimal. The kernel of f' is the simple module $S(b)$, the kernel of h is the simple module $S(a)$ and the kernel of f is $P(b)$.

First, let $C = S(b) \oplus S(c)$, thus $\tau C = S(a) \oplus S(b)$ and both f' and h are right C-determined, whereas f is not right C-determined.

Second, let $C = Q(b) \oplus S(c)$, thus $\tau C = P(b) \oplus S(b)$. Then both f and f' are right C-determined, whereas h is not right C-determined.

Let $f : X \to Y$ and $f' : X' \to Y$ such that (f, f') is a pair of neighbors. We say that (f, f') is *of type* C_0 (or better of type $[S(C_0)]$) where C_0 is an indecomposable direct summand of C, provided there is a map $\phi : C_0 \to X'$ such that $f'\phi$ does not factor through f. Such a summand C_0 must exist, since otherwise f' would factor through f, due to the fact that f is right C_0-determined. The following proposition shows that C_0 is uniquely determined.

Proposition 6.1 *If (f, f') is a pair of C-neighbors of type C_0, then $\eta_{CY}(f')/\eta_{CY}(f)$ is isomorphic to the simple $\Gamma(C)$-module $S(C_0) = \operatorname{top}\operatorname{Hom}(C, C_0)$. Thus, the type of a pair of C-neighbors is well-defined.*

Thus, if (f, f') is a pair of C-neighbors of type C_0, we may write $\mathbf{type}_C(f, f') = [S(C_0)] \in K_0(\Gamma(C))$.

Proof Let $\phi : C_0 \to X'$ be a map such that $f'\phi$ does not factor through f. We obtain a homomorphism of $\Gamma(C)$-modules

$$\operatorname{Hom}(C, f'\phi) : \operatorname{Hom}(C, C_0) \to \operatorname{Hom}(C, Y)$$

which maps into $f' \operatorname{Hom}(C, X')$ (since the image consists of the maps $f'\phi\psi$ with $\psi : C \to C_0$), We claim that $\operatorname{Hom}(C, f'\phi)$ does not map into $f \operatorname{Hom}(C, X)$. Assume, for the contrary, that $\operatorname{Hom}(C, f'\phi)$ maps into $f \operatorname{Hom}(C, X)$. Choose $m : C_0 \to C$ and $e : C \to C_0$ with $em = 1$. By assumption, the element $\operatorname{Hom}(C, f'\phi)(e) = f'\phi e$ belongs to $f \operatorname{Hom}(C, X)$, thus there is $\phi' : C \to X$ with $f'\phi e = f\phi'$ and therefore

$$f'\phi = f'\phi em = f\phi' m$$

shows that $f'\phi$ factor through f, a contradiction.

Thus, the image of $\operatorname{Hom}(C, f'\phi)$ is a $\Gamma(C)$-submodule of $\operatorname{Im}\operatorname{Hom}(C, f')$ which is not contained in $\operatorname{Im}(C, f')$ and which is an epimorphic image of the projective module $\operatorname{Hom}(C, C_0)$. Since we know that $\operatorname{Im}\operatorname{Hom}(C, f)$ is a maximal submodule of $\operatorname{Im}\operatorname{Hom}(C, f')$, it follows that

$$\operatorname{Im}\operatorname{Hom}(C, f')/\operatorname{Im}\operatorname{Hom}(C, f') \simeq \operatorname{top}\operatorname{Im}\operatorname{Hom}(C, f'\phi) \simeq \operatorname{top}\operatorname{Hom}(C, C_0).$$

This is what we wanted to show. □

We note the following: *If (f, f') is a pair of C-neighbors and $f = f'h$, then h may be neither injective nor surjective.* Let us exhibit examples with $f' = 1_Y$.

Example 6 As in the examples 5, let Δ be the linearly directed quiver of type $\mathbb{A}_3$ and take now as Λ the path algebra of Δ modulo the zero relation $\alpha\beta$.

Let $Y = P(c)$ and $C = S(b)$, then

$$^{S(b)}[\to P(c)\rangle \quad \leftrightarrow \quad \mathcal{S}\operatorname{Hom}(S(b), P(c))$$

are lattices with precisely two elements: in $^{S(2)}[\to P(3)\rangle$, there is the right equivalence class of the identity map $f' = 1_{P(c)}$ as well as the right equivalence class of any non-zero map $f : P(b) \to P(c)$. Note that f is right minimal and right $S(b)$-determined, and it is neither mono nor epi.

Now consider a right minimal right C-determined map f ending in Y. As we have mentioned, we want to attach to f an element $\mathbf{type}_C(f) \in K_0(\Gamma(C))$.

Proposition 6.2 *Let C, Y be Λ-modules. Let f be right minimal and right C-determined map ending in Y with maximal right C-factorization $h_1, \ldots, h_t$. Write $f_i = h_1 \cdots h_i$, for $0 \le i \le t$. Then*

$$\boxed{\mathbf{type}_C\, f = \sum_{i=1}^{t} \mathbf{type}_C(f_i, f_{i-1})}$$

is a well-defined element of $K_0(\Gamma(C))$ and we have

$$\boxed{\mathbf{type}_C(f) = \mathbf{dim}\operatorname{Hom}(C,Y)/\eta_{CY}(f) = \mathbf{d} - \mathbf{dim}\,\eta_{CY}(f),}$$

with $\mathbf{d} = \mathbf{dim}\operatorname{Hom}(C, Y)$.

Proof Unter the Auslander bijection η_{CY}, the chain

$$[f\rangle = [f_t\rangle < [f_{t-1}\rangle < \cdots < [f_0\rangle$$

is mapped to a chain of submodules

$$\eta_{CY}(f) = \eta_{CY}(f_t) \subset \eta_{CY}(f_{t-1}) \subset \cdots \subset \eta_{CY}(f_0) = \operatorname{Hom}(C, Y)$$

with simple factors $\eta_{CY}(f_{i-1})/\eta_{CY}(f_i)$, thus we obtain in this way a composition series of $\mathrm{Hom}(C, Y)/\eta_{CY}(f)$. The Jordan–Hölder theorem for $\mathrm{Hom}(C, Y)/\eta_{CY}(f)$ asserts that this yields the dimension vector $\mathbf{dim}\,\mathrm{Hom}(C, Y)/\eta_{CY}(f)$, independent of the choice of the composition series. □

For any $\Gamma(C)$-dimension vector $\mathbf{e}$, let us denote by ${}^C[\to Y\rangle^{\mathbf{e}}$ the set of all elements $[f\rangle$ in ${}^C[\to Y\rangle$ such that $\mathbf{type}_C(f) = \mathbf{e}$. Note that ${}^C[\to Y\rangle^{\mathbf{e}}$ is non-empty only in case $\mathbf{e} \leq \mathbf{dim}\,\mathrm{Hom}(C, Y)$, thus we obtain a decomposition

$$\boxed{{}^C[\to Y\rangle = \bigsqcup_{\mathbf{e}} {}^C[\to Y\rangle^{\mathbf{e}}}$$

into a finite number of disjoint subsets.

Proposition 6.3 *The Auslander bijection η_{CY} yields a bijection*

$$\boxed{{}^C[\to Y\rangle^{\mathbf{e}} \xrightarrow{1:1} \mathcal{S}_{\mathbf{d}-\mathbf{e}}\,\mathrm{Hom}(C, Y)}$$

for every $\Gamma(C)$-dimension vector $\mathbf{e}$.

Assume now that k is an algebraically closed field and that Λ and Γ are k-algebra. If M is a Γ-module and $\mathbf{e}$ a dimension vector for Γ, we write $\mathbb{G}_{\mathbf{e}}M$ instead of $\mathcal{S}_{\mathbf{e}}M$. Note that $\mathbb{G}_{\mathbf{e}}M$ is in a natural way an algebraic variety, it is called a *quiver Grassmannian.* Namely, all the Γ-modules with dimension vector $\mathbf{e}$ have the same k-dimension, say e (if $\mathbf{e} = \sum_{[S]} e_S[S]$, then $e = \sum_{[S]} e_S \dim_k S$). Denote by $\mathbb{G}_e({}_kM)$ the usual Grassmannian of all e-dimensional subspaces U of the vector space ${}_kM$. Using Plücker coordinates one knowns that $\mathbb{G}_e({}_kM)$ is a closed subset of a projective space, thus $\mathbb{G}_e({}_kM)$ is a projective variety. Now $\mathbb{G}_{\mathbf{e}}M$ is a subset of $\mathbb{G}_e({}_kM)$ defined by the vanishing of some polynomials (which express the fact that we consider submodules U with a fixed dimension vector), thus also $\mathbb{G}_{\mathbf{e}}M$ is an algebraic variety and indeed a projective variety (but usually not even connected).

Proposition 6.3 can be reformulated as follows:

Proposition 6.4 *The Auslander bijection η_{CY} yields a bijection*

$$\boxed{{}^C[\to Y\rangle^{\mathbf{e}} \xrightarrow{1:1} \mathbb{G}_{\mathbf{d}-\mathbf{e}}\,\mathrm{Hom}(C, Y)}$$

for every $\Gamma(C)$-dimension vector $\mathbf{e}$.

In particular, we see that the set ${}^C[\to Y\rangle^{\mathbf{e}}$ is a projective variety: these *Auslander varieties* (as they should be called) furnish an important tool for studying the right equivalence classes of maps ending in a given module. As we have mentioned at the end of Sect. 2, the study of the set of right minimal maps ending in a fixed module Y can be separated nicely into that of the local symmetries described by the right automorphism groups and that of the global directedness given by the right factorization lattice. Auslander's first theorem describes the right factorization lattice as the filtered union of the right C-factorization lattices, and, as we now see, these right C-factorization lattices are finite disjoint unions of (transversal) subsets which are projective varieties, the Auslander varieties.

Remark It seems that quiver Grassmannians first have been studied by Schofield [42] and Crawley-Boevey [10] in order to deal with generic properties of quiver representations. In 2006, Caldero and Chapoton [9] observed that quiver Grassmannians can be used effectively in order to analyse the structure of cluster algebras as introduced by Fomin and Zelevinsky. Namely, it turns out that cluster variables can be described using the Euler characteristic of quiver Grassmannians. In this way quiver Grassmannians are now an indispensable tool for studying cluster algebras and quantum cluster algebras. We should add that quiver Grassmannians were also used (at least implicitly) in the study of quantum groups, see for example the calculation of Hall polynomials in [34]. A large number of papers is presently devoted to special properties of quiver Grassmannians.

There is the famous assertion that any projective variety is a quiver Grassmannian, see the paper [29] by Reineke (answering in this way a question by Keller) as well as blogs by Le Brujn [26] (with a contribution by Van den Bergh) and by Baez [6]. Actually, the construction as proposed by Van den Bergh in Le Bruyn's blog is much older, it has been mentioned explicitly already in 1996 by Hille [21] dealing with moduli spaces of thin representations (see the example at the end of that paper), and it can be traced back to earlier considerations of Huisgen-Zimmermann dealing with moduli spaces of serial modules, even if they were published only later (see [22], Theorem G, but also [8], Corollary B, and [12], Example 5.4). It follows from Proposition 4.6 above that given a strictly wild algebra Λ and any projective variety V, there are Λ-modules C, Y and a dimension vector $\mathbf{e}$ such that ${}^{C}[\to Y\rangle^{\mathbf{e}}$ is isomorphic to V. We will show in [39] that this holds true for all controlled wild algebras.

7 Maps of right C-length 1

By definition, a right minimal right C-determined map f has right C-length 1 provided f is not invertible and given any factorization $f = f'h$ with f' right minimal right C-determined, then one of the maps f', h is invertible. Let us denote by ${}^{C}[\to Y\rangle^{1}$ the set of right equivalence classes of the maps ending in Y which have right C-length 1.

Warning: an irreducible map f is of course right minimal, but if f is irreducible and right C-determined, we may have $|f|_C > 1$. For example, consider the Kronecker quiver, take $C = {}_{\Lambda}\Lambda$. The irreducible map $f : P_0 \to P_1$ has $|f|_C = 2$ (note the factorizazion $P_0 \subset \operatorname{rad} P_1 \subset P_1$).

Here is an immediate consequence of Proposition 5.4.

Corollary 7.1 *Let $f : X \to Y$ be right minimal and right C-determined. Then $|f|_C = 1$ if and only if $\eta_{CY}(f)$ is a maximal $\Gamma(C)$-submodule of* $\operatorname{Hom}(C, Y)$.

If we denote by $\mathcal{S}_m \operatorname{Hom}(C, Y)$ the set of maximal submodules of $\operatorname{Hom}(C, Y)$, then the restriction of η_{CY} furnishes a bijection

$$\boxed{{}^{C}[\to Y\rangle^{1} \xrightarrow{1:1} \mathcal{S}_m \operatorname{Hom}(C,Y).}$$

In order to analyze maps of right C-length 1, we will need the following lemma.

Lemma 7.2 *Assume that $f : X \to Y$ and $f' : X' \to Y$ are epimorphisms with $f = f'h$. Then we have the following commutative diagram with exact rows:*

$$\begin{array}{ccccccccc} 0 & \longrightarrow & K' & \longrightarrow & X' & \xrightarrow{f'} & Y & \longrightarrow & 0 \\ & & \uparrow{\scriptstyle h'} & & \uparrow{\scriptstyle h} & & \| & & \\ 0 & \longrightarrow & K & \longrightarrow & X & \xrightarrow{f} & Y & \longrightarrow & 0 \end{array}$$

If f is right minimal and h' is a split epimorphism, then also f' is right minimal.

Remark Observe that it is not enough to assume that h' is an epimorphism. As an example, take the indecomposable injective Kronecker module $X = Q_1$ of length 3, let K be a submodule of length 2, and $K' = K/\operatorname{soc}$. Then $Q_1 \to Q_1/K$ is right minimal. But the induced sequence is just the short exact sequence $K/\operatorname{soc} \to Q_1/\operatorname{soc} \to Y$ which splits.

Proof Denote the kernel of h' by K'', thus we can assume that $K = K' \oplus K''$ such that h' is the canonical projection $K \to K'$ with kernel K''. Assume that $X' = U \oplus V$, where U is contained in the kernel of f', thus $U \subseteq K'$. Since $X' = X/K''$, there are submodules U', V' of X' both containing K'' such that $U' + V' = X'$ and $U' \cap V' = K''$, with $U = U'/K''$ and $V = V'/K''$.

Consider $U'' = U' \cap K'$, this is a submodule of the kernel K of f. Also, $U' = K'' + U''$ (using the modular law). Thus we have $U'' \cap V' = U' \cap K' \cap V' \subseteq K' \cap K'' = 0$ and $U'' + V' = U'' + V' + K'' = U' + V' = X$. This shows that U'' is a direct summand of X which is contained in the kernel of f. Since f is right minimal, we see that $U'' = 0$. Since $U' = K'' + U'' = K''$, it follows that $U = 0$. □

Corollary 7.3 *Let C be a module. Let $f : X \to Y$ be a right minimal right C-determined epimorphism with $|f|_C = 1$. Then the kernel of f is indecomposable.*

Proof Let $K = \tau C$. The kernel of f has to be non-zero, thus assume it is decomposable, say equal to $K_1 \oplus K_2$ with non-zero modules $K_1, K_2 \in \operatorname{add} K$. The canonical projection $p_1 : K_1 \oplus K_2 \to K_1$ yields a commutative diagram with exact rows:

$$\begin{array}{ccccccccc} 0 & \longrightarrow & K_1 \oplus K_2 & \longrightarrow & X & \xrightarrow{f} & Y & \longrightarrow & 0 \\ & & \downarrow{\scriptstyle p_1} & & \downarrow{\scriptstyle p_1'} & & \| & & \\ 0 & \longrightarrow & K_1 & \longrightarrow & X' & \xrightarrow{f'} & Y & \longrightarrow & 0 \end{array}$$

Since f is right minimal and p_1 is a split epimorphism, lemma 7.2 asserts that f' is right minimal. Since f' is also right C-determined, we see that $|f|_C \geq 2$, a contradiction. □

Proposition 7.4 *Let C be indecomposable and non-projective and let $K = \tau C$. If*

$$\epsilon : \quad 0 \longrightarrow K^t \longrightarrow X \xrightarrow{f} Y \longrightarrow 0$$

is an exact sequence, then $[f\rangle$ belongs to ${}^C[\to Y\rangle^1$ if and only if $t = 1$ and the equivalence class $[\epsilon]$ is a non-zero element of the $\Gamma(K)$-socle of $\mathrm{Ext}^1(Y, K)$.

Proof First, assume that $[f\rangle$ belongs to ${}^C[\to Y\rangle^1$. According to Corollary 7.3, we must have $t = 1$. Also, since f is not split epi, we see that $[\epsilon]$ is a non-zero element of $\mathrm{Ext}^1(Y, K)$. Assume that $[\epsilon]$ does not belong to the $\Gamma(K)$-socle of $\mathrm{Ext}^1(Y, K)$. Then there is a nilpotent endomorphism ϕ of K such that the induced exact sequence does not split. Thus there is a commutative diagram with exact rows

$$\begin{array}{ccccccccc} 0 & \longrightarrow & K & \longrightarrow & X & \xrightarrow{f} & Y & \longrightarrow & 0 \\ & & \downarrow{\scriptstyle \phi} & & \downarrow{\scriptstyle \phi'} & & \| & & \\ 0 & \longrightarrow & K & \longrightarrow & X' & \xrightarrow{f'} & Y & \longrightarrow & 0 \end{array}$$

with the upper row being the sequence ϵ. Since the lower sequence does not split, the map f' is right minimal. The kernel shows that f' is also C-determined. Since ϕ' is not invertible, the factorization $f = f'\phi'$ shows that (ϕ', f') is a C-factorization of f of length at least 2, thus $|f|_C \geq 2$, a contradiction. This shows that $[\epsilon]$ belongs to the $\Gamma(K)$-socle of $\mathrm{Ext}^1(Y, K)$.

Conversely, assume that $t = 1$ and $[\epsilon]$ is a non-zero element of the $\Gamma(K)$-socle of $\mathrm{Ext}^1(Y, K)$. Now f is right minimal, right C-determined, and not an isomorphism, thus $|f|_C \geq 1$. Assume that $|f|_C \geq 2$, thus there is a C-factorization (h_1, h_2) of f. Since $f = h_2 h_1$ is surjective, also h_2 is surjective. Since h_2 is right minimal, right C-determined and not invertible, its kernel has to be of the form K^t for some $t \geq 1$. Thus there is a commutative diagram with exact rows of the following form

$$\begin{array}{ccccccccc} 0 & \longrightarrow & K & \longrightarrow & X & \xrightarrow{f} & Y & \longrightarrow & 0 \\ & & \downarrow{\scriptstyle h_1'} & & \downarrow{\scriptstyle h_1} & & \| & & \\ 0 & \longrightarrow & K' & \longrightarrow & X' & \xrightarrow{h_2} & Y & \longrightarrow & 0 \end{array}$$

where again the upper row is ϵ. Write $h_1' = (\phi_1, \ldots, \phi_t)$ with endomorphisms $\phi_i : K \to K$. If all ϕ_i belong to the radical of $\Gamma(K)$, then all the sequences induced from ϵ by the maps ϕ_i split, thus also the lower sequence splits, since it is induced from ϵ by h_1'. Thus, at least one of the maps ϕ_i has to be invertible and therefore h_1' is a split monomorphism.

If $t > 1$, then the lower sequence splits off a sequence $0 \to K \xrightarrow{1} K \to 0 \to 0$, but this means that h_2 is not right minimal. Thus $t = 1$. But then h_1' is an automorphism, thus h_1 is invertible, a contradiction. This shows that $|f|_C = 1$. □

We say that an epimorphism f is *epi-irreducible,* provided for any factorization $f = f'f''$ with f'' a proper epimorphism, the map f' is a split epimorphism (the dual concept of mono-irreducible maps has been considered in [35]).

Proposition 7.5 *Let C be indecomposable, non-projective and let $K = \tau C$. If $f : X \to Y$ is an epi-irreducible epimorphism with kernel $K = \tau C$, then f belongs to ${}^{C}[\to Y\rangle^{1}$.*

Proof According to Proposition 7.4, we have to show that the given exact sequence

$$\epsilon : \quad 0 \longrightarrow K \longrightarrow X \xrightarrow{f} Y \longrightarrow 0$$

belongs to the $\Gamma(K)$-socle of $\mathrm{Ext}^1(Y, K)$. Thus, let ϕ be a non-invertible endomorphism of K, write it in the form $\phi = \phi'\phi''$ with ϕ'' epi and ϕ' mono. Consider the exact sequence induced from ϵ by ϕ''

$$\begin{array}{ccccccccc} 0 & \longrightarrow & K & \longrightarrow & X & \xrightarrow{f} & Y & \longrightarrow & 0 \\ & & \downarrow{\scriptstyle \phi''} & & \downarrow{\scriptstyle f''} & & \| & & \\ 0 & \longrightarrow & K & \longrightarrow & X' & \xrightarrow{f'} & Y & \longrightarrow & 0 \end{array}$$

Since ϕ'' is a proper epimorphism, also f'' is a proper epimorphism, thus f' is a split epimorphism. But this implies that also the exact sequence induced from ϵ by ϕ splits. □

We will need some basic facts concerning the Gabriel–Roiter measure of finite length modules, see [35]. The Gabriel–Roiter measure of a module M will be denoted by $\gamma(M)$. We recall that any indecomposable module M which is not simple has a Gabriel–Roiter submodule M', this is a certain indecomposable submodule of M and the embedding $M' \to M$ is called a *Gabriel–Roiter inclusion.* Recall that a Gabriel–Roiter inclusion $M' \to M$ is mono-irreducible: this means that for any proper submodule M'' of M with $M' \subseteq M''$, the inclusion $M' \subseteq M''$ splits. As a consequence, given any nilpotent endomorphism f of M/M', the sequence induced from $0 \to M' \to M \to M/M' \to 0$ using f splits. Also, it follows that the cokernel M/M' of a Gabriel–Roiter inclusion is indecomposable (and not projective).

Of course, we may use duality and consider a Gabriel–Roiter submodule U of DM, the corresponding projection $M = D^2M \to DU$ will be called a *co-Gabriel–Roiter projection.* By duality, a co-Gabriel–Roiter projection is an epi-irreducible epimorphism with (non-injective) indecomposable kernel.

Corollary 7.6 *Let M be an indecomposable module which is not simple and let $f : M \to Y$ be a co-Gabriel–Roiter projection, say with kernel K. Let $C = \tau^{-}K$. Then f is right minimal, right C-determined and $|f|_C = 1$, thus $[f\rangle$ belongs to ${}^{C}[\to Y\rangle^{1}$.*

Remark If M is indecomposable and not simple, we also may consider a Gabriel–Roiter submodule U of M, say with projection $p : M \to M/U = Y'$ and consider $C' = \tau^{-}U$. Then p is right minimal and right C'-determined, however in general there is not a fixed number t such that $[p\rangle$ belongs to ${}^{C'}[\to Y'\rangle^{t}$ or to ${}^{C'}[\to Y'\rangle_{t}$. A typical example is example 8 presented in the next section. The two modules $P(b)$ and $\tau^{-}S(a)$ both have $P(a)$ as a Gabriel–Roiter submodule with factor module $S(b)$. Let $C' = \tau^{-}P(a)$. The projection $P(b) \to S(b)$ belongs to ${}^{C'}[\to S(b)\rangle^{2} = {}^{C'}[\to S(b)\rangle_{0}$ whereas the projection $\tau^{-}S(a) \to S(b)$ belongs to ${}^{C'}[\to S(b)\rangle^{1} = {}^{C'}[\to S(b)\rangle_{1}$.

8 Epimorphisms in $[\to Y\rangle$

The set of right equivalence classes $[f\rangle$ where f is an epimorphism is obviously a coideal of the lattice $[\to Y\rangle$, we denote it by $[\to Y\rangle_{\text{epi}}$. Since the pullback of an epimorphism is again an epimorphism, we see that $[\to Y\rangle_{\text{epi}}$ is closed under meets. Also, for any module C, the subset ${}^C[\to Y\rangle_{\text{epi}}$ of ${}^C[\to Y\rangle$ consisting of the right equivalence classes of all right C-determined epimorphisms ending in Y is a coideal which is closed under meets. Since ${}^C[\to Y\rangle$ is a lattice of finite height, we see that ${}^C[\to Y\rangle_{\text{epi}}$ has a unique minimal element, say $[f_0\rangle$ and our first aim will be to describe $\eta_{CY}(f_0)$.

Before we deal with this question, let us point out in which way the projectivity or non-projectivity of indecomposable direct summands of C are related to the fact that right minimal right C-determined morphisms f are mono or epi. If f is a monomorphism, then f is right minimal and right Λ-determined (see Corollary 3.8), thus right C-determined for some projective module C. Conversely, if C is projective, then any right minimal, right C-determined morphism is a monomorphism (see Corollary 3.7). Namely, if K is an indecomposable direct summand of the kernel of f, where f is right minimal, then K is not injective and $\tau^- K$ is a direct summand of any module C such that f is right C-determined. Of course, since K is not injective, $\tau^- K$ is an indecomposable non-projective module. One should be aware that a morphism may be right C-determined for some module C without any indecomposable projective direct summand, without being surjective.

Example 7 As in example 6, we take as Λ the path algebra of the linearly directed quiver Δ of type $\mathbb{A}_3$, modulo the zero relation $\alpha\beta$. Again, let $Y = P(c)$ and $C = S(b)$. As we have mentioned already, the non-zero maps $f : P(b) \to P(c)$ are not surjective, but right $S(b)$-determined, and, of course, $S(b)$ is not projective. (As we will see below, it is essential for this feature that the kernel of f has injective dimension at least 2; for a general discussion of maps which are not surjective, but right C-determined by a module C without any indecomposable projective direct summands, we refer to [38]).

The submodule $\operatorname{Hom}(C, \mathcal{P}, Y)$ **of** $\operatorname{Hom}(C, Y)$. We denote by $\operatorname{Hom}(C, \mathcal{P}, Y)$ the set of morphisms $C \to Y$ which factor through a projective module. Note that $\operatorname{Hom}(C, \mathcal{P}, Y)$ is a $\Gamma(C)$-submodule of $\operatorname{Hom}(C, Y)$.

Proposition 8.1 *Assume that $f : X \to Y$ is right C-determined. Then f is surjective if and only if $\eta_{CY}(f) \supseteq \operatorname{Hom}(C, \mathcal{P}, Y)$.*

Proof One direction is a trivial verification: First assume that f is surjective. Let h belong to $\operatorname{Hom}(C, \mathcal{P}, Y)$, thus $h = h_2 h_1$ where $h_1 : C \to P)$ and $h_2 : P \to Y$ with P projective. Since f is surjective and P is projective, there is $h_2' : P \to X$ such that $h_2 = f h_2'$. Thus shows that $h = h_2 h_1 = f h_2' h_1$ belongs to $f \operatorname{Hom}(C, X) = \eta_{CY}(f)$.

The converse is more interesting, here we have to use that f is right C-determined. We assume that $\eta_{CY}(f) \supseteq \operatorname{Hom}(C, \mathcal{P}, Y)$. Let $p : P(Y) \to Y$ be a projective cover of Y. Consider an arbitrary morphism $\phi : C \to P(Y)$. The composition $p\phi$ belongs to $\operatorname{Hom}(C, \mathcal{P}, Y)$, thus to $\eta_{CY}(f) = f \operatorname{Hom}(C, X)$. Since f is right C-determined, it follows that p itself factors through f, say $p = fp'$ for some $p' : P(Y) \to X$. Now the composition $fp' = p$ is surjective, there f has to be surjective. □

Let us denote by ${}^{C}[\to Y\rangle_{\mathrm{epi}}$ the subset of ${}^{C}[\to Y\rangle$ given by all elements $[f\rangle$ with f an epimorphism.

Proposition 8.2 *The restriction of the Auslander bijection η_{CY} yields a poset isomorphism*

$$ {}^{C}[\to Y\rangle_{epi} \longrightarrow \mathcal{S}(\mathrm{Hom}(C,Y)/\mathrm{Hom}(C,\mathcal{P},Y)) = \mathcal{S}\underline{\mathrm{Hom}}(C,Y) $$

such that the following diagram commutes:

$$\begin{array}{ccc} {}^{C}[\to Y\rangle_{epi} & \longrightarrow & \mathcal{S}\underline{\mathrm{Hom}}(C,Y) \\ \downarrow & & \downarrow \\ {}^{C}[\to Y\rangle & \xrightarrow{\eta_{CY}} & \mathcal{S}\,\mathrm{Hom}(C,Y) \end{array}$$

Here, the vertical maps are the canonical inclusions.

Proof It is well-known that given a module M and a submodule M', then the lattice of submodules of the factor module M/M' is canonically isomorphic to the lattice of the submodules U of M satisfying $M' \subseteq U$. This is the vertical map on the right. □

More generally, dealing with a morphism f which is right C-determined, we can recover the image of f as follows:

Proposition 8.3 *Let $f : X \to Y$ be right C-determined. Then one recovers the image of f as the largest submodule Y' of Y (with inclusion map $u : Y' \to Y$) such that $u\,\mathrm{Hom}(C,\mathcal{P},Y') \subseteq f\,\mathrm{Hom}(C,X)$.*

Proof Let Y' be the image of f with inclusion map u and $uf' = f$ (with f' surjective). First of all, we show that $u\,\mathrm{Hom}(C,\mathcal{P},Y') \subseteq f\,\mathrm{Hom}(C,X)$. Let $\phi' : C \to \Lambda$ and $\phi'' : \Lambda \to Y'$ (the maps $\phi''\phi'$ obtained in this way generate $\mathrm{Hom}(C,\mathcal{P},Y')$ additively). We want to show that $u\phi''\phi$ factors through f. Since $f' : X \to Y'$ is surjective, there is $\psi : \Lambda \to X$ such that $\phi'' = f'\psi$ (since Λ is projective). Thus $u\phi''\phi' = uf'\psi\phi' = f\psi\phi'$. Thus $u\phi''\phi'$ factors through f.

On the other hand, let $u'' : Y'' \to Y$ be a submodule of Y such that $u''\,\mathrm{Hom}(C,\mathcal{P},Y'') \subseteq f\,\mathrm{Hom}(C,X)$. Let $p : P(Y'') \to Y''$ be a projective cover. Consider the map $f' = u''p : P(Y'') \to Y$. It has the property that for all maps $\phi : C \to P(Y'')$ the composition $f'\phi$ factors through f (namely $f'\phi = u''p\phi$ belongs to $u''\,\mathrm{Hom}(C,\mathcal{P},Y'') \subseteq f\,\mathrm{Hom}(C,X)$). But f is right C-determined, thus we conclude that f' factors through α, say $f' = f\phi'$ for some $\phi' : C \to P(Y'')$. Thus the image Y'' of f' is contained in the image Y' of f. This is what we wanted to prove. □

We recover in this way Proposition 8.1. Namely, if f is surjective, then Y is the image of f, thus Y is one of the submodule Y' with $u\,\mathrm{Hom}(C,\mathcal{P},Y') \subseteq f\,\mathrm{Hom}(C,X)$, thus $\mathrm{Hom}(C,\mathcal{P},Y) \subseteq f\,\mathrm{Hom}(C,Y)$.

Conversely, if $\mathrm{Hom}(C,\mathcal{P},Y) \subseteq f\,\mathrm{Hom}(C,Y)$, then Y is one of the submodules Y' with $u\,\mathrm{Hom}(C,\mathcal{P},Y') \subseteq f\,\mathrm{Hom}(C,X)$ and therefore the image of f contains Y, thus is equal to Y. This shows: *If f is right C-determined, then f is surjective if and only if* $\operatorname{Hom}(C, \mathcal{P}, Y) \subseteq f \operatorname{Hom}(C, X)$.

Corollary 8.4 *Let C, Y be modules.*

(a) *All maps in ${}^{C}[\to Y\rangle$ are epimorphisms if and only of* $\operatorname{Hom}(C, \mathcal{P}, Y) = 0$.

(b) $\operatorname{Hom}(C, \mathcal{P}, Y) = \operatorname{Hom}(C, Y)$ *if and only if the only element $[f\rangle$ in ${}^{C}[\to Y\rangle$ with f surjective is the right equivalence class of the identity map $Y \to Y$, if and only if any surjective map ending in Y with kernel in* add τC *splits.*

Kernels with injective dimension at most 1.

Proposition 8.5 *Let K be a module and $C = \tau^{-} K$. The following conditions are equivalent:*

(i) *The injective dimension of K is at most* 1.
(ii) *If Y is any module, then all maps in ${}^{C}[\to Y\rangle$ are epimorphisms.*
(iii) *We have* $\operatorname{Hom}(C, \mathcal{P}, Y) = 0$ *for all modules Y.*

Proof Recall from [33], 2.4 that K has injective dimension at most 1 if and only if $\operatorname{Hom}(C, \Lambda) = 0$. Thus, if K has injective dimension at most 1 and Y is an arbitrary module, then $\operatorname{Hom}(C, \mathcal{P}, Y) = 0$, this shows that (i) implies (iii). Conversely, assume the condition (iii), thus $\operatorname{Hom}(C, \mathcal{P}, Y) = 0$ for all modules Y. If the injective dimension of K would be at least 2, then $\operatorname{Hom}(C, \Lambda) \neq 0$. But $\operatorname{Hom}(C, \mathcal{P}, \Lambda) = \operatorname{Hom}(C, \Lambda)$. This contradiction shows that (iii) implies (i). For the equivalence of (ii) and (iii) see Corollary 8.4(a). □

Corollary 8.6 *Let Λ be hereditary and C a module without any indecomposable projective direct summand. Then any right C-determined morphism $f : X \to Y$ is an epimorphism.*

Proof Let $K = \tau C$. Since C has no indecomposable projective direct summand, it follows that $C = \tau^{-} K$. Since Λ is hereditary, the injective dimension of any module is at most 1. Since the injective dimension of K is at most 1, it follows from the proposition that all right C-determined maps are epimorphisms. □

Of course, we also can show directly that $\operatorname{Hom}(C, \mathcal{P}, Y) = 0$. Namely, let $g : C \to Y$ be in $\operatorname{Hom}(C, \mathcal{P}, Y)$. Then $g = g_2 g_1$ with $g_1 : C \to P$, where P is a projective module. The image P' of g_1 is a submodule of P, thus, since Λ is hereditary, the module P' is also projective. Thus, we have a surjective map $C \to P'$ with P' projective. Such a map splits. This shows that P' is isomorphic to a direct summand of C. It follows that $P' = 0$ and therefore $g = 0$.

Riedtmann–Zwara degenerations. Recall that M' is a Riedtmann–Zwara degeneration of M if and only if there is an exact sequence of the form

$$0 \to K \to K \oplus M \to M' \to 0,$$

or, equivalently, if and only if there is an exact sequence of the form

$$0 \to M' \to M \oplus L \to L \to 0;$$

(in both sequences we can assume that the maps $K \to K$ and $L \to L$, respectively, are in the radical).

In terms of the Auslander bijection, we may deal with these data in several different ways: namely, we may look at the right equivalence classes of both $[M \to M'\rangle$ and $[K \oplus M \to M'\rangle$ in $[\to M'\rangle$ as well as at the right equivalence class $[M' \to M\rangle$ in $[\to M\rangle$. In case we deal with $[K \oplus M \to M'\rangle$, one should be aware that this map $[K \oplus M \to M'\rangle$ is the join of the two maps $[K \to M'\rangle$ and $[M \to M'\rangle$ in $[\to M'\rangle$.

In addition, we also may concentrate on the possible maps $K \to K$ and $L \to L$ (sometimes called steering maps).

When dealing with epimorphisms in ${}^{C}[\to Y\rangle$, Riedtmann–Zwara degenerations play a decisive role, as the following proposition shows:

Proposition 8.7 *Let $f : X \to Y$ and $f' : X' \to Y$ be epimorphisms with isomorphic kernels. If $[f\rangle \leq [f'\rangle$, then X' is a Riedtmann–Zwara degeneration of X.*

Proof Let $h : X \to X'$ with $f = f'h$. Let $u : K \to X$ and $u' : K \to X'$ be the kernel maps. Since $f = f'h$, there is $h' : K \to K$ such that $u'h' = hu$, thus we deal with the following commutative diagram with exact sequences:

$$\begin{array}{ccccccccc} 0 & \longrightarrow & K & \xrightarrow{u} & X & \xrightarrow{f} & Y & \longrightarrow & 0 \\ & & \downarrow{\scriptstyle h'} & & \downarrow{\scriptstyle h} & & \| & & \\ 0 & \longrightarrow & K & \xrightarrow{u'} & X & \xrightarrow{f'} & Y & \longrightarrow & 0 \end{array}$$

The diagram shows that the lower exact sequence is induced from the upper one by h'. But this means that the following sequence is exact:

$$0 \longrightarrow K \xrightarrow{\begin{bmatrix} u \\ -h' \end{bmatrix}} K \oplus X \xrightarrow{\begin{bmatrix} h & u' \end{bmatrix}} X' \longrightarrow 0$$

This is a Riedtmann–Zwara sequence, thus X' is a Riedtmann–Zwara degeneration of X. □

Example 8 Let Λ be given by the quiver with one loop α at the vertex a and an arrow $b \to a$, with the relation $\alpha^2 = 0$.

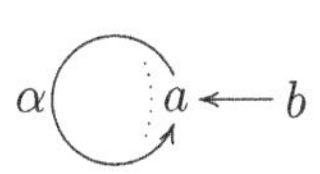

Let $K = P(a)$ and $Y = S(b)$, thus $\dim \operatorname{Ext}^1(Y, K) = 2$. The module $C = \tau^- K$ is of length 4 with socle a and top $b \oplus b$, thus $\dim \operatorname{Hom}(C, Y) = 2$. Note that $\Gamma(C) = \operatorname{End}(C)^{\mathrm{op}} = k[t]/t^2$ and that $\operatorname{Hom}(C, Y)$ as a $\Gamma(C)$-module is cyclic.

The universal cover of the Auslander–Reiten quiver of Λ looks as follows:

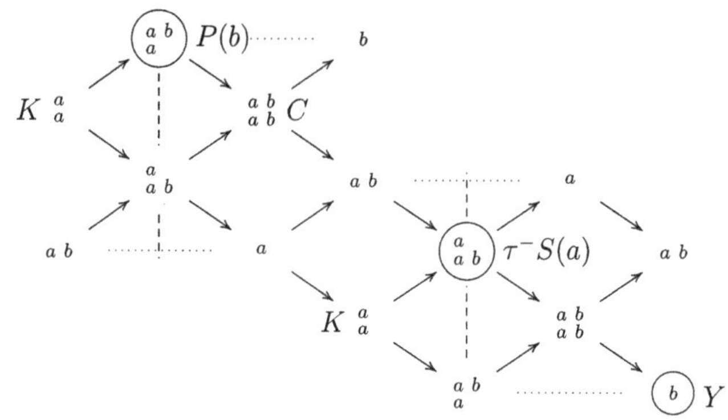

The region in-between the dashed lines is a fundamental domain of the Auslander–Reiten quiver of Λ; the Auslander–Reiten quiver of Λ is obtained by identifying these lines in order to form a Moebius strip.

The encircled vertices in the Auslander–Reiten quiver yield ${}^{C}[\to\ Y\rangle$. Here is ${}^{C}[\to\ Y\rangle$ as well as $\mathcal{S}\,\mathrm{Hom}(C, Y)$:

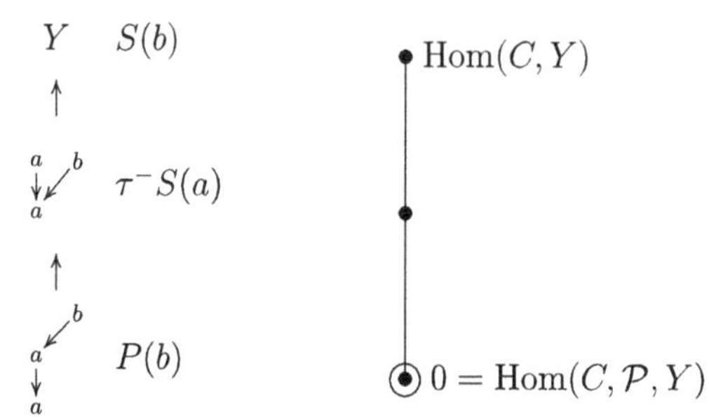

9 Modules K with semisimple endomorphism ring

We start with a well-known characterization of such modules.

Lemma 9.1 *Let K be a module. The following conditions are equivalent:*

(i) *The endomorphism ring of K is semisimple.*
(ii) *There are pairwise orthogonal bricks $K_1, \ldots, K_n$ such that* $\mathrm{add}\,K = \mathrm{add}\,\{K_1, \ldots, K_n\}$.

Proposition 9.2 *Let K be a module with semisimple endomorphism ring. Let $f = f'h$, where f, f' are right minimal epimorphisms with kernels in* add K*, starting at the same module X. Then h is surjective and its kernel belongs to* add K.

Proof By assumption, there is a commutative diagram with exact rows

$$\begin{array}{ccccccccc} 0 & \longrightarrow & K & \xrightarrow{u} & X & \xrightarrow{f} & Y & \longrightarrow & 0 \\ & & \downarrow{\scriptstyle h'} & & \downarrow{\scriptstyle h} & & \| & & \\ 0 & \longrightarrow & K' & \xrightarrow{u'} & X & \xrightarrow{f'} & Y & \longrightarrow & 0 \end{array}$$

with inclusion maps u, u' such that both K, K' belong to add K. Since K, K' belong to add K and the endomorphism ring of K is semisimple, there is a submodule K'' of K' such that $K' = \mathrm{Im}(h') \oplus K''$. Let us denote by $u'' : K'' \to K'$ the inclusion map and by $q' : K' \to K'/\mathrm{Im}(h') = K''$ the canonical projection, thus $q'u'' = 1_{K''}$. Now $q' : K' \to K''$ is the cokernel of h', and we can identify K'' and we can complete the diagram above by inserting the cokernels of h and h' as follows:

$$\begin{array}{ccccccccc} 0 & \longrightarrow & K & \xrightarrow{u} & X & \xrightarrow{f} & Y & \longrightarrow & 0 \\ & & \downarrow{\scriptstyle h'} & & \downarrow{\scriptstyle h} & & \| & & \\ 0 & \longrightarrow & K' & \xrightarrow{u'} & X & \xrightarrow{f'} & Y & \longrightarrow & 0 \\ & & \downarrow{\scriptstyle q'} & & \downarrow{\scriptstyle q} & & & & \\ & & K'' & = & K'' & & & & \\ & & \downarrow & & \downarrow & & & & \\ & & 0 & & 0 & & & & \end{array}$$

Since $qu' = q'$, we see that $qu'u'' = q'u'' = 1_{K''}$, thus K'' is a direct summand of X which lies inside the kernel of f'. Since we assume that f' is right minimal, it follows that $K'' = 0$, thus h' is surjective. Therefore also h is surjective.

On the other hand, the kernel of h can be identified with the kernel of h', and again using that K, K' belong to add K and that the endomorphism ring of K is semisimple, we see that the kernel of h belongs to add K. □

Proposition 9.3 *Let K be a module and assume that the endomorphism ring of K is semisimple. Let $C = \tau^- K$ and assume that* $\mathrm{Hom}(C, \mathcal{P}, Y) = 0$. *If $f : X \to Y$ is right minimal and right C-determined, then f is surjective and*

$$\boxed{|f|_C = \mu(\mathrm{Ker}(f))}$$

where $\mu(M)$ is the Krull–Remak–Schmidt number (the number of direct summands when M is written as a direct sum of indecomposable modules.

Also, $\eta_{CY}^{-1}(0)$ is given by the universal extension from below

$$0 \to K' \to X \to Y \to 0$$

with $K' \in \mathrm{add}\, K$.

Proof Since we assume that $\mathrm{Hom}(C, \mathcal{P}, Y) = 0$, all right minimal right C-determined maps $f : X \to Y$ are surjective, and the kernel of such a map is in add K. Let $|f|_C = n$ and consider a (maximal) chain

$$X_n \xrightarrow{h_n} X_{n-1} \xrightarrow{h_{n-1}} \cdots \xrightarrow{h_2} X_1 \xrightarrow{h_1} X_0 = Y$$

of non-invertible maps such that the compositions $f_t = h_1 \cdots h_t$ for $1 \le t \le n$ are right minimal and right C-determined. Now all the maps f_t are right minimal epimorphisms with kernels in add K, thus, according to proposition 7.2, also the maps h_t are epimorphisms with kernels K_t in add K. Since we assume that the endomorphism ring of K is semisimple, we see that the kernel of f_t is just $\bigoplus_{i=1}^{t} K_t$. Now let $K_t = K' \oplus K''$ be a direct decomposition with K' indecomposable, and let $h : X_t \to X_t/K'$ be the canonical projection. Since K' is contained in the kernel of h_t, we can factor h_t through h and obtain a map $h'_t : X_t/K' \to X_{t-1}$ with kernel X'' such that $h_t = h'_t h$. Altogether we have obtained a refinement

$$X_n \xrightarrow{h_n} X_{n-1} \xrightarrow{h_{n-1}} \cdots \longrightarrow X_t \xrightarrow{h} X_t/K' \xrightarrow{h'_t} X_{t-1} \longrightarrow \cdots \xrightarrow{h_2} X_1 \xrightarrow{h_1} X_0 = Y$$

with $h_t = h'_t h$. Let $f' = f_{t-1}h'_t = h_1 \cdots h_{t-1}h'_t$. We apply Proposition 9.1 to f_t and f'. Note that $f_t = f_{t-1}h_t = f_{t-1}h'_t h = f'h$ and the corresponding map from the kernel of f_t to the kernel of f' is just the split epimorphism $K_t \to K_t/K'$. Thus 9.1 asserts that f' is right minimal (and of course also right C-determined). The maximality of the chain $(h_1, \ldots, h_t)$ implies that h'_t has to be invertible, thus $K'' = 0$. This shows that K_t is indecomposable. Altogether, we see that $\mu(\mathrm{Ker}(f_t)) = t$, thus $\mu(\mathrm{Ker}(f)) = n = |f|_C$.

The last assertion is obvious: if

$$0 \longrightarrow K' \longrightarrow X \xrightarrow{f} Y \longrightarrow 0$$

is the universal extension from below with $K' \in \mathrm{add}\, K$, then $[f\rangle$ belongs to ${}^{C}[\to Y\rangle$ and any other extension of Y from below with kernel in add K is induced from it. Thus $[f\rangle$ has to be the zero element of the lattice ${}^{C}[\to Y\rangle$. □

Corollary 9.4 *Let K be a module with injective dimension at most* 1 *and assume that the endomorphism ring of K is semisimple. Let $C = \tau^{-}K$. If $f : X \to Y$ is right minimal and right C-determined, then f is surjective and $|f|_C = \mu(\mathrm{Ker}(f))$. Also, $\eta_{CY}^{-1}(0)$ is given by the universal extension from below using modules in* add K.

Proof Combine Proposition 8.5 and Proposition 9.3. □

Example 9 Consider the 3-subspace quiver

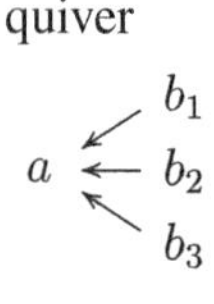

We consider the indecomposable modules which are neither projective, nor injective: these are the modules $N(i) = \tau Q(b_i)$, for $1 \le i \le 3$ and $M = \tau Q(a)$.

Let $Y = Q(a)$ and $C = N(1) \oplus N(2) \oplus N(3)$. Then ${}^{C}[\to Q(a)\rangle$ looks as follows:

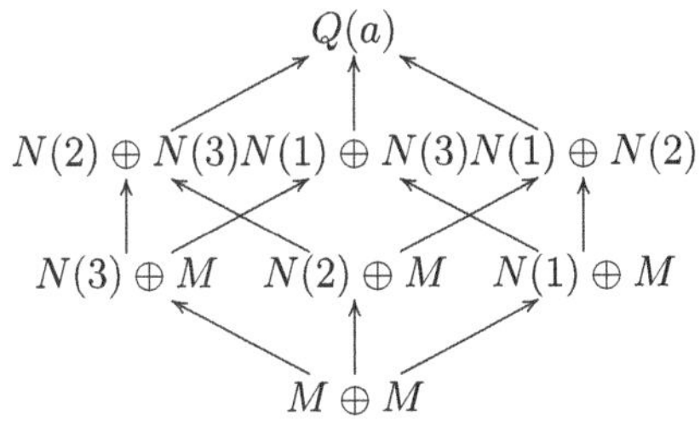

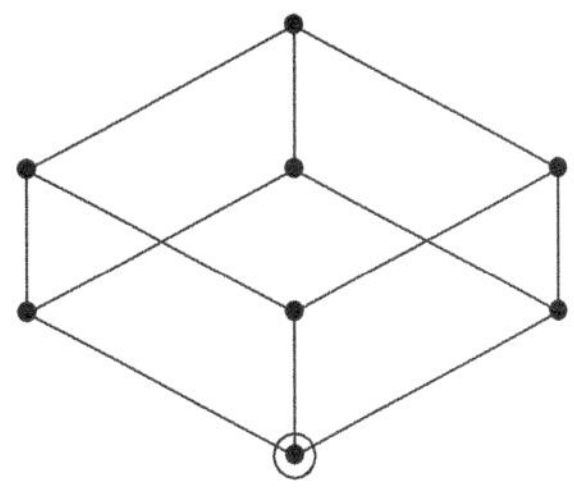

Note that here we have $K = \tau C = P(B_1) \oplus P(b_2) \oplus P(b_3)$.

10 Comparison with Auslander–Reiten theory

Let $C = Y$ be indecomposable and consider the Auslander bijection for $C = Y$:

$$^{Y}[\to Y\rangle \longleftrightarrow \mathcal{S}\,\mathrm{Hom}(Y, Y).$$

The subspace $\mathrm{rad}(Y, Y)$ of $\mathrm{Hom}(Y, Y)$ on the right corresponds to the right almost split map ending in Y. Two possible cases have to be distinguished:

If $Y = P$ is projective, then we get a morphism which is right determined by a projective module, thus we must get a monomorphism. Of course, what we obtain is just the embedding of the radical $\mathrm{rad}\, P$ into P.

If Y is not projective, then we get an epimorphism with kernel in $\mathrm{add}\, \tau Y$. Actually, we get the epimorphism of the Auslander–Reiten sequence ending in Y, thus the kernel is precisely τY.

What we see is that *the minimal right almost split map ending in Y is a waist in* $^{Y}[\to Y\rangle$ and it just corresponds to the waist $\mathrm{rad}(Y, Y) \subset \mathrm{End}(Y)$.

More generally, we have:

Proposition 10.1 *Let Y be indecomposable and Y a direct summand of C, and consider*

$$^{C}[\to Y\rangle \longleftrightarrow \mathcal{S}\,\mathrm{Hom}(C, Y).$$

Then $\mathcal{S}\,\mathrm{Hom}(C, Y)$ is a local module (with maximal submodule $\mathrm{Im}\,\mathrm{Hom}(C, g)$, *where g is minimal right almost split ending in Y).*

Proof Let $\tau Y \longrightarrow \mu Y \xrightarrow{g} Y$ be the Auslander–Reiten sequence ending in Y. Then the right-equivalence class $[g\rangle$ of g belongs to $^{C}[\to Y\rangle$ and every map $X \to Y$ which is not a split epimorphism, factors through g. This means that every element of $^{C}[\to Y\rangle$ different from the identity map $Y \to Y$ is less or equal to $[g\rangle$. This means that $^{C}[\to Y\rangle$ has a unique maximal submodule, namely $\mathrm{Hom}(C, g)$. □

The Auslander–Reiten formula $\mathrm{Ext}^1(Y, K) \simeq D\underline{\mathrm{Hom}}(\tau^- K, Y)$. The Auslander bijection provides a bijection between the submodules of $\underline{\mathrm{Hom}}(\tau^- K, Y)$ with the right equivalence classes of surjective maps $M \to Y$ with kernel in $\mathrm{add}(K)$ (namely, the submodules of $\underline{\mathrm{Hom}}(\tau^- K, Y)$ just correspond bijectively to the submodules of

$\mathrm{Hom}(\tau^- K, Y)$ which contain $\mathrm{Hom}(\tau^- K, \mathcal{P}, Y)$). Consider the following triangle:

$$\begin{array}{ccccc}
(f,g) & & \{K \xrightarrow{f} M \xrightarrow{g} Y \mid \text{short exact}\} & & \\
\downarrow & & \downarrow & \searrow^{\; g \mapsto g\,\mathrm{Hom}(\tau^- K, M)} & \\
[f,g] & & \mathrm{Ext}^1(Y,K) & & \\
\downarrow & & \downarrow & & \\
[g\rangle & & {}^{\tau^- K}[\to Y\rangle_{\mathrm{epi}} & \xrightarrow{\eta_{\tau^- K, Y}} & \mathcal{S}\underline{\mathrm{Hom}}(\tau^- K, Y)
\end{array}$$

Here, the two left hand columns concern the canonical way of attaching to a short exact sequence $\epsilon = (f, g)$ the corresponding element $[\epsilon] = [f, g]$ in Ext^1; if (f, g) and (f', g') are short exact sequences with $[f, g] = [f', g']$ in Ext^1, then the maps f, f' are right equivalent, thus $[f, g] \mapsto [g\rangle$ is a well-defined map $\mathrm{Ext}^1(Y, K) \to {}^{\tau^- K}[\to Y\rangle_{\mathrm{epi}}$ and we obtain a commutative triangle as shown.

Having another look at the left columns of the triangle, the reader should obverse that the split short exact sequence which yields the zero element of Ext^1 gives the unit element of the lattice ${}^{\tau^- K}[\to Y\rangle_{\mathrm{epi}}$, namely the identity map $1 : Y \to Y$.

Proposition 10.2 *Let $g : M \to Y$ be surjective with kernel K and $C = \tau^- K$. Then an element $h \in \mathrm{Hom}(C, Y)$ belongs to the set $\eta_{C,Y}(g)$ if and only if the induced sequence $h_*([f, g])$ splits. Also, the set $\eta_{C,Y}(g)$ is a $\Gamma(C)$-submodule of* $\mathrm{Hom}(C, Y)$, *where* $\Gamma(C) = \mathrm{End}(C)^{op}$.

Proof Given an element $h \in \mathrm{Hom}(C, Y)$, the induced sequence $h_*([f, g])$ splits if and only if h factors through g. But we have already noted that $\eta_{C,Y}(g)$ is the set of all elements $h \in \mathrm{Hom}(C, Y)$ which factor through g. And we know that $\eta_{C,Y}(g)$ is a $\Gamma(C)$-submodule of $\mathrm{Hom}(C, Y)$. □

Now let us invoke the Auslander–Reiten formula: $\mathrm{Ext}^1(Y, K) \simeq D\underline{\mathrm{Hom}}(\tau^- K, Y)$. Here we use that we deal with an artin algebra Λ, thus the center k of Λ is a (commutative) artinian ring and D is the duality functor $\mathrm{mod}\, k \to \mathrm{mod}\, k$ given by a minimal cogenerator of $\mathrm{mod}\, k$.

There are the following two horizontal bijections as well as the vertical map on the left:

$$\begin{array}{ccccc}
[f,g] & \mathrm{Ext}^1(Y,K) & \simeq & D\underline{\mathrm{Hom}}(\tau^- K, Y) \\
\downarrow & \downarrow & & \vdots \\
[g\rangle & {}^{\tau^- K}[\to Y\rangle_{\mathrm{epi}} & \xrightarrow{\eta_{\tau^- K, Y}} & \mathcal{S}\underline{\mathrm{Hom}}(\tau^- K, Y)
\end{array}$$

We have inserted a dashed arrow on the right which corresponds to the composition of the three given maps

$$D\underline{\mathrm{Hom}}(\tau^- K, Y) \longrightarrow \mathrm{Ext}^1(Y, K) \longrightarrow {}^{\tau^- K}[\to Y\rangle_{\mathrm{epi}} \longrightarrow \mathcal{S}\underline{\mathrm{Hom}}(\tau^- K, Y)$$

It seems to be of interest to describe in detail this composition! One may conjecture that here one attaches to a linear map $\alpha \in D\underline{\mathrm{Hom}}(\tau^- K, Y)$ the largest Γ-submodule of $\underline{\mathrm{Hom}}(\tau^- K, Y)$ lying in the kernel of α.

Let us now assume that K is indecomposable so that add K consists of direct sums of copies of K. Given a short exact sequence $K \xrightarrow{f} M \xrightarrow{g} Y$, the map g is right minimal if and only if the sequence does not split. On the other hand, given $[g\rangle \in {}^{\tau^- K}[\to Y\rangle_{\mathrm{epi}}$ with g right minimal, the kernel of g is the direct sum of say t copies of K and $t \leq 1$ just means that $[g\rangle$ is obtained from a short exact sequence $K \xrightarrow{f} M \xrightarrow{g} Y$. It follows that under the assumption that K is indecomposable, it is easy to identify the elements of $\mathcal{S}\,\mathrm{Hom}(\tau^- K, Y)$ which are images of the elements of $\mathrm{Ext}^1(Y, K)$ under $\eta_{\tau^- K, Y}$.

Comparison. In which way are the Auslander bijections better than the Auslander–Reiten formula? What is the advantage of the Auslander bijection compared to the Auslander–Reiten formula?

1) We do not only deal with the set $\underline{\mathrm{Hom}}(C, Y)$ but with all of $\mathrm{Hom}(C, Y)$. To extend such a bijection as given by the Auslander–Reiten formula to a larger setting should always be of interest. But also note that the set $\underline{\mathrm{Hom}}(C, Y)$ depends on the module category which we consider, not just on the modules themselves.
2) The duality is replaced by a covariant bijection.
3) The usual Auslander–Reiten picture concerns indecomposable modules, and almost split sequences, thus indecomposable modules and irreducible maps. In the language of the Auslander bijection, we only deal with $C = Y$ indecomposable and only with the submodule $\mathrm{rad}(C, C) \subset \mathrm{Hom}(C, C)$, whereas
   - we should not restrict to indecomposable modules,
   - and not to the condition $C = Y$,
   - and we want to deal with all submodules of $\mathrm{Hom}(C, Y)$, not just the radical subspace.

   Concerning the Auslander–Reiten-theory, there is an essential difference whether C is projective or not. If C is projective, we obtain an inclusion map, whereas if C is not projective, then we obtain an extension. —This feature dominates also the Auslander bijections: First of all, there is the extreme case of C being projective, then we consider submodules (and we consider arbitrarily ones, not just the radicals of the indecomposable projective modules). In general we deal with extensions ending in a submodule Y' of Y; if C is a generator, then we deal with all possible extensions of submodules of Y from below using modules in add τC.
4) The Auslander–Reiten theory only deals with the factor category of mod Λ modulo the infinite radical. The Auslander bijection takes care of **any** morphism.
5) Families of modules do not play any role in the Auslander–Reiten theory. As we will see soon, families of modules are an essential features in the frame of the Auslander bijections.

II. Families of modules

11 The modules present in ${}^C[\to Y\rangle$ are of bounded length

We say that a module M is *present in* ${}^C[\to Y\rangle$ provided there exists a right minimal map $f : M \to Y$ which is right C-determined; similarly, we say that M is *present in* ${}^{C}[\to Y\rangle^{t}$ provided there exists a right minimal map $f : M \to Y$ which is right C-determined and $|f|_C = t$.

Proposition 11.1 *There is a constant* $\lambda = \lambda(\Lambda)$ *such that for any pair of modules* C, Y *with* $C \neq 0$ *and any right minimal right* C*-determined map* $X \to Y$, *we have*

$$\boxed{|X| \le \lambda |C|^2 |Y|.}$$

Proof Write $C = \bigoplus C_i$ with indecomposable direct summands C_i and let $K_i = \tau C_i$. Note that $|K_i| \le d^2|C_i|$, where $d = |{}_\Lambda\Lambda|$ (see for example [30]). Of course, there is also the weaker bound $|K_i| \le d^2|C|$.

Let Λ be an artin k-algebra, where k is a commutative artinian ring. Since there are only finitely many simple modules and since $\mathrm{Ext}^1(Y, X)$ is a k-module of finite length for all Λ-modules X, Y of finite length, the length of the k-modules $\mathrm{Ext}^1(S, S')$, where S, S' are simple Λ-modules, is bounded, thus let e be the maximum.

The long exact Hom-sequences imply that the length of the k-module $\mathrm{Ext}^1(Y, X)$ is bounded by $e|X||Y|$, for any Λ-modules X, Y of finite length.

Now, let $f : X \to Y$ be right minimal and right C-determined. There is a short exact sequence $K' \to X \to Y'$ with Y' a submodule of Y, such that the module K' belongs to add τC and such that f is the composition of $X \to Y'$ and the inclusion map $Y' \to Y$. With f also the map $X \to Y'$ is right minimal.

Write $K' = \bigoplus_{i=1}^{s} K_i^{t_i}$. The Ext-Lemma of [35] asserts that t_i is bounded by the k-length of $\mathrm{Ext}^1(Y', K_i)$, since the map $X \to Y'$ is right minimal. Thus

$$t_i \le e|K_i||Y'| \le d^2e|C_i||Y'| \le d^2e|C_i||Y|,$$

and therefore $\sum t_i \le d^2e|C||Y|$. Thus

$$|K'| = \sum t_i|K_i| \le \sum t_i d^2|C| = d^2|C|\sum t_i \le d^4e|C|^2|Y|,$$

and therefore

$$|X| = |K'| + |Y'| \le d^4e|C|^2|Y| + |Y| \le d^4e|C|^2|Y| + |C|^2|Y| = (d^4e+1)|C|^2|Y|$$

(here we use that $C \neq 0$). Thus, let $\lambda = d^4e + 1$. □

There is the following converse:

Proposition 11.2 *Let* $Q = (D\Lambda)^b$. *Then any module of length at most* b *is present in* ${}^{\Lambda}[\to Q\rangle$.

Proof Let M be a module of length at most b, thus the socle soc M of M is a semisimple module of length at most b and therefore a submodule of $Q = (D\Lambda)^b$. It follows that M itself can be embedded into Q. Such an embedding $f : M \to Q$ is right minimal and right Λ-determined, thus M is present in ${}^{\Lambda}[\to Q\rangle$. □

Corollary 11.3 *Let $\mathcal{M}$ be a set of modules. The modules in $\mathcal{M}$ are of bounded length if and only if there exist modules C, Y such that any module M in $\mathcal{M}$ is present in ${}^{C}[\to Y\rangle$.*

Proof If all the modules in $\mathcal{M}$ are present in ${}^{C}[\to Y\rangle$, then they have to be of length at most $\lambda(\Lambda)|C|^2|Y|$, thus of bounded length, see Proposition 11.1. Conversely, if the modules in $\mathcal{M}$ are of length at most b, then they are present in ${}^{\Lambda}[\to Q\rangle$ where $Q = (D\Lambda)^b$, according to Proposition 11.2. □

Remark **Modules versus Morphisms.** As we have seen, given any infinite set $\mathcal{M}$ of **modules** of bounded length, there are modules C, Y such that all the modules in $\mathcal{M}$ are present in ${}^{C}[\to Y\rangle$. On the other hand, given modules X, Y, one cannot expect that the right equivalence classes $[f\rangle$ of all (or at least infinitely many) non-zero **morphisms** $f : X \to Y$ belong to some ${}^{C}[\to Y\rangle$, since the kernels of these maps f may belong to infinitely many isomorphism classes.

12 Minimal infinite families

Recall that a Krull–Remak–Schmidt category $\mathcal{C}$ is said to be *finite* provided the number of isomorphism classes of indecomposable objects in $\mathcal{C}$ is finite, otherwise $\mathcal{C}$ is said to be *infinite.*

Let $\mathcal{M}$ be a family of modules. We say that $\mathcal{M}$ is *minimal infinite* provided add $\mathcal{M}$ is infinite whereas add $\mathcal{M}'$ is finite, where $\mathcal{M}'$ is the set of modules M' which are proper submodules or proper factor modules of modules in $\mathcal{M}$.

Lemma 12.1 *If $\mathcal{M}$ is a minimal infinite family of modules (not necessarily of the same length), then there is an infinite subset $\mathcal{N} \subseteq \mathcal{M}$ which consists of pairwise non-isomorphic indecomposable modules of fixed length.*

Of course, $\mathcal{N}$ is again minimal infinite.

Proof Since add $\mathcal{M}$ is infinite, there is a sequence of modules $M_i \in \mathcal{M}$ with $i \in \mathbb{N}$ such that M_i does not belong to add$\{M_1, \dots, M_{i-1}\}$ for all i. Write $M_i = N_i \oplus N_i'$ such that N_i is indecomposable and does not belong to add$\{M_1, \dots, M_{i-1}\}$. It follows that the modules $N_1, N_2, \dots$ are indecomposable and pairwise non-isomorphic. Let I be the set of natural numbers such that $N_i' \neq 0$. If $i \in I$, then N_i is a proper submodule of M_i, thus belongs to add $\mathcal{M}'$. Since add $\mathcal{M}'$ is finite and the modules N_i with $i \in I$ are indecomposable and pairwise non-isomorphic, it follows that I is finite. Let $\mathcal{N}$ be the set of modules N_i with $i \notin I$, then $\mathcal{N}$ is an infinite subset of $\mathcal{M}$ and consists of pairwise non-isomorphic indecomposable modules.

It is easy to see that *the modules in a minimal infinite family of indecomposable modules are of bounded length.* Namely, according to [36] (as well as [37]), any indecomposable module M of length at least 2 has an indecomposable proper submodule of length at least $\frac{1}{pq}|M|$, where p is the maximal length of an indecomposable projective module, q the maximal length of an indecomposable injective module. Thus, if the modules in $\mathcal{N}$ are indecomposable, but not of bounded length, then we find indecomposable submodules of modules in $\mathcal{N}$ which are of arbitrarily large length.

Thus, assume that the modules in $\mathcal{N}$ are of length at most b. Then there are infinitely many modules in $\mathcal{N}$ of fixed length b', for some $1 \leq b' \leq b$. □

Proposition 12.2 *Let $\mathcal{M}$ be a minimal infinite family. Then there are indecomposable non-projective modules C, Y and infinitely many indecomposable modules $M_i \in \mathcal{M}$ with short exact sequences*

$$0 \to \tau C \longrightarrow M_i \xrightarrow{f_i} Y \to 0$$

such that f_i is a co-Gabriel–Roiter projection. All these equivalence classes $[f_i\rangle$ belong to ${}^C[\to Y\rangle^1$.

Proof According to Lemma 12.1, we can assume that all the modules in $\mathcal{M}$ are indecomposable and of fixed length b. Let $\mathcal{M}'$ be the set of modules M' which are proper submodules or proper factor modules of modules in $\mathcal{M}$. By assumption, there are only finitely many isomorphism classes of modules in $\mathcal{M}'$. Since there are only finitely many simple modules, we must have $b \geq 2$, thus any $M \in \mathcal{M}$ has a co-Gabriel–Roiter factor-module $Q_M = M/K_M$; here, K_M is a submodule of M. Note that K_M is a proper non-zero submodule, that Y_M is a proper factor module of M, and that both modules K_M and $Y_M = M/K_M$ are indecomposable. Of course, both K_M and Y_M belong to $\mathcal{M}'$. Since there are (up to isomorphism) only finitely many pairs (K, Y) in $(\mathcal{M}')^2$, it follows that there is a pair (K, M) such that there are infinitely many pairwise non-isomorphic indecomposable modules M with $K = K_M$ and $Y = Y_M$. The exact sequences $0 \to K \to M \to Y \to 0$ with M indecomposable show that K cannot be injective, thus $C = \tau^- K$ is again indecomposable. This completes the proof of the proposition.

It remains to observe that $[f_i\rangle$ belongs to ${}^C[\to Y\rangle^1$, but this has been shown in Corollary 7.6. □

The lattices ${}^C[\to Y\rangle$ which arise in this way can have arbitrarily large height, as the following examples show.

Example 10 Let Λ be the n-Kronecker algebra, this is the path algebra of the quiver

$$a \overset{\longleftarrow}{\underset{\longleftarrow}{\vdots}} b$$

with n arrows. Let $Y = S(b)$ and $C = \tau^- S(a)$. Then C has dimension vector $(n^2 - 1, n)$, thus $\dim \mathrm{Hom}(C, Y) = n$. Since $\mathrm{End}(C) = k$, we see that $\mathcal{S}\,\mathrm{Hom}(C, Y)$ is of the form $\mathbb{G}(n)$ as exhibited in Sect. 19.7. The family of modules which we are interested in are the indecomposable modules M of length 2. Given such a module M, there is an exact sequence

$$0 \to S(a) \longrightarrow M \xrightarrow{f_M} S(b) \to 0,$$

with f_M a co-Gabriel–Roiter projection. As we will see in the next section, this family of maps f_M is a cofork.

13 Forks and coforks

We call a family of maps $(g_i : X \to M_i)_{i\in I}$ a *fork* provided for any finite subset $J \subseteq I$, the map $g_J = (g_i)_{i\in J} : X \to \bigoplus_{i\in J} M_i$ is left minimal. The dual notion will be that of a *cofork,* this is a set of maps $(f_i : M_i \to Y)_{i\in I}$ such that the direct sum map $f^J = (f_i)_{i\in J} : \bigoplus_{i\in J} M_i \to Y$ with J any finite subset of I is right minimal.

Lemma 13.1 *Let $g_i : X \to M_i$ with $i \in I$ be a family of non-zero maps with indecomposable modules M_i. Then $(g_i : X \to M_i)_i$ is a fork if and only if $g_i \notin \sum_{j\neq i} \operatorname{Hom}(M_j, M_i)g_j$ for all $i \in I$.*

Proof First, assume that there is some i with $g_i \in \sum_{j\neq i} \operatorname{Hom}(M_j, M_i)g_j$, say there is the subset $\{1, 2, \dots, t\} \subseteq I$ such that $g_1 \in \sum_{j=2}^{t} \operatorname{Hom}(M_j, M_1)g_j$. Thus, for $2 \le j \le t$ there are maps $p_j : M_j \to M_1$ such that $g_1 = \sum_{j=2}^{t} p_j g_j$. We want to show that the map $g_J : X \to M = \bigoplus_{i=1}^{t} M_i$ is not left minimal. Let N be the kernel of the map

$$(-1, p_2, \dots, p_t) : \bigoplus_{i=1}^{t} M_i \longrightarrow M_1.$$

It is easy to check that M is the direct sum of M_1 and N. But the equality $-g_1 + \sum_{j=2}^{t} p_j g_j$ shows that the image of g_J is contained in N. Since the image of g_J is contained in a proper direct summand of M, we see that g_J is not left minimal.

Conversely, assume that $g_i \notin \sum_{j\neq i} \operatorname{Hom}(M_j, M_i)g_j$ for all $i \in I$. We want to show that for any finite subset $J \subseteq I$, the map $g_J : X \to \bigoplus_{i\in J} M_i$ is left minimal. If J is empty, then we deal with the zero map $X \to 0$ which of course is left minimal. If J consists of the single element i, then we deal with $g_i : X \to M_i$. Since $g_i \neq 0$ and M_i is indecomposable, the map g_i is left minimal. Thus we can assume that J contains $t \ge 2$ elements, say $J = \{1, 2, \dots, t\}$. Assume that the map $g_J : X \to M$ with $M = \bigoplus_{i=1}^{t} M_i$ is not left minimal. Then there is a proper direct decomposition $M = N \oplus N'$ such that the image of g_J is contained in N. We can assume that N' is indecomposable, thus isomorphic to some M_i, say to M_1. Let $p : M \to N' = M_1$ be the projection with kernel N, write $p = (p_i)_i$ with $p_i : M_i \to M_1$. Note that p_1 has to be an isomorphism. Replacing any p_i by $(p_1)^{-1}p_i$, we can assume that $p_1 = 1$. Since the image of g_J is contained in the kernel N of p, we have $\sum_{i=1}^{t} p_i g_i = 0$, thus $g_1 = -\sum_{i=2}^{t} p_i g_i$. This completes the proof. □

We may use forks and coforks in order to construct inductively families of modules:

Proposition 13.2 *Let $(f_i : M_i \to Y)_{i\in I}$ be an infinite cofork, with indecomposable modules M_i which have the same Gabriel–Roiter measure γ_0. Let*

$$\mathcal{K} = \operatorname{add}\{\operatorname{Ker}(f^J) \mid J \subseteq I,\ |J| < \infty\}.$$

Then $\mathcal{K}$ is infinite, all indecomposable modules K in $\mathcal{K}$ are cogenerated by $\{M_i \mid i \in I\}$ and have Gabriel–Roiter measure $\gamma(K) < \gamma_0$.

Before we give the proof, let us note the following consequence: either $\mathcal{K}$ will contain indecomposable modules of arbitrarily large length, or else the indecomposable modules in $\mathcal{K}$ are of bounded length. In the latter case, there is an infinite subset of indecomposable modules K_i in $\mathcal{K}$ all of which have the same Gabriel–Roiter measure $\gamma_1 < \gamma_0$.

Proof First, assume that $\mathcal{K} = \text{add } K$ for some module K, let $C = \tau^- K$. For any natural number t, we may choose a subset J of I of cardinality t. Then the module $M_J = \bigoplus_{i \in J} M_i$ is present in ${}^C[\to Y\rangle$. But the modules which are present in ${}^C[\to Y\rangle$ are of bounded length, whereas $|M_J| \geq t$. This contradiction shows that $\mathcal{K}$ is infinite.

Let K be an indecomposable module in $\mathcal{K}$, say a direct summand of the kernel K_J of $f^J : M_J \to Y$. Assume that $\gamma(K) \geq \gamma_0$. This implies that the inclusion map $K \subseteq K_J \subset M_J$ splits. But this contradicts the fact that the map $f^J : M_J \to Y$ is right minimal. □

Let us present two different ways for obtaining forks:

Proposition 13.3 (Gabriel–Roiter forks) *Let M_i with $i \in I$ be pairwise non-isomorphic indecomposable modules of fixed length with isomorphic Gabriel–Roiter submodule U, say with embeddings $u_i : U \to M_i$. Then the family $(u_i : U \to M_i)_{i \in I}$ is a fork.*

Proof The maps $u_i : U \to M_i$ are non-zero maps and the modules M_i are indecomposable. Thus, if the family $(u_i : U \to M_i)_i$ is not a fork, then Lemma 13.1 asserts that there is some i with $u_i \in \sum_{j \neq 1} \text{Hom}(M_j, M_1) f_j$, thus we can assume that there is the subset $\{1, 2, \ldots, t\} \subseteq I$ and maps $p_j : M_j \to M_1$ for $2 \leq j \leq t$ such that $u_1 = \sum_{j=2}^{t} p_j u_j$.

Let $M'' = \bigoplus_{j=2}^{t} M_j$. For $2 \leq j \leq t$, consider the submodule $p_j u_j(U)$ of M_1. It is a proper submodule of M_1, thus $\gamma(p_j u_j(U)) \leq \gamma(U)$, since U is a Gabriel–Roiter submodule of M_1. On the other hand, the image of $(p_j u_j)_{j=2}^{t} : U \to M''$ is contained in $\bigoplus_{j=2}^{t} p_j u_j(U)$. Since $u_1 = \sum_{j=2}^{t} p_j u_j$, this map $(p_j u_j)_{j=2}^{t}$ is injective, thus U is a submodule of $\bigoplus_{j=2}^{t} p_j u_j(U)$. It follows that $\gamma(U) \leq \max \gamma(p_j u_j(U))$. Thus there is some s with $2 \leq s \leq t$ such that $\gamma(U) = \gamma(p_s u_s(U))$. Since U is indecomposable, U has to be a direct summand of $p_s u_s(U)$. But $p_s u_s(U)$ is a factor module of U, thus $p_s u_s : U \to p_s u_s(U)$ is an isomorphism. As a consequence, u_s is a split monomorphism. But this is impossible, since u_s is the inclusion of a Gabriel–Roiter submodule. □

Let us add the dual assertion.

Proposition 13.4 (co-Gabriel–Roiter coforks) *Let M_i with $i \in I$ be pairwise non-isomorphic indecomposable modules of fixed length with isomorphic co-Gabriel–Roiter factor modules Y, say with projections $v_i : M_i \to Y$. Then the family $(v_i : M_i \to Y)_{i \in I}$ is a cofork.*

Remark Let us stress the following: Let M_i with $i \in I$ be pairwise non-isomorphic indecomposable modules with Gabriel–Roiter submodules $U_i \subset M_i$ and assume that the modules M_i/U_i are isomorphic, say to Y, with projection maps $f_i : M_i \to Y$. Then the family $(f_i : M_i \to Y)_{i \in I}$ is not necessarily a cofork. Namely, consider again example 8, and look at the projections $f : P(b) \to S(b)$ and $f' : \tau^- S(a) \to S(b)$. As we mentioned already, the kernels of both maps are Gabriel–Roiter submodules. Since there is a factorization $f = f'h$, the map $[f, f'] : P(b) \oplus \tau^- S(a) \to S(b)$ is not right minimal.

Here is a consequence of Proposition 13.4 and Corollary 7.6.

Proposition 13.5 *Let M_i with $i \in I$ be pairwise non-isomorphic indecomposable modules with isomorphic co-Gabriel–Roiter factor modules Y, say with projections $v_i : M_i \to Y$ and assume that also the kernels are isomorphic, say isomorphic to K. Let $C = \tau^- K$. Then $(v_i : M_i \to Y)_{i \in I}$ is a cofork which belongs to ${}^C[\to Y\rangle^1$.*

A proof similar to Proposition 13.3 shows that starting with any infinite family $\mathcal{M}$ of indecomposable modules with fixed length, there are infinite forks consisting of maps $S \to M$, where S is a simple module and $M \in \mathcal{M}$. A fork $(g_i : S \to M_i)_i$ with S simple will be called a *simple fork*. Similarly, a cofork $(f_i : M_i \to S)_i$ with S simple is called a *simple cofork*.

Proposition 13.6 *Let M_i with $i \in I$ be pairwise non-isomorphic indecomposable modules with fixed Gabriel–Roiter measure. Then there exists for every index $i \in I$ a simple submodule S_i of M_i, say with inclusion map $u_i : S_i \to M_i$, such that for every simple module S and $I(S) = \{i \in I \mid S_i = S\}$ the family $(u_i : S \to M_i)_{i \in I(S)}$ is a fork.*

Proof Let us assume that the modules in $\mathcal{M}$ have length b. Of course, $b > 1$. Consider a module M_i with $i \in I$. It is not cogenerated by the remaining modules, thus the intersection of the kernels of all maps $\phi : M_i \to M_j$ with $j \neq i$ is non-zero. Let S_i be a simple submodule of M_i which is contained in this intersection (thus $\phi(S_i) = 0$ for all maps $\phi : M_i \to M_j$ with $j \neq i$. Denote by $u_i : S_i \to M$ the inclusion map, thus $\mathrm{Hom}(M_i, M_j)u_i) = 0$ for all $j \neq i$. For any simple module S, let $I(S) = \{i \in I \mid S_i = S\}$.

In order to see that $(u_i : S \to M_i)_{i \in I(S)}$ is a fork, we have to show the following: For any finite subset J of I, say $J = \{1, 2, \ldots, t\}$, the map $u^J : (u_i)_i : S \to \bigoplus_{i=1}^t M_i$ is left minimal. Again, this is clear for $t = 1$, since the maps u_i are non-zero and the modules M_i are indecomposable. Thus we can assume that $t \geq 2$. If the map u^J is not left minimal, then, up to permutation of the indices, there are maps $p_i : M_i \to M_1$ with $2 \leq i \leq t$ such that $u_1 = \sum_{i=2}^t p_i u_i$. However, by construction, $\mathrm{Hom}(M_i, M_1)u_i = 0$ for $i \neq 1$, thus all the summands $p_i u_i$ with $2 \leq i \leq t$ are zero. Since $u_1 \neq 0$, we obtain a contradiction. □

Corollary 13.7 (Simple forks) *Let $\mathcal{M}$ be an infinite set of pairwise non-isomorphic indecomposable modules of fixed length. Then there exists an infinite simple fork $(u_i : S \to M_i)_i$ such that all M_i belong to $\mathcal{M}$.*

Proof Since the modules in $\mathcal{M}$ are of bounded length, only finitely many Gabriel–Roiter measures occur, thus there is an infinite subset $\mathcal{M}' \subseteq \mathcal{M}$ which consists of indecomposable modules with fixed Gabriel–Roiter measure. Now we apply Proposition 13.6. Since there are only finitely many simple modules, one of the forks $(u_i : S \to M_i)_{i \in I(S)}$ has to be infinite. □

Again we add the dual assertion.

Corollary 13.8 (Simple coforks) *Let $\mathcal{M}$ be an infinite set of pairwise non-isomorphic indecomposable modules of fixed length. Then there exists an infinite simple cofork $(v_i : M_i \to S)_i$ such that all M_i belong to $\mathcal{M}$.*

The setting developed here allows to provide a proof of the following result which first was established in [35]. Note that this result strengthens the assertion of the first Brauer–Thrall conjecture [41].

Corollary 13.9 (First Brauer–Thrall conjecture) *Let $\mathcal{M}$ be an infinite set of indecomposable modules of a fixed length. Then there are indecomposable modules of arbitrarily large length which are cogenerated by modules in $\mathcal{M}$.*

Proof Let $\mathcal{M}_0 = \mathcal{M}$ and apply Corollary 13.8. Thus, there is a simple cofork $(v_i : M_i \to S)_i$ such that all M_i belong to $\mathcal{M}_0$. Let $\mathcal{K} = \operatorname{add}\{\operatorname{Ker}(v^J) \mid J \subseteq I,\ |J| < \infty\}$. According to Proposition 13.2 we know that $\mathcal{K}$ is infinite, that all indecomposable modules K in $\mathcal{K}$ are cogenerated by $\{M_i \mid i \in I\}$ and that they have Gabriel–Roiter measure $\gamma(K) < \gamma_0$. Now either the indecomposable modules in $\mathcal{K}$ are of unbounded length, then we are done. Or else they are of bounded length: then we find in $\mathcal{K}$ an infinite set $\mathcal{M}_1$ of indecomposable modules having the same Gabriel–Roiter measure, say γ_1 and $\gamma_1 < \gamma_0$. Inductively, we construct a sequence of sets of indecomposable modules

$$\mathcal{M}_0, \mathcal{M}_1, \ldots, \mathcal{M}_i$$

such that the modules in $\mathcal{M}_i$ are cogenerated by $\mathcal{M}_{i-1}$ and have fixed Gabriel–Roiter measure $\gamma_i < \gamma_{i-1}$, for $i \geq 1$. The procedure stops in case there are indecomposable modules of unbounded length which are cogenerated by $\mathcal{M}_i$, and then these modules are cogenerated by $\mathcal{M}$. Otherwise the procedure can be continued indefinitely. But then we have constructed infinitely many sets $\mathcal{M}_i$ of indecomposable modules. Since the modules in $\mathcal{M}_i$ have Gabriel–Roiter measure γ_i and the measures γ_i are pairwise different, the modules in $\bigcup_i \mathcal{M}_i$ cannot be of bounded length. Also, the modules in any $\mathcal{M}_i$ are cogenerated by $\mathcal{M}$. This completes the proof. □

14 The Kronecker algebra

Throughout this section, Λ will be the Kronecker algebra as introduced already in Example 2. It is a very important artin algebra and a clear understanding of its module category mod Λ seems to be of interest.

For all pairs C, Y of indecomposable Λ-modules, we are going to describe the lattice ${}^C[\to Y\rangle$ as well as all the modules present in ${}^C[\to Y\rangle^1$, this is the subset of ${}^C[\to Y\rangle$ of elements of right C-length 1.

Let us recall the structure of the category mod Λ (see for example [33], or [5], section VIII.7). There are the preprojective and the preinjective modules, modules without an indecomposable direct summand which is preprojective or preinjective are said to be regular. For any Λ-module M, its defect is defined by $\delta(M) = \dim \operatorname{Hom}(M, Q_0) - \dim \operatorname{Hom}(P_0, M)$. Any indecomposable preprojective Λ-module has defect -1, the indecomposable preinjective modules have defect 1, all the regular modules have defect 0. There are countably many indecomposable preprojective modules, they are labeled P_i, and also countably many indecomposable preinjective modules, they are labeled Q_i; both P_i and Q_i have length $2i + 1$. The indecomposable regular modules are those modules which belong to stable Auslander–Reiten components, and all these components are stable tubes of rank 1. The full subcategory $\mathcal{R}$ of all regular modules is abelian. By definition, the simple regular modules are the regular modules which are simple objects in this subcategory. Given any indecomposable regular module R, its endomorphism ring $\operatorname{End}(R)$ is a commutative ring (namely a ring of the form $k[T]/\langle f\rangle$, where $k[T]$ is the polynomial ring in one variable T with coefficients in k and f is a power of an irreducible polynomial) and $\dim R = 2 \dim \operatorname{End}(R)$.

As we have mentioned, we are interested in pairs C, Y of indecomposable Λ-modules such that a family of modules is present in ${}^{C}[\to Y\rangle$. It turns out that only the case of C being preprojective, Y being preinjective is relevant, as the following proposition shows.

Proposition 14.1 *Let Λ be the Kronecker algebra and C, Y indecomposable Λ-modules. If C is preprojective or preinjective, then ${}^{C}[\to Y\rangle$ is a projective geometry. If C is regular, then ${}^{C}[\to Y\rangle$ is a chain.*

By definition, a projective geometry $\mathbb{G}(d)$ over the field k is the lattice of subspaces of the k-space of dimension d. The chain $\mathbb{I}(d)$ is the set of integers i with $0 \le i \le d$ with the usual ordering. The labels have been chosen in such a way that the height of $\mathbb{G}(d)$ as well as of $\mathbb{I}(d)$ is just d.

The following table provides the precise data: Here, an indecomposable regular module of regular length t and with regular socle R is denoted by $R[t]$.

row	C	Y		${}^{C}[\to Y\rangle$
1)	P_i	P_j	$(i \le j)$	$\mathbb{G}(j-i+1)$
2)	P_i	$R[t]$		$\mathbb{G}(\frac{1}{2}\dim R[t])$
3)	P_i	Q_j		$\mathbb{G}(i+j)$
4)	$R[s]$	$R[t]$		$\mathbb{I}(\min(s,t))$
5)	$R[s]$	Q_j		$\mathbb{I}(\frac{1}{2}\dim R[s])$
6)	Q_i	Q_j	$(i \ge j)$	$\mathbb{G}(i-j+1)$

Let us stress that in row 4), the regular modules C, Y are supposed to belong to the same tube, namely to the tube containing a fixed simple regular module R. For all pairs C, Y of indecomposable Kronecker modules which are not contained in the table, one has $\operatorname{Hom}(C, Y) = 0$, thus ${}^{C}[\to Y\rangle$ consists of a single element.

Proof of proposition 14.1 First, let us calculate $\dim \operatorname{Hom}(C, Y)$. The reflection functors of [7] yield

$$\begin{aligned}
\mathrm{Hom}(P_i, P_j) &= \dim \mathrm{Hom}(P_0, P_{j-i}) = j - i + 1\\
\mathrm{Hom}(P_i, R[t]) &= \dim \mathrm{Hom}(P_0, R[t]) = \tfrac{1}{2} \dim R[t]\\
\mathrm{Hom}(P_i, Q_j) &= \dim \mathrm{Hom}(P_0, Q_{j+i}) = i + j\\
\mathrm{Hom}(R[s], Q_j) &= \dim \mathrm{Hom}(R[s], Q_0) = \tfrac{1}{2} \dim R[s]\\
\mathrm{Hom}(Q_i, Q_j) &= \dim \mathrm{Hom}(Q_{i-j}, Q_0) = i - j + 1
\end{aligned}$$

As we have mentioned, we have $\dim R = 2 \dim \mathrm{End}(R)$. It follows that

$$\dim \mathrm{Hom}(R[s], Q_j) = \tfrac{1}{2} \dim R[s] = s \dim \mathrm{End}(R).$$

Finally,

$$\dim \mathrm{Hom}(R[s], R[t]) = \min(s, t) \dim \mathrm{End}(R).$$

In case $C = P_i$ or Q_j one has $\mathrm{End}(C) = k$, thus Auslander's Second Theorem asserts that ${}^{C}[\to Y\rangle$ is of the form $\mathbb{G}(\dim \mathrm{Hom}(C, Y))$. This yields the rows 1), 2), 3) and 6) of the table.

It remains to look at the rows 4) and 5), thus we assume now that $C = R[s]$ and $Y = R[t]$ or $Y = Q_j$. We show that $\mathrm{Hom}(C, Y)$ is a cyclic $\Gamma(C)$-module, thus we have to find an element $g \in \mathrm{Hom}(C, Y)$ such that $g\,\mathrm{End}(C) = \mathrm{Hom}(C, Y)$, or, equivalently, such that $g(\mathrm{rad}\,\mathrm{End}(C))^{r-1} \neq 0$, where r is the length of $\mathrm{Hom}(C, Y)$ (as a $\Gamma(C)$-module). Note that $\Gamma(C)$ is a local ring, thus there is a unique simple $\Gamma(C)$-module S. Since S has k-dimension $\dim \mathrm{End}(R)$, the calculations above show that $r = s$ in case $Y = Q_j$ and $r = \min(s, t)$ in case $Y = R[t]$. On the other hand, $(\mathrm{rad}\,\mathrm{End}(C))^{r-1}$ is generated by any endomorphism of $R[s]$ with image $R[s-r+1]$. Thus let $p : R[s] \to R[s-r+1]$ be the canonical projection, $u : R[s-r+1] \to R[s]$ the canonical inclusion, then up generates $(\mathrm{rad}\,\mathrm{End}(C))^{r-1}$. Our aim is to exhibit $g : R[s] \to Y$ such that $gup \neq 0$.

First, let $Y = R[t]$ and $s \leq t$. Then $r = \min(s, t) = s$ and $R[s-r+1] = R[1] = R$. Let $g : R[s] \to R[t]$ be the canonical inclusion, thus $gu : R \to R[t]$ is an inclusion, in particular non-zero, and therefore also $gup \neq 0$.

Second, let $Y = R[t]$ and $s > t$. Then $r = \min(s, t) = t$ and $R[s-r+1] = R[s-t+1]$. Let $g : R[s] \to R[t]$ be the canonical projection. Then $gu : R[s-t+1] \to R[t]$ has image R (since the kernel of g is $R[s-t]$). This shows that gu is non-zero, and therefore also $gup \neq 0$.

Finally, we have to deal with the case $Y = Q_j$. Take a non-zero map $g' : R \to Q_j$. Since $\mathrm{Ext}^1(R[s]/R, Q_j) = 0$, there exists $g : R[s] \to Q_j$ such that $gu = g'$. Since $gu \neq 0$, it follows that also $gup \neq 0$.

Thus, always we have found $g : R[s] \to Y$ such that $gup \neq 0$. As a consequence, $\mathrm{Hom}(C, Y)$ is a cyclic $\Gamma(C)$-module. Since $\Gamma(C)$ is a local uniserial ring, it follows that $\mathcal{S}\,\mathrm{Hom}(C, Y)$ is of the form $\mathbb{I}(r)$, where r is the length of $\mathrm{Hom}(C, Y)$. According to our calculations, $r = \min(s, t)$ in case $Y = R[t]$ and $r = \frac{1}{2} \dim R[s]$ in case $Y = Q_j$.

Thus we have verified the assertions presented in the table. If the pair C, Y does not occur in the table, then it is well-known that $\mathrm{Hom}(C, Y) = 0$, thus $\mathcal{S}\,\mathrm{Hom}(C, Y)$

consists of a single element and therefore is of the form $\mathbb{I}(0) = \mathbb{G}(0)$. This completes the proof. □

The following assertion which has been shown in the proof will be of further interest:

Lemma 14.2 *Any non-zero map $g' : R \to Q_j$ can be extended to a map $g : R[t] \to Q_j$, and any such $g : R[t] \to Q_j$ generates the $\Gamma(C)$-module* $\mathrm{Hom}(R[t], Q_j)$.

Remark We have mentioned in Sect. 4 that both sides of the Auslander bijection concern maps with target Y, but that they invoke these maps in quite different ways. A nice illustration seems to be Proposition 14.1. The map $g : R[t] \to Q_j$ constructed there is a generator of the maximal submodule of $\mathrm{Hom}(R[t], Q_j)$ and is used in the proof of Lemma 14.2 in order to show that $\mathcal{S}\,\mathrm{Hom}(R[t], Q_j)$ is of the form $\mathbb{I}(d)$ for some d. On the other hand, in Proposition 14.9, we will consider the right equivalence class $[g\rangle$ as an element of $[\to Q_j\rangle$.

Proposition 14.3 *Let Λ be the Kronecker algebra, let C, Y be indecomposable Λ-modules and M an indecomposable direct summand of a module present in ${}^C[\to Y\rangle$.*

If C, Y both are preprojective, also M is preprojective. If C, Y both are preinjective, also M is preinjective. If both C, Y belong to the tube $\mathcal{T}$, also M belongs to $\mathcal{T}$.

If C is preprojective and Y belongs to the tube $\mathcal{T}$, then M is preprojective or belongs to $\mathcal{T}$. If C belongs to the tube $\mathcal{T}$ and Y is preinjective, then M is preinjective or belongs to the tube $\mathcal{T}$.

Proof First, assume that Y is preprojective. If $f : X \to Y$ is right minimal, then X has to be preprojecive. Thus, for any module C, all the modules present in ${}^C[\to Y\rangle$ are preprojective.

Second, assume that Y is regular, say belonging to the tube $\mathcal{T}$. Again, assume that $f : X \to Y$ is right minimal. Then X is the direct sum of a preprojective module and a module in $\mathcal{T}$. Thus, for any module C, all the modules present in ${}^C[\to Y\rangle$ are direct sums of preprojective modules and modules in $\mathcal{T}$.

Finally, assume that Y is preinjective and C is regular or preinjective. Let $f : X \to Y$ be right minimal and right C-determined. Since C has no indecomposable projective direct summand and Λ is hereditary, we see that f is surjective and its kernel is in add K with $K = \tau C$. Now, if C is preinjective, then also K is preinjective and X as an extension of a preinjective module by a preinjective module is preinjective again. On the other hand, if C is regular, say belonging to the tube $\mathcal{T}$, then also K belongs to $\mathcal{T}$. Since X is an extension of a module in $\mathcal{T}$ by a preinjective module, it is the direct sum of a module in $\mathcal{T}$ and a preinjective module. This completes the proof. □

Remark Let us stress that the cases $C = P_i, Y = R[t]$ and $C = R[s], Y = Q_j$ are not at all dual (as the consideration of $\mathrm{Hom}(P_i, R[t])$ and $\mathrm{Hom}(R[t], Q_j)$ could suggest), but are of completely different nature. The reason is that we always consider $\mathrm{Hom}(C, Y)$ as a $\Gamma(C)$-module!

Of course, we have already seen that ${}^C[\to Y\rangle$ is in the first case a projective geometry, in the second case a chain. But also if we look at the different layers, we encounter clear differences. In the chain case, all the elements of ${}^C[\to Y\rangle$ are given by short exact sequences of the form $(R[s] \to M \to Q_j)$, thus by elements of

$\operatorname{Ext}^1(Q_j, R[s])$. In this case, we may interpret ${}^C[\to Y\rangle$ as a display of the various orbits in $\operatorname{Ext}^1(Q_j, R[s])$ with respect to the action of the automorphism group of $R[s]$. In contrast, in the projective geometry case, only the elements of ${}^C[\to Y\rangle$ of right C-length at most 1 are given by short exact sequences of the form $(P_i \to M \to R[t])$, whereas for the elements of right C-length at least 2, we need short exact sequences of the form $(P_i{}^a \to M \to R[t])$ with $a \geq 2$.

In the proof of Proposition 14.3, we have seen that for $C = P_i, Y = R[t]$, the modules M present in ${}^C[\to Y\rangle$ are direct sums of preprojective modules and modules in the tube which contains $\mathcal{T}$, and that for $C = R[s]$, $Y = Q_j$ the modules present in ${}^C[\to Y\rangle$ are direct sums of preinjective modules and modules in $\mathcal{T}$. However, in the first case the number of indecomposable preprojective direct summands of M may be large, whereas in the second case there is just one direct summand of M which is indecomposable preinjective.

Preprojective C, preinjective Y. Let us focus now the attention on ${}^C[\to Y\rangle$, where C is indecomposable preprojective and Y is indecomposable preinjective. Here is the general behavior as seen in Proposition 14.1:

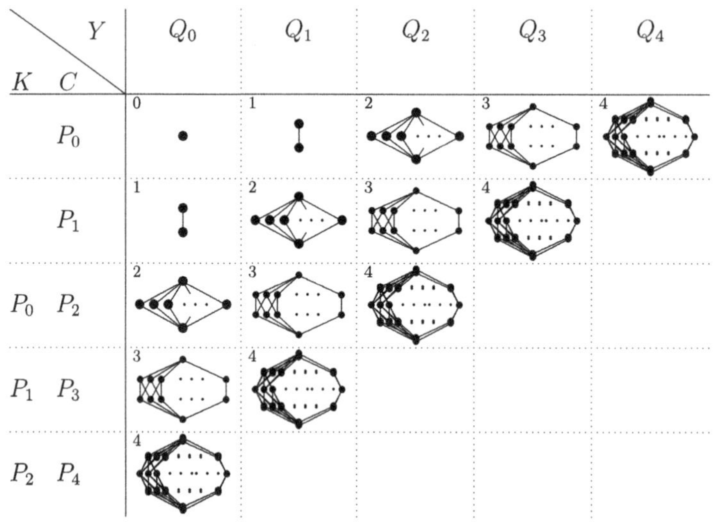

We want to know which modules are present in ${}^C[\to Y\rangle^1$. As we will show, these are certain regular modules.

Lemma 14.4 *Let Λ be the Kronecker algebra and M a regular module. The following conditions are equivalent.*

(i) *If $M = M' \oplus M''$, then* $\operatorname{Ext}^1(M', M'') = 0$.
(ii) *The regular socle of M is multiplicity-free.*
(ii*) *The regular top of M is multiplicity-free.*
(iii) *$M = \bigoplus_{i=1}^n M_i$, with indecomposable modules M_i belonging to pairwise different tubes.*
(iv) $\operatorname{End}(M)$ *is commutative.*

Proof The equivalence of (ii) and (iii), and dually of (ii*) and (iii) is straight forward. The implication (iii) $\Longrightarrow$ (i) follows from the fact that $\operatorname{Ext}^1(M_i, M_j) = 0$ in case M_i, M_j are regular and belong to different Auslander–Reiten components. The converse implication (i) $\Longrightarrow$ (iii) follows from the fact that $\operatorname{Ext}^1(M_i, M_j) \neq 0$ in case M_i, M_j are nonzero regular and belong to the same Auslander–Reiten component. In order to see the implication (iii) $\Longrightarrow$ (iv), one should be aware that for $M = \bigoplus_{i=1}^n M_i$, with indecomposable modules M_i belonging to pairwise different tubes, one has $\operatorname{End}(M) = \prod_i \operatorname{End}(M_i)$ and that $\operatorname{End}(R)$ is commutative for any indecomposable regular module R. Conversely, in order to show the implication (iv) $\Longrightarrow$ (iii), assume that M is regular and assume that $M = M' \oplus M'' \oplus M''$ with M', M'' non-zero modules belonging to some tube. Then $\operatorname{Hom}(M', M'') \neq 0$, thus there is a non-zero homomorphism $\phi : M' \to M''$ and we may consider this as an endomorphism of M by setting ϕ to be zero on $M'' \oplus M'''$. Let $e' : M \to M$ be the projection of M to M' with kernel $M'' \oplus M'''$. Then $\phi e' = \phi \neq 0$, whereas $e'\phi = 0$. This shows that $\operatorname{End}(M)$ is not commutative. □

A regular Kronecker-module M will be said to be *strongly regular* provided the equivalent conditions of the Lemma are satisfied. Let $\mathcal{R}(i)$ be the set of isomorphism classes of strongly regular modules of length i. Note that $\mathcal{R}(i)$ is empty in case i is odd or negative, and $\mathcal{R}(0)$ has just one element, namely the isomorphism class of the zero module.

As we have mentioned, we want to see in which way families of modules may be present in ${}^C[\to Y\rangle^1$, with C indecomposable preprojective, and Y indecomposable preinjective. Here is the description of these sets.

Proposition 14.5 *Let Λ be the Kronecker algebra, C indecomposable preprojective, Y indecomposable preinjective.*

If $C = P(S)$ for some simple module S, then ${}^C[\to Y\rangle^1$ may be identified with the set of inclusion maps $X \to Y$ such that the socle of Y/X is equal to S.

If $C = \tau^- K$ for some indecomposable module K, then ${}^C[\to Y\rangle^1$ consists of the right equivalence classes of surjective maps $X \to Y$ with kernel K.

Proof See Corollary 3.7 and Corollary 7.3. □

Given a morphism $f : X \to Y$, let $\sigma(f) = [X]$, the isomorphism class of the source X of f. We study the function σ defined on ${}^C[\to Y\rangle^1$ with values in the set of isomorphism classes of modules. The main result of this section is the following description of ${}^C[\to Y\rangle^1$:

Proposition 14.6 *Let Λ be the Kronecker algebra, C indecomposable preprojective, Y indecomposable preinjective. Then σ is a bijection*

$$\boxed{\sigma : {}^C[\to Y\rangle^1 \longrightarrow \mathcal{R}(i) \quad \textit{with} \quad i = |C| + |Y| - 4.}$$

For the proof, we need some preliminary considerations.

Lemma 14.7 *Let Λ be the Kronecker algebra, let $0 \to U \to X \to Y \to 0$ be a non-split exact sequence with Y indecomposable preinjective, U preprojective. Then no indecomposable direct summand of X is preinjective.*

Proof Let $X = X' \oplus X''$ with X' indecomposable. Assume that X' is preinjective. Denote the map $X \to Y$ by f and let f' be its restriction to X'. Then $f' \neq 0$, since otherwise X' is a direct summand of the kernel of f, thus equal to K, so that the sequence splits. Non-zero maps between indecomposable preinjective modules are surjective, thus f' is surjective. Of course, f' is not an isomorphism, since otherwise the sequence would split. Let U' be the kernel of f'. As we see, $U' \neq 0$. We have the following commutative diagram with exact rows:

$$\begin{array}{ccccccccc} 0 & \longrightarrow & U' & \longrightarrow & U & \longrightarrow & X'' & \longrightarrow & 0 \\ & & \| & & \downarrow & & \downarrow & & \\ 0 & \longrightarrow & U' & \longrightarrow & X' & \xrightarrow{f'} & Y & \longrightarrow & 0 \end{array}$$

Now U' is a submodule of U, thus it is preprojective. Since $U' \neq 0$, we must have $\delta(U') \leq -1$, where δ denotes the defect. It follows that $\delta(X') = \delta(U') + \delta(Y) \leq 0$, since $\delta(Y) = 1$. This contradicts our assumption that X' is preinjective. □

Lemma 14.8 *Let X be regular, Y indecomposable preinjective. Then the following conditions are equivalent:*

(i) *X is strongly regular,*
(ii) *There exists $f : X \to Y$ such that the kernel of f does not contain a simple regular submodule.*

Any map $f : X \to Y$ with no simple regular submodule in its kernel is a monomorphism or an epimorphism. If $f : X \to Y$ and $f' : X' \to Y$ are maps with no simple regular submodule in the kernel, and X, X' are isomorphic, then f, f' are right equivalent.

Proof First, we show that (ii) implies (i). Assume that there is $f : X \to Y$ such that the kernel of f does not contain a simple regular submodule of X. In order to show that X is strongly regular, we show that its regular socle is multiplicity-free. Assume, for the contrary, that X has a submodule U such that $U = R \oplus R$ with R simple regular. Let f_1, f_2 be the restrictions of f to $R \oplus 0$ or $0 \oplus R$, respectively. Since no simple regular submodule of X is contained in the kernel of f, we see that both f_1, f_2 are non-zero maps. According to Lemma 14.2, there is a map $h : R \to R$ such that $f_1 = f_2 h$. But then $R' = \{(-h(x), x) \mid x \in R\}$ belongs to the kernel of f. Of course, R' is isomorphic to R, thus R' is a simple regular submodule of X which belongs to the kernel of f, a contradiction.

Conversely, assume that $X = \bigoplus_i R_i[t_i]$ is a strongly regular module, with pairwise different simple regular modules R_i. According to Lemma 14.2, there is a map $f_i : R_{[}t_i] \to Y$ such that the restriction of f_i to R_i is non-zero. Since the simple regular submodules R_i are pairwise non-isomorphic, these are the only simple regular submodules of X, thus no simple regular submodule of X lies in the kernel of $f = (f_i)_i : X \to Y$,

Now assume that $f : X \to Y$ is a map such that the kernel of f does not contain a simple regular submodule. Assume that f is not an epimorphism. The image of f is a factor module of X, thus the direct sum of a regular and a preinjective module. But Y has no non-zero proper submodule which is preinjective. Thus, the image of f is regular. The kernel of a map between regular modules is regular, thus either a monomorphism, or it contains a simple regular submodule. Since the latter is not possible, we see that f has to be a monomorphism.

Finally, assume that $f : X \to Y$ and $f' : X' \to Y$ are maps with no simple regular submodule in the kernel. Let $g : X \to X'$ be an isomorphism. Write $X = \bigoplus R_i[t_i]$ with pairwise non-isomorphic simple regular modules R_i. Let f_i and f_i' be the restriction of f and $f'g$, respectively, to $R_i[t_i]$. Since R_i is not in the kernel of f, the restriction of f_i to R_i is non-zero. According to Lemma 14.2, there is a map $h_i : R_i \to R_i$ such that $f_i = f_i'h_i$. But also the restriction of f_i' to R_i is non-zero, thus R_i is not in the kernel of h_i and therefore h_i is an automorphism. Let $h = (h_i)_i : X \to X$. This is an automorphism of X with $f = f'gh$. Since $f = f'gh$ and gh is an isomorphism, we see that f, f' are right equivalent. □

Proposition 14.9 *Let Y be indecomposable preinjective. The maximal submodules of Y are pairwise non-isomorphic and these are, up to isomorphism, all the strongly regular modules of length $|Y| - 1$.*

The kernels of the non-zero maps $Y \to Q_1$ are pairwise non-isomorphic and these are, up to isomorphism, all the strongly regular modules of length $|Y| - 3$.

Proof Let X be a maximal submodule of Y. Then Y/X is simple injective, thus $\delta(X) = \delta(Y) - \delta(Y/X) = 0$. Since Y has no proper non-zero preinjective submodule, we see that X has to be regular. According to Lemma 14.8, the inclusion map $X \to Y$ shows that X is even strongly regular, and of course of length $|Y| - 1$. Conversely, if X is strongly regular and of length $|Y| - 1$, Lemma 14.8 yields a monomorphism $X \to Y$. Now assume that two maximal submodules X, X' are isomorphic, let $f : X \to Y$ and $f' : X' \to Y$ be the inclusion maps. Then, according to Lemma 14.8, f, f' are right equivalent, thus $X = X'$.

In the same way, we consider the kernels X of the non-zero maps $Y \to Q_1$. Clearly, all non-zero maps $Y \to Q_1$ are surjective, thus $\delta(X) = 0$ and again, X has to be regular, and according to Lemma 14.8 even strongly regular, of course of length $|Y| - 3$. Conversely, if X is strongly regular and of length $|Y| - 1$, Lemma 14.8 yields a monomorphism $X \to Y$. The factor module Y/X is of length 3 and has the same composition factors as Q_1. But the only factor module of Y of length 3 with the same composition factors as Q_1 is Q_1 itself, thus X is the kernel of a non-zero map $Y \to Q_1$. Finally, we use again Lemma 14.8 in order to see that isomorphic kernels of non-zero maps $Y \to Q_1$ are actually identical. □

Remark It should be stressed that for $C = P_0$ and $C = P_1$, the inverse η_{CY}^{-1} of the Auslander bijection can be seen very well:

If M is a Kronecker module, we may write M in the form

$$M = (M_a, M_b;\ \alpha : M_b \to M_a, \beta : M_b \to M_a)$$

where M_a, M_b are vector spaces and M_α, M_β are linear maps. We can identify M_a with $\mathrm{Hom}(P_0, M)$ and M_b with $\mathrm{Hom}(P_1, M)$, since $P_0 = P(a)$, $P_1 = P(b)$.

Consider the case $C = P_1$. Given a maximal submodule U of $\mathrm{Hom}(C, Y)$, we may interpret it as a maximal submodule of $Y_b = \mathrm{Hom}(C, Y)$, and we may consider the submodule ΛU of Y generated by U, let $u : \Lambda U \to Y$ be the inclusion map. Then $[u : \Lambda U \to Y\rangle$ belongs to ${}^C[\to Y\rangle_1$ and $\eta_{CV}(u) = U$. This shows that $[u\rangle = \eta_{CY}^{-1}(U)$.

Similarly, for $C = P_0$, starting with a maximal subspace V_a of the vector space $Y_a = \mathrm{Hom}(C, Y)$, we may construct $V_b = Y_\alpha^{-1}(V_a) \cap Y_\beta^{-1}(V_a)$, then $V = (V_a, V_b; Y_\alpha|V_b, Y_\beta|V_b)$ is a submodule of Y, say with inclusion map $v : V \to Y$. Then $[v : V \to Y\rangle$ belongs to ${}^C[\to Y\rangle_1$ and $\eta_{CV}(v) = V_a$. This shows that $[v\rangle = \eta_{CY}^{-1}(V_a)$.

Proposition 14.10 *Let Λ be the Kronecker algebra and Y indecomposable preinjective, C indecomposable preprojective. Let $d = |C| + |Y| - 4$. Let X be any module.*

There is $f : X \to Y$ right minimal, right C-determined with $|f|_C = 1$, if any only if the isomorphism class of X belongs to $\mathcal{R}(d)$.

If $f : X \to Y$ and $f' : X' \to Y$ are right minimal, right C-determined maps with $|f|_C = |f'|_C = 1$. Then f, f' are right equivalent if and only if X, X' are isomorphic.

Proof First, consider the case when C is not projective, thus $|C| \geq 5$, therefore $d > |Y|$.

Since Λ is hereditary and C has no indecomposable projective direct summand, any right C-determined map $f : X \to Y$ is surjective and its kernel belongs to add K, where $K = \tau C$. If we assume that $|f|_C = 1$, then the kernel of f has to be equal to K. It follows that $\delta(X) = \delta(Y) + \delta(K) = 0$. Since all indecomposable submodules of X are preprojective or regular, we see that X has to be regular. Of course, its length is just d. Note that no simple regular submodule R of X can be contained in the kernel of f, since by assumption the kernel of f is K, thus preprojective. The Lemma 14.8 shows that X is strongly regular, thus the isomorphism class of X belongs to $\mathcal{R}(d)$.

Conversely, assume that X is a strongly regular module of length $|X| = d$. According to the Lemma 14.8 there exists a morphism $f : X \to Y$ such that its kernel contains no simple regular module. Since $|X| = d > |Y|$, the map f cannot be a monomorphism, thus it is an epimorphism. Since the kernel of f does not contain a simple regular module, it is the direct sum of indecomposable preprojective modules. Since the defect of the kernel is $\delta(X) - \delta(Y) = -1$, we see that the kernel of f is indecomposable preprojective. The length of the kernel is $|X| - |Y| = |C| - 4 = |\tau C|$, thus the kernel of f is isomorphic to τC. Altogether, we have shown: if the isomorphism class of X belongs to $\mathcal{R}(d)$, then there is $f : X \to Y$ right minimal, right C-determined with $|f|_C = 1$.

Finally, assume that $f : X \to Y$ and $f' : X' \to Y$ are right minimal, right C-determined maps with $|f|_C = |f'|_C = 1$. Then X, X' are strongly regular. If these maps are right equivalent, then clearly X, X' are isomorphic. Conversely, assume that X, X' are isomorphic. In order to show that f, f' are right equivalent, we may assume that $X' = X$. Lemma 14.8 asserts that f, f' are right equivalent.

Now assume that C is projective, thus $|C| \leq 3$, therefore $d < |Y|$. In case $C = P_1$, we have to consider the submodules X of Y with $Y/X = Q_0$, in case $C = P_0$, we have to consider the submodules X of Y with $Y/X = Q_1$. This has been done in the previous proposition. □

Proof of proposition 14.6 The proposition is an immediate consequence of Proposition 14.5 and Proposition 14.10. □

Corollary 14.11 *Let* Λ *be the Kronecker algebra, C indecomposable preprojective, Y indecomposable preinjective. Then* $\sigma\eta_{CY}^{-1}$ *yields a bijection*

$$\sigma\eta_{CY}^{-1} : \mathcal{S}_m \operatorname{Hom}(C,Y) \longrightarrow \mathcal{R}(i) \quad \textit{with} \quad i = |C| + |Y| - 4.$$

Note that $\Gamma(C) = k$, thus $\operatorname{Hom}(C, Y)$, considered as a $\Gamma(C)$-module is just a vector space, thus $\mathcal{S}_m \operatorname{Hom}(C, Y)$ is the set of the maximal subspaces of a vector space, and therefore a projective space. We see that *we obtain a parameterization of the set* $\mathcal{R}(2i)$ *of all strongly regular Kronecker modules of length 2i by the projective space* $\mathbb{P}^i$.

Remark This parameterization of $\mathcal{R}(d)$ extends the well-known description of the geometric quotient of the "open sheet", when dealing with conjugacy classes of $(n \times n)$-matrices with coefficients in a field, see for example Kraft [23].

Let us mention some details: Let $R = k[T]$ be the polynomial ring in one variable T with coefficients in the field k, and let us assume that k is algebraically closed. We consider R as the path algebra of the quiver with one vertex and one loop; in this way, the R-modules of dimension n are just pairs (V, ϕ), where V is a k-space of dimension n and $\phi : V \to V$ an endomorphism of V, or, after choosing a basis of V, we just deal with $(n \times n)$-matrices with coefficients in k. Isomorphism of R-modules translates to equivalence (or conjugacy) of matrices. The assertions of Lemma 14.4 can be reformulated in this context, *the properties mentioned there characterize just the cyclic R-modules of finite length.* Note that the category of $k[T]$-modules of finite length is equivalent to the full subcategory $\mathcal{R}'$ of all regular Kronecker modules without eigenvalue ∞ (to be precise: let us denote the two arrows of the Kronecker quiver by α, β and let $\mathcal{T}_\infty$ be the Auslander–Reiten component which contains the indecomposable regular representation M with $M_\alpha = 0$. The subcategory $\mathcal{R}'$ consists of all regular representations with no indecomposable direct summand in $\mathcal{T}_\infty$, or equivalently, it is the full subcategory of all representations M such that M_α is bijective. Under the equivalence of the category of finite-dimensional $k[T]$-modules and $\mathcal{R}'$ $k[T]$-module N corresponds to the representation M such that $M_\alpha : N \to N$ is the identity map, and $M_\beta : N \to N$ the multiplication by T). A representation M in $\mathcal{R}'$ is strongly regular if and only if M corresponds under this equivalence to a cyclic $k[T]$-module. Thus, one may be tempted to call the strongly regular Kronecker modules "cyclic" modules, but this would be in conflict with standard terminology.

Remark **"Modules determined by morphisms".** A bijection of two sets can always be read in two different directions. This survey is concerned with the Auslander bijection

$$\eta_{CY} : {}^{C}[\to Y\rangle \longrightarrow \mathcal{S}\operatorname{Hom}(C, Y),$$

In the previous sections, the focus was going from left to right: Any right minimal morphism $f \in {}^{C}[\to Y\rangle$ yields under η_{CY} a submodule of $\operatorname{Hom}(C, Y)$ and is uniquely determined by this submodule, this is the philosophy of saying that morphisms as elements of $[\to Y\rangle)$ are determined by modules.

The considerations in the present section point into the reverse direction: we use sets of morphisms as convenient indices for parameterizing isomorphism classes of modules. We recall from Sect. 7 that the restriction of η_{CY} furnishes a bijection

$$\eta_{CY} : {}^{C}[\to Y\rangle^{1} \longrightarrow \mathcal{S}_m \operatorname{Hom}(C, Y),$$

thus we can use the right hand set in order to parametrize the left hand set.

Our interest lies in the special case where maps $f : M \to Y$ and $f' : M' \to Y$ of right C-length 1 are right equivalent only in case M and M' are isomorphic. In this case, the maximal submodule $\eta_{CY}(f : M \to Y)$ (thus a set of morphisms) uniquely determines the module M. In addition, we will assume that C is a brick, or at least that C is indecomposable and $\operatorname{rad}\operatorname{End}(C)$ annihilates $\operatorname{Hom}(C, Y)$. In this case, $\mathcal{S}_m \operatorname{Hom}(C, Y)$ is a projective space (namely the projective d-space over the division ring $\operatorname{End}(C)/\operatorname{rad}\operatorname{End}(C)$, provided $\operatorname{Hom}(C, Y)$ is a module of length $d + 1$).

Let us determine $\eta_{CY}^{-1}(0)$ for $C = P_i,\ Y = Q_j$ with $i, j \in \mathbb{N}_0$. Note that

$$\dim \operatorname{Hom}(P_{i-1}, Q_j) = i + j - 1,$$

where $P_{-1} = 0$. Thus, the universal map from add P_{i-1} to Q_j is of the form $P_{i-1}^{i+j-1} \to Q_j$ (for $j = 0$, it is a map of the form $P_{i-1}^{i-1} \to Q_0$).

Proposition 14.12 *Let Λ be the Kronecker algebra. Let $C = P_i,\ Y = Q_j$ with $i, j \in \mathbb{N}_0$. Let $f : P_{i-1}^{i+j-1} \to Q_j$ be the universal map from* add P_{i-1} *to Q_j. Then*

$$\boxed{\eta_{CY}^{-1}(0) = [f : P_{i-1}^{i+j-1} \to Q_j\rangle.}$$

Proof For $i = 0$, the intersection of the kernels of the maps $Q_j \to Q_1$ is zero. For $i = 1$, the module $P_0 = P_{i-1}$ is simple projective, thus, for $j \geq 1$, the universal map $f : P_0^j \to Q_j$ is just the embedding of the socle of Q_j into Q_j and the socle is the intersection of the kernels of the maps $Q_j \to Q_0$. For $i = 1$ and $j = 0$, the intersection of the kernels of the maps $Q_j \to Q_0$ is zero, but also P_{i-1}^{i+j-1} is zero.

Assume now that $i \geq 2$. We have

$$\dim \operatorname{Ext}^1(Q_j, P_{i-2}) = \dim \operatorname{Hom}(P_i, Q_j) = \dim \operatorname{Hom}(P_{i+j}, Q_0) = i + j,$$

thus the universal extension of Q_j from below using copies of P_{i-2} looks as follows

$$0 \to P_{i-2}^{i+j} \to X \xrightarrow{f} Q_j \to 0$$

with a module X such that $\operatorname{Ext}^1(X, X) = 0$. The dimension vector of X is

$$\mathbf{dim}\, X = (i + j)\, \mathbf{dim} P_{i-2} + \mathbf{dim}\, Q_j = (i + j - 1)\, \mathbf{dim} P_{i-1}.$$

Since also $\mathrm{Ext}^1(P_{i-1}, P_{i-1}) = 0$, it follows that $X = P_{i-1}^{i+j-1}$. Since f is right minimal and $\dim \mathrm{Hom}(P_{i-1}, Q_j) = i + j - 1$, we see that f has to be the universal map from add P_{i-1} to Q_j. □

Example 11 We consider the special case of the Kronecker modules $C = P_4$, $Y = Q_0$. Note that $\mathbf{dim}\, Y = (0, 1)$ and $\mathbf{dim} P_4 = (5, 4)$, thus $\tau C = P_2$ has dimension vector $(3, 2)$ and $\mathrm{Hom}(\tau C, Y) = 2$. Here is a sketch of ${}^C[\to Y\rangle$. In any layer ${}^C[\to Y\rangle_t$ with $1 \le t \le 3$, we indicate the elements $[f : X \to Y\rangle$ in the form $P_2^t \to \mathbf{dim}\, X \to Y$. The map in the layer $t = 0$ has been described in Proposition 14.12.

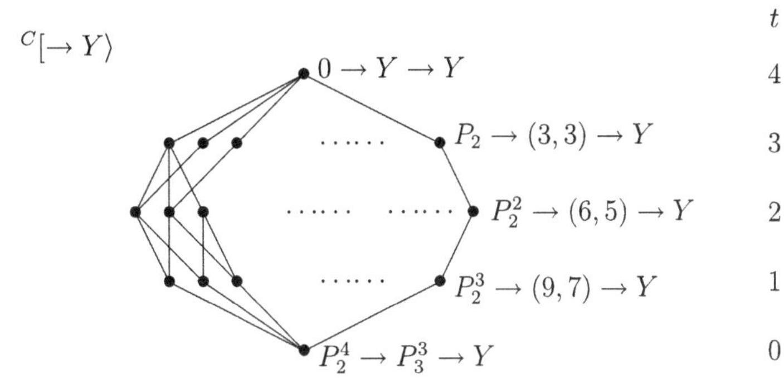

Let us exhibit some of the short exact sequences $P_1^t \to X \to Y$.

For $(3, 3)$, the module X must be a strongly regular module, thus, there are three different kinds: direct sums of three pairwise non-isomorphic indecomposable modules of length 2, direct sums of an indecomposable module M of length 4 and an indecomposable module M' of length 2 such that $\mathrm{Hom}(M, M') = 0$, and finally indecomposable modules of length 6.

For $(6, 5)$ and $(9, 7)$, we deal with direct sums of preprojective and regular modules. We consider the case $\mathbf{dim}\, X = (6, 5)$, thus $t = 2$, in detail. Let $X = X_a, X_b; X_\alpha, X_\beta)$ be a Kronecker module with a submodule U isomorphic to P_2^2 such that X/U is isomorphic to Q_0.

We start with a basis of U_b and add an element x in order to obtain a basis of X_b, thus $X_a = U_a$, and $X_b = U_b \oplus \langle x\rangle$. What we have to describe are the elements $\alpha(x)$ and $\beta(x)$ in U_a, and we have to provide a decomposition of X into indecomposables. In this way, we also will see that the map $X \to Q_0$ with kernel U is right minimal.

Since U is isomorphic to P_2^2, we can exhibit U as follows: U_a has a basis $v_0, \dots, v_5$ and U_b has a basis u_1, u_2, u_4, u_5 and $\alpha(u_i) = v_{-1}$ and $\beta(u_i) = v_i$. In order to describe X, we have to discuss possible values for $\alpha(x)$ and $\beta(x)$. Four different cases will be of interest. In the first three cases, let $\alpha(x) = v_2$.

(1) If we define $\beta(x) = v_3$, then clearly $X = P_5$.
(2) If we define $\beta(x) = v_4$, then we get a decomposition of X as follows: The elements $u_1, u_2, x, u_5, v_0, v_1, v_2, v_4, v_5$ yield a submodule of the form P_4, the elements $x - u_4, v_2 - v_3$ also yield a submodule (since $\beta(x - u_4) = 0$). These two submodules provide a direct decomposition.
(3) If we define $\beta(x) = v_5$, then we see that the elements $u_1, u_2, x; v_0, v_1, v_2, v_5$ yield a submodule of the form P_3. The elements $u_2 - u_4, x - u_5, v_1 - v_3, v_2 - v_5$

yield a 4-dimensional indecomposable submodule, and we obtain in this way a direct decomposition.

(4) Finally, let $\alpha(x) = v_1, \beta(x) = v_4$. Then we get a submodule of X with basis $u_1, x, u_5, v_0, v_1, v_4, v_5$ which is of the form P_3 as well as two indecomposable submodules of length 2, namely with basis $x - u_1, v_4 - v_2$ and with basis $x - u_4, u_1 - u_3$.

In this way, we obtain short exact sequences

$$0 \to P_2^2 \to X \to Q_0 \to 0$$

such that X is of the form P_5, of the form $P_4 \oplus R$, of the form $P_3 \oplus R''$, or finally of the form $P_3 \oplus R \oplus R'$, where R, R' are both regular of length 2, and R'' is regular of length 4.

15 Lattices of height at most 2

The height of $\mathcal{S}\operatorname{Hom}(C, Y)$ is the length of the $\Gamma(C)$-module $\operatorname{Hom}(C, Y)$. If Λ is a k-algebra, k is algebraically closed, and C is multiplicity-free (what we can assume), then the height of $\mathcal{S}\operatorname{Hom}(C, Y)$ is the k-dimension of $\operatorname{Hom}(C, Y)$.

Our main interest will concern the height 2 lattices which are not distributive, since this is the first time that one may encounter infinite families. Here is a discussion of the lattices of height at most 2, in general.

Height 0. The lattice $\mathcal{S}M$ has height 0 if and only if $M = 0$. Thus, to say that $\mathcal{S}\operatorname{Hom}(C, Y)$ has height zero means that $\operatorname{Hom}(C, Y) = 0$.

Height 1. The lattice $\mathcal{S}M$ has height 1 if and only if M is a simple module. Thus, in our case $M = \operatorname{Hom}(C, Y)$, we deal with a simple $\Gamma(C)$-module. Note that $\operatorname{Hom}(C, Y)$ is a simple $\Gamma(C)$-module if any only if there is a right minimal, right C-determined map $f : X \to Y$ which is not an isomorphism, such that for any right minimal, right C-determined map $f' : X' \to Y$ which is not an isomorphism, there is an isomorphism $h : X \to X'$ such that $f = f'h$.

Three cases should be noted:

(1) f may be an epimorphism. For example, take the path algebra of the quiver of type $\mathbb{A}_2$ as exhibited in example 1. Let $C = Y = S(b)$ (thus $\tau C = S(a)$). Then the epimorphism $f : P(b) \to S(b) = Y$ is, up to isomorphism, the only right minimal, right C-determined map ending in Y which is not an isomorphism.

(2) f may be a monomorphism. To obtain an example, take again the path algebra of the quiver of type $\mathbb{A}_2$. Let $C = S(a)$ and $Y = P(b)$, thus now C is projective. The monomorphism $f : S(a) \to P(b) = Y$ is, up to isomorphism, the only right minimal, right C-determined map ending in Y which is not an isomorphism.

(3) f is neither epi nor mono. As an example, take the radical square zero algebra with the linearly oriented quiver of type $\mathbb{A}_3$, see example 6. Let $Y = P(c)$ and $C = S(b)$ (thus $\tau C = S(a)$). The non-zero map $f : P(b) \to P(c)$ is, up to right equivalence, the only right minimal, right C-determined map ending in Y which is not an isomorphism.

Height 2. A lattice of height 2 may be either a chain (thus of the form $\mathbb{I}(2)$) or not. If the lattice is not a chain, the lattice still may be distributive (case III) or not (case IV). The submodule lattices $\mathcal{S}M$ of type III and IV occur for a semisimple module M of length 2; in case M is the direct sum of two non-isomorphic simple modules, we deal with case III, otherwise M is the direct sum of two isomorphic simple modules and then we deal with case IV. Since the lattices we are interested are submodule lattices (here of a module of length 2), we distinguish also in the case of a serial module M of length 2, whether the two composition factors are isomorphic (case I) or not (case II).

Altogether, we see that for a lattice $\mathcal{S}\,\mathrm{Hom}(C, Y)$ of height 2, there are the following four cases:

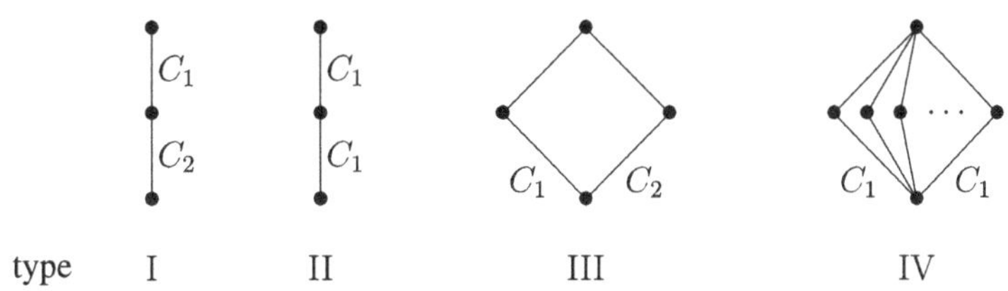

here, the labels C_1, C_2 concern the type of the corresponding pair of neighbors: C_1, C_2 are non-isomorphic indecomposable modules.

Type I. *Example 12* Take the linearly oriented quiver of type $\mathbb{A}_3$ as discussed in Examples 5. Let $Y = S(3)$ and $C = Q(2) \oplus S(3)$. Then $\Gamma(C)$ is the path algebra of the quiver of type $\mathbb{A}_2$ and $\mathrm{Hom}(C, Y)$ is the indecomposable $\Gamma(C)$ module of length 2. The lattices ${}^C[\to Y\rangle$ and $\mathcal{S}\,\mathrm{Hom}(C, Y)$ look as follows:

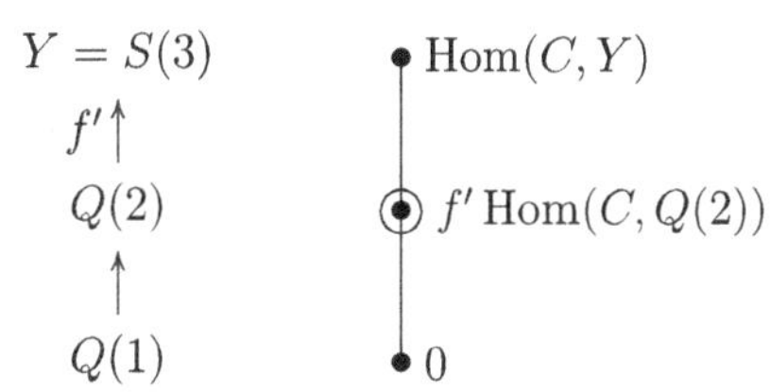

Type II. In the last section, we have seen such examples for the Kronecker algebra, namely: if C, Y are modules with C indecomposable, such that ${}^C[\to Y\rangle$ is of the form $\mathbb{I}(2)$, then we must be in type II.

Another example has been presented already in Sect. 7 when dealing with Riedtmann–Zwara degenerations, namely example 9. There, we have chosen (non-projective) indecomposable modules C and Y such that $\mathrm{Hom}(C, Y)$ was a cyclic module of length 2. There, an additional module Y was considered with an epimorphism $M' \to Y$ such that the composition $M \to M' \to Y$ is non-zero. Note that this procedure fits into the consideration of families $\mathcal{M} = \{M_i \mid i \in I\}$ of modules which is based on dealing with fixed morphisms $f_i : M_i \to Y$ for some module Y.

Type III. *Example 13* Take the quiver of type $\mathbb{A}_3$ with two sinks:

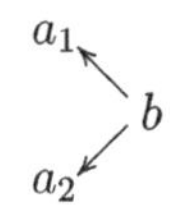

and let Λ be its path algebra. Here is the Auslander–Reiten quiver of Λ:

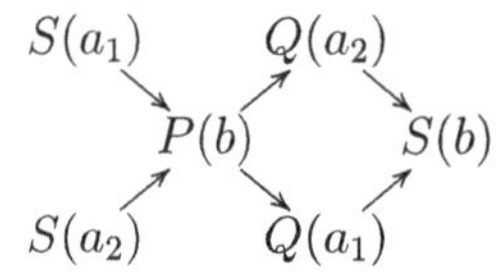

Let $Y = S(b)$ and $C = Q(a_1) \oplus Q(a_2)$. As usual, let us show both ${}^{C}[\to Y\rangle$ as well as $\mathcal{S}\,\mathrm{Hom}(C, Y)$:

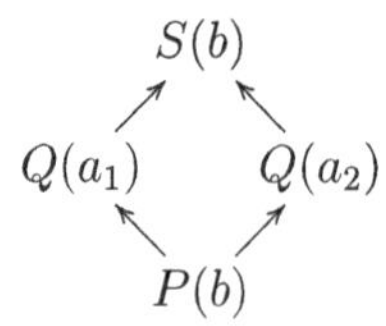

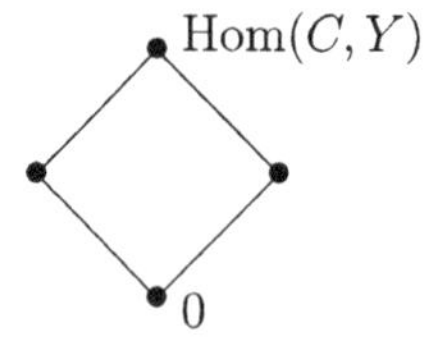

Example 14 Here is a second example of type III. In contrast to the previous example, here the two incomparable right minimal maps $f : X \to Y$ and $f' : X' \to Y$ have the property that the modules X and X' are isomorphic. We consider the Kronecker algebra (see Sect. 11). Let $Y = S(b)$ and $C = R \oplus R'$, where R, R' are non-isomorphic regular modules of length 2. We may consider R, R' as submodules of $Q(a)$ and we denote by $f : Q(a) \to S(b)$ the projection with kernel R, by $f' : Q(a) \to S(b)$ the projection with kernel R'. The pullback of f and f' is the preinjective module Q_2 (with dimension vector (2, 3)).

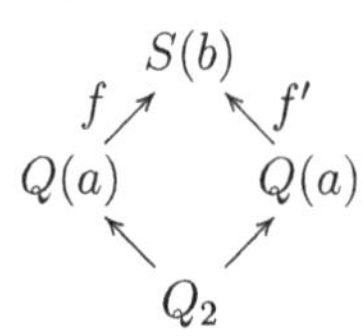

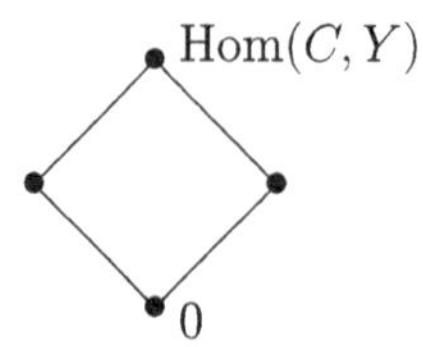

Type IV. Let C, Y be a pair of modules such that ${}^{C}[\to Y\rangle$ is of type IV. As we will see, and this is our main concern, the behaviour of the modules present in ${}^{C}[\to Y\rangle^{1}$ may be quite different.

Example 15 **where the modules present in ${}^{C}[\to Y\rangle^{1}$ are all isomorphic.** As we have seen in Sect. 11, there are many such examples for the Kronecker algebra, thus let Λ be the Kronecker algebra, let $P_0, P_1, P_2, \ldots$ be the indecomposable preprojective modules, and $Q_0, Q_1, Q_2, \ldots$ the indecomposable preinjective modules, with both P_i and Q_i being of length $2i + 1$.

First, let $C = P_i$ and $Y = P_{i+1}$, for some $i \geq 0$, thus $\dim \mathrm{Hom}(C, Y) = 2$ and therefore $\mathcal{S}\,\mathrm{Hom}(C, Y)$ is of the form IV.

If $i = 0$, then we deal with the lattice of submodules U of P_1 such that the socle of P_1/U is generated by P_0. Such a submodule is either 0 or simple (thus of the form

P_0) or equal to P_1:

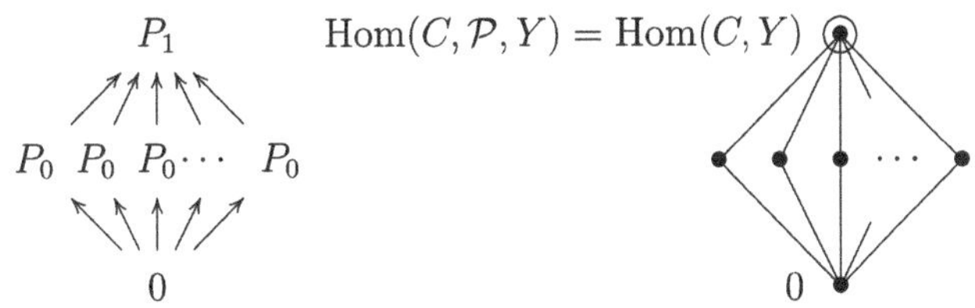

If $i = 1$, then we deal with the lattice of all submodules U of P_2 which contain the socle of P_2 (these are just the submodules U of P_2 such that the socle of P_2/U is generated by P_1). Note that such a submodule is either the socle itself, thus isomorphic to ${P_0}^3$, or of the form $N = P_0 \oplus P_1$, or P_2 itself.

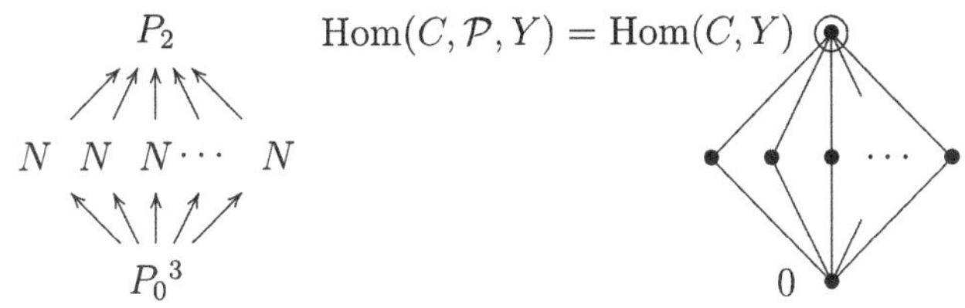

with $N = P_0 \oplus P_1$

Whereas for the cases $i = 0, 1$ all the maps shown in the lattice ${}^C[\to Y\rangle$ are inclusion maps, the maps exhibited in ${}^C[\to Y\rangle$ for $i \geq 2$ are all epimorphisms, see Proposition 6.6 and Lemma 6.8.

So let us assume that $i \geq 2$. A map $f : X \to Y$ is right P_i-determined if and only if its kernel is a direct sum of copies of P_{i-2}. Since $\mathrm{Ext}^1(P_{i+1}, P_{i-2}) = D\,\mathrm{Hom}(P_i, P_{i+1})$ is a two-dimensional vector space, we see that we deal with short exact sequences of the form

$$0 \to P_{i-2} \to N \to P_{i+1} \to 0$$

$$0 \to {P_{i-2}}^2 \to M \to P_{i+1} \to 0$$

such that the maps $N \to P_{i+1}$ and $M \to P_{i+1}$ are right minimal. It follows easily that all these modules N have to be of the form $P_{i-1} \oplus P_i$, and M has to be of the form ${P_{i-1}}^3$. Thus, the lattice ${}^C[\to Y\rangle$ looks as follows:

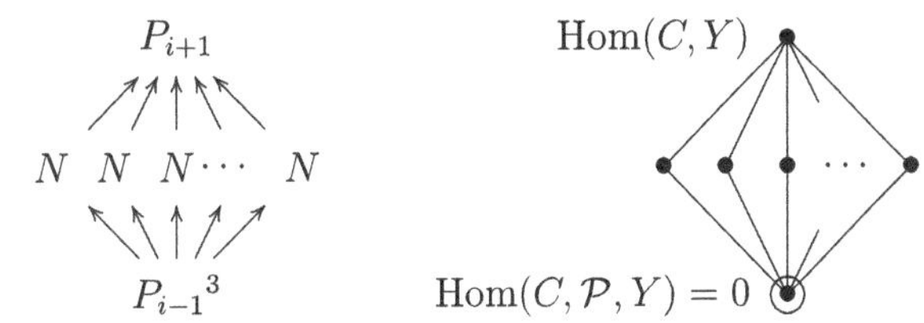

with $N = P_{i-1} \oplus P_i$

Let us describe the right minimal maps $N = P_1 \oplus P_2 \to P_3$ in detail. We have $\dim \mathrm{Hom}(P_1 \oplus P_2, P_3) = 3$, but actually only the homomorphisms $P_2 \to P_3$ matter and $\mathrm{Hom}(P_2, P_3) = 2$. Why only the homomorphisms $P_2 \to P_3$ matter? We need an epimorphism $f : P_1 \oplus P_2 \to P_3$, write it as $f = [f_1, f_2]$ with $f_1 : P_1 \to P_3$ and

$f_2 : P_2 \to P_3$. In order that f is an epimorphism, the following two conditions have to be satisfied:

(1) The restriction f_2 of f to P_2 has to be non-zero.
(2) The image $f_1(P_1)$ is not contained in $f_2(P_2)$, or, equivalently (since P_1 is projective) f_1 does not factor through f_2.

If two such maps $[f_1, f_2]$ and $[f_1', f_2']$ are given with both f_2 and f_2' non-zero, and $f_1(P_1) \not\subset f_2(P_2)$ as well as $f_1'(P_1) \not\subset f_2'(P_2)$, then $[f_1, f_2]$ is right equivalent to $[f_1', f_2']$ if and only if f_2 is right equivalent to f_2', if and only if there is a scalar $c \in k^*$ such that $f_2' = cf_2$.

It follows that the existence of the one-parameter family of right minimal maps $N \to P_3$ comes from the fact that $\dim \operatorname{Hom}(P_2, P_3) = 2$.

Observe that the lattices ${}^C[\to Y\rangle$ for $C = P_i, Y = P_{i+1}$ and all $i \geq 0$ have the same form, provided we set $P_{-1} = 0$,

Example 16 **where the modules in** ${}^C[\to Y\rangle^1$ **are pairwise non-isomorphic.** Again, we deal with the Kronecker algebra. Let $Y = Q_0$ and $C = P_2$, thus $\dim \operatorname{Hom}(C, Y) = 2$ and $K = \tau C = P_0$. The right minimal right C-determined morphisms ending in Y are epimorphisms with kernel in add K. Here are the lattices in question:

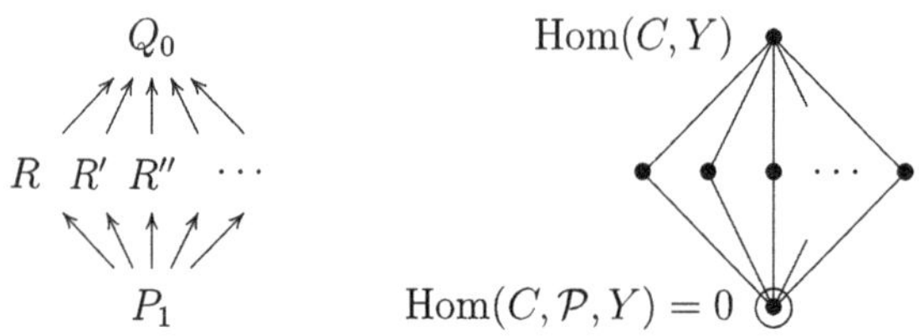

with pairwise non-isomorphic indecomposable representations $R, R', R'', \ldots$ of length 2.

Example 17 **where the modules in** ${}^C[\to Y\rangle^1$ **belong to a finite number of isomorphism classes with a fixed dimension vector.**

Take the 3-subspace quiver as considered in example 9. Let $Y = Q(a)$, and C the maximal indecomposable module, thus $C = \tau Q(a)$ and $\dim \operatorname{Hom}(C, Y) = 2$. Then $\mathcal{S} \operatorname{Hom}(C, Y)$ is the non-distributive lattice of height 2, and the elements of height 1 form a $\mathbb{P}_1$-family.

This $\mathbb{P}_1$-family in $\mathcal{S} \operatorname{Hom}(C, Y)$ contains three special elements, namely the three subspaces which are generated by the composition of irreducible maps $C \to \tau^- P(b_i)$ and $\tau^- P(b_i) \to Y$, for $1 \leq i \leq 3$. The remaining elements of the $\mathbb{P}_1$-family are generated by the surjective maps $C \to Y$.

Correspondingly, in ${}^C[\to Y\rangle_1$, there are the right equivalence classes $[f_i\rangle$ of surjective maps $f_i : M(i) \to Y$, where $M(i) = P(b_i) \oplus \tau^- P(b_i)$ for $1 \leq i \leq 3$; the remaining elements of the $\mathbb{P}_1$-family are the right equivalence classes of the surjective maps $C \to Y$. Note that all the modules $M(1), M(2), M(3)$ as well as C have the same dimension vector, namely $2\,\begin{smallmatrix}1\\1\\1\end{smallmatrix}$.

The zero element of the lattice ${}^C[\to Y\rangle$ is the projective cover $P(Y) \to Y$. In the submodule lattice $\mathcal{S} \operatorname{Hom}(C, Y)$, the zero element is $\operatorname{Hom}(C, \mathcal{P}, Y) = 0$.

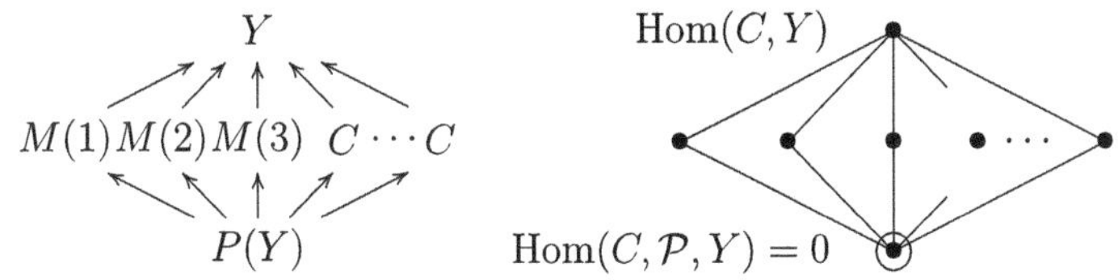

Example 18 **where the modules in ${}^C[\to Y\rangle^1$ form a family of modules with varying dimension vectors.** Again, we take the 3-subspace quiver as considered in example 9. Now let $C = P(a)$ and $Y = \tau Q(a)$ the maximal indecomposable module.

Since C is the projective module $P(a)$, the right minimal right C-determined maps $f : X \to Y$ are the inclusions maps of submodules X of Y such that the socle of Y/X is a direct sum of copies of $S(a)$. The indecomposable submodules of Y are $P(b_1), P(b_2), P(b_3)$ as well as submodules isomorphic to $S(a)$. In order that the socle of Y/X is a direct sum of copies of $S(a)$, either $X = Y$ (and $Y/X = 0$), or $X = 0$ (and $Y/X = Y$), or X has to be indecomposable. If X is an indecomposable proper submodule of Y, then the socle of Y/X is either isomorphic to $S(0)$ or else X is contained in one of the modules $P(b_1), P(b_2), P(b_3)$. It follows that the 1-parameter family in the middle of ${}^C[\to Y\rangle$ consists of the indecomposable proper submodules X of Y such that the socle of Y/X is isomorphic to $S(a)$, thus either X is one of $P(b_1), P(b_2), P(b_3)$, or else X is a simple submodule of Y not contained in $P(b_1), P(b_2), P(b_3)$.

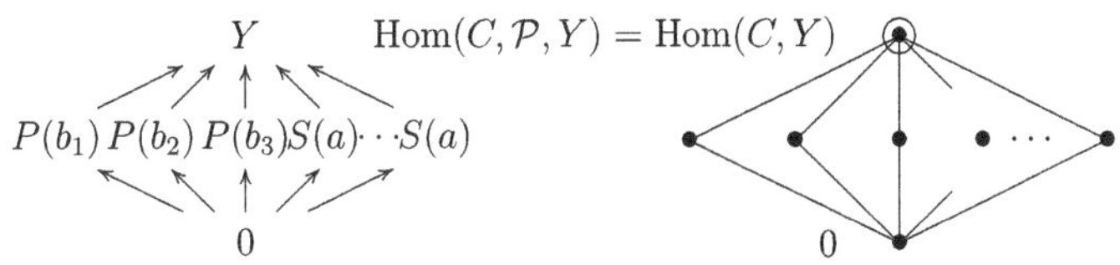

Example 19 **Another example where the modules in ${}^C[\to Y\rangle^1$ form a family of modules with varying dimension vectors.** Take the one-point extension Λ of the Kronecker algebra using a regular module R of length 2, say $R = R_\infty$

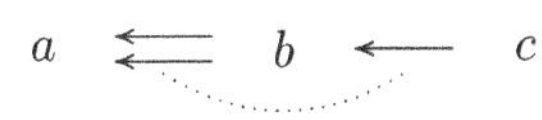

The vertex c is the extension vertex and $\operatorname{rad} P(c) = R_\infty$. The regular Kronecker modules of length 2 different from R_∞ will be denoted by R_λ with $\lambda \in k$.

Let $K = S(a)$ and $C = \tau^- K$, this is the indecomposable Kronecker-module P_2, its dimension vector (as a Λ-module) is $(3, 2, 0)$. Let $Y = P(c)/S(a)$, its dimension vector is $(0, 1, 1)$. We have $\dim \operatorname{Hom}(C, Y) = 2$. Since $\operatorname{End}(C) = k$, the submodule lattice $\mathcal{S}\operatorname{Hom}(C, Y)$ is just the projective line. Under η_{SY} we obtain a $\mathbb{P}_1$-family of right minimal maps ending in Y, namely those with the following short exact sequences:

$$K \longrightarrow P(c) \xrightarrow{p} Y$$
$$K \longrightarrow R_\lambda \xrightarrow{f'_\lambda} S(b) \quad \text{for } \lambda \in k$$

Here is, on the left, ${}^C[\to Y\rangle$, and, on the right, $\mathcal{S}\operatorname{Hom}(C, Y)$

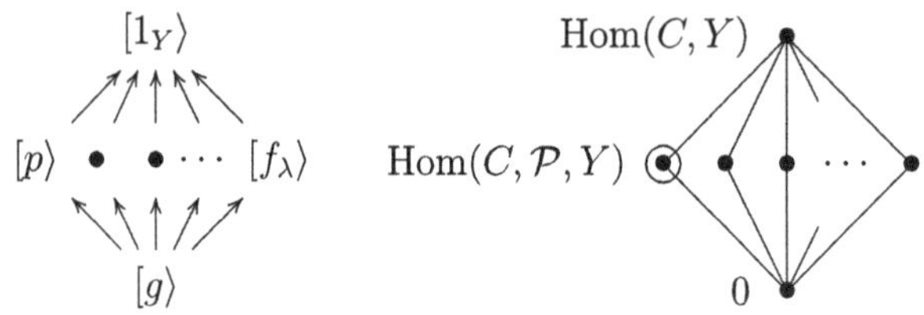

The 0-subspace of $\mathrm{Hom}(C, Y)$ corresponds to a map $g : P(b) \to Y$ with image $S(b)$, namely to the projective presentation of $S(b)$

$$K^2 \longrightarrow P(b) \xrightarrow{g'} S(b).$$

It should be noted that the short exact sequence

$$K \longrightarrow R_\infty \xrightarrow{f'_\infty} S(b)$$

yields the map $R_\infty \xrightarrow{f'_\infty} S(b) \subset Y$ which we denote by f_∞ and which is not C-determined. Namely, there is the following commutative diagram

$$\begin{array}{ccc} \mathrm{rad}\, P(c) & \xrightarrow{\iota} & P(c) \\ \downarrow & & \downarrow p \\ R_\infty & \xrightarrow{f_\infty} & Y \end{array}$$

with inclusion map ι. Since p does not factor through f_∞, we see that p almost factors through f_∞, thus the theory asserts that $P(c)$ has to belong to any determiner of f_∞ and therefore f_∞ cannot belong to ${}^{C}[\to Y\rangle$.

As we have noted, $P(c)$ has to belong to any determiner of f_∞. Let us add $P(c)$ to C and consider the Auslander bijection for $C \oplus P(c)$ and Y. We have $\dim \mathrm{Hom}(C \oplus P(c), Y) = 3$. Note that the endomorphism ring of $C \oplus P(c)$ is hereditary of type $\mathbb{A}_2$ and the submodule structure of $\mathrm{Hom}(C \oplus P(c), Y)$ looks as follows:

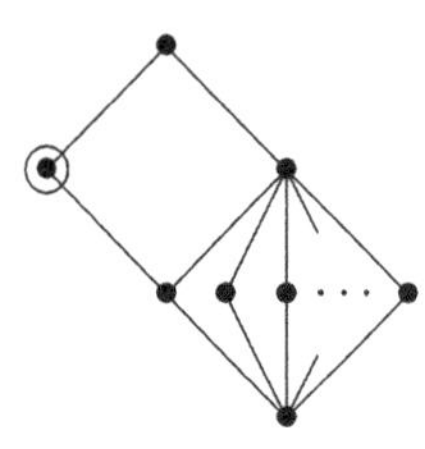

For the proof, we only have to verify that a non-trivial map $C \to P(c)$ does not annihilate the module $\mathrm{Hom}(C \oplus P(c), Y)$. The encircled vertex is the submodule $\mathrm{Hom}(C \oplus P(c), \mathcal{P}, Y)$.

The corresponding diagram in ${}^{C \oplus P(c)}[\to Y\rangle$ looks as follows; here, we write its elements as short exact sequences ending in a submodule of Y:

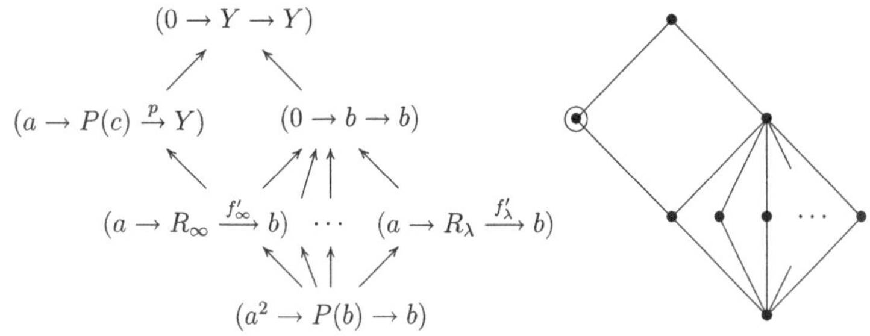

We may label the lines of the Hasse diagram of $\mathcal{S}\,\mathrm{Hom}(C \oplus P(c), Y)$ by the corresponding type, thus either by C or by $P(c)$

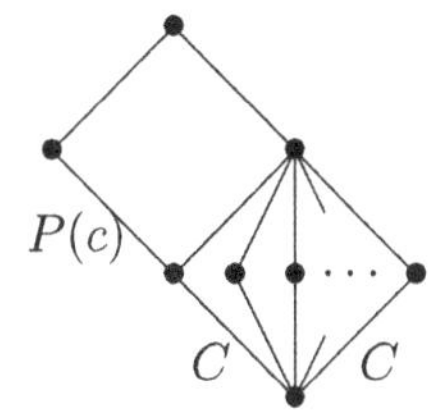

We should add that conversely, we can recover $\mathcal{S}\,\mathrm{Hom}(C, Y)$ from $\mathcal{S}\,\mathrm{Hom}(C \oplus P(c), Y)$ by deleting the shaded part:

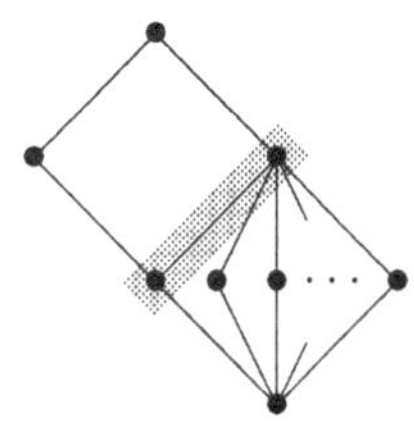

III. Special cases

16 The module C being a generator

The special case $C = \Lambda$ has been discussed already at the end of Sect. 4. In this case, $\mathcal{S}\,\mathrm{Hom}(C, Y)$ is just the lattice of all submodules of Y.

Proposition 16.1 *Let C be a generator. Then $f : X \to Y$ is right C-determined if and only if the intrinsic kernel of f is in* $\mathrm{add}\,\tau C$.

Proof According to Theorem 3.4, the map f is right C-determined if and only if $C(f) \in \mathrm{add}\,C$. Thus, assume that $C(f) \in \mathrm{add}\,C$. Let K be an indecomposable direct summand of the intrinsic kernel of f. By definition of $C(f)$, we know that $\tau^- K$ belongs to $\mathrm{add}\,C(f)$, and K is not injective. Thus $K = \tau\tau^- K$ belongs to $\mathrm{add}\,\tau C(f) \subseteq \mathrm{add}\,C$. This shows that the intrinsic kernel of f belongs to $\mathrm{add}\,\tau C$.

Conversely, assume that the intrinsic kernel belongs to $\mathrm{add}\,\tau C$, in particular any indecomposable direct summand of the intrinsic kernel K belongs to $\mathrm{add}\,\tau C$, thus $\tau^- K$ belongs to $\mathrm{add}\,\tau^-\tau C \subseteq \mathrm{add}\,C$. Also, since C is a generator, any indecomposable projective module belongs to add C. As a consequence, $C(f)$ belongs to add C. This completes the proof. □

Everyone admits that the concept of being determined is not very intuitive, however in the special case when C is a generator (and this is often the only important case), one knows: The maps in ${}^C[\to Y\rangle$ can be described by the exact sequences

$$0 \to K \to X \xrightarrow{e} Y' \to 0$$

where K is in add τC and Y' is a submodule of Y say with inclusion map $m : Y' \to Y$, the map in ${}^C[\to Y\rangle$ in question is the composition me. To repeat: *If C is a generator, a right minimal is right C-determined if and only if its kernel belongs to* add τC.

This means: for C a generator, the set ${}^C[\to Y\rangle$ can be visualized very well. Unfortunately, this seems to be difficult in general, but the notion of right determination just allows to have the prolific bijection ${}^C[\to Y\rangle \leftrightarrow \mathcal{S}\,\mathrm{Hom}(C, Y)$.

Proposition 16.2 *Let C be a generator and $u : Y' \to Y$ a monomorphism. Then $f \mapsto uf$ defines an embedding $u_* : {}^C[\to Y'\rangle \to {}^C[\to Y\rangle$. The image of this embedding is an ideal of the lattice ${}^C[\to Y\rangle$ and the following diagram commutes:*

$$\begin{array}{ccc} {}^C[\to Y\rangle & \xrightarrow{\eta_{CY}} & \mathcal{S}\,\mathrm{Hom}(C,Y) \\ u_*\uparrow & & \uparrow \mathcal{S}\,\mathrm{Hom}(C,u) \\ {}^C[\to Y'\rangle & \xrightarrow{\eta_{CY'}} & \mathcal{S}\,\mathrm{Hom}(C,Y') \end{array}$$

Proof Let $f : X \to Y'$ be right C-determined. Since the intrinsic kernel of f and of uf are the same, also uf is right C-determined by 14.1. If $f \preceq f'$ are maps ending in Y', then there is a map h such that $f = f'h$, therefore $uf = uf'h$, thus $uf \preceq uf'$. This shows that $f \mapsto uf$ yields a map $u_* : {}^C[\to Y'\rangle \to {}^C[\to Y\rangle$. In order to see that u_* is an embedding, let us assume that f, f' are maps ending in Y such that $uf \preceq uf'$. Then there is h with $uf = uf'h$, but this implies that $f = f'h$ (since u is a monomorphism), and therefore $f \preceq f'$. In order to see that the image of u_* is an ideal, let f be right C-determined and ending in Y' and g right C-determined ending in Y such that $g \preceq u_*(f)$. We want to show that g is in the image of u_*. But $g \preceq uf$ means that there is h with $g = ufh$, thus $g = u_*(fh)$.

It remains to show that the diagram commutes. Let $f : X \to Y'$ be right C-determined. Then we have $\eta_{CY'}(f) = f\,\mathrm{Hom}(C, X)$. The map $\mathrm{Hom}(C, u)$ send any $\phi \in \mathrm{Hom}(C, Y')$ to $\mathrm{Hom}(C, u)(\phi) = u\phi$, thus $\mathcal{S}\,\mathrm{Hom}(C, u)$ sends $f\,\mathrm{Hom}(C, X)$ to $uf\,\mathrm{Hom}(C, X)$. On the other hand, we also have $\eta_{CY}(u_*(f)) = \eta_{CY}(uf) = uf\,\mathrm{Hom}(C, X)$. □

If $u : Y' \to Y$ is a monomorphism, but C is not a generator, then given $[f\rangle \in {}^C[\to Y'\rangle$, the right equivalence class of the composition uf usually will not belong to ${}^C[\to Y\rangle$.

Example 20 Take the quiver of type $\mathbb{A}_2$ as considered in example 1 and take $C = Y = S(2)$ and $Y' = 0$ with inclusion map $u : 0 \to S(2)$. The zero map $0 \to 0$ yields an element of ${}^C[\to Y'\rangle$, but u is not right C-determined, since $P(2)$ almost factors through u, so that $P(2)$ has to be a direct summand of $C(u)$ and therefore of any right determiner of u.

17 The case of $[\to Y\rangle$ being of finite height

Here we consider modules Y such that there are only finitely many indecomposable modules M_i with $\operatorname{Hom}(M_i, Y) \neq 0$. Let M be the direct sum of these modules and consider the Auslander bijection

$$^{M}[\to Y\rangle \leftrightarrow \mathcal{S}\operatorname{Hom}(M, Y),$$

it maps $\bigoplus_{i,j} M_i^{n_i} \xrightarrow{(f_{ij})} Y$ to the $\Gamma(M)$-submodule of $\operatorname{Hom}(M, Y)$ generated by the maps f_{ij} (considered as maps $M \to M_i \to Y$ where the map $M \to M_i$ is the canonical projection onto the direct summand).

Proposition 17.1 *Let Y be a module.*
(a) *If $[\to Y\rangle$ is of height h, then the number of isomorphism classes of indecomposable modules M_i with* $\operatorname{Hom}(M_i, Y) \neq 0$ *is at most h.*
(b) *If $M_1, \dots, M_t$ are all the indecomposable modules M_i with* $\operatorname{Hom}(M_i, Y) \neq 0$, *one from each isomorphism class, and $M = \bigoplus_i^t M_i$, then $[\to Y\rangle = {}^{M}[\to Y\rangle$ and any module C with $[\to Y\rangle = {}^{C}[\to Y\rangle$ satisfies $M \in \operatorname{add} C$.*

Proof Let $M_1, \dots, M_t$ be pairwise non-isomorphic indecomposable modules with $\operatorname{Hom}(M_i, Y) \neq 0$, and let $M = \bigoplus_i^t M_i$. Let $\Gamma(M) = \operatorname{End}(M)^{\mathrm{op}}$. Then the indecomposable projective $\Gamma(M)$-modules $P(i) = \operatorname{Hom}(M, M_i)$ are pairwise non-isomorphic, and $\operatorname{Hom}_{\Gamma(M)}(P(i), \operatorname{Hom}(M, Y)) \neq 0$, thus the $\Gamma(M)$-module $\operatorname{Hom}(M, Y)$ has length at least t. Now, according to the Auslander bijection, the poset $\mathcal{S}M$ is isomorphic to the subposet ${}^{M}[\to Y\rangle$ of $[\to Y\rangle$. The length of $\mathcal{S}M$ is at least t, the length of $[\to Y\rangle$ is h. This shows that $t \leq h$. This shows (a).

In order to show (b), we recall from Proposition 3.9 that given a map f ending in Y, any indecomposable direct summand C_0 of $C(f)$ satisfies $\operatorname{Hom}(C_0, Y) \neq 0$, thus C_0 is in $\operatorname{add} M$ and therefore $C(f) \in \operatorname{add} M$. Thus any map $f : X \to Y$ is right M-determined and therefore $[\to Y\rangle = {}^{M}[\to Y\rangle$.

On the other hand, let us assume that $[\to Y\rangle = {}^{C}[\to Y\rangle$ for some module C. Using again 3.9, we can assume that any indecomposable direct summand C_i of C satisfies $\operatorname{Hom}(C_i, Y) \neq 0$. Thus we can assume that C is a direct summand of M. But if C is a proper direct summand of M, say $C = \bigoplus_{i=1}^{t'} M_i$ with $t' < t$, then

$$\begin{aligned} |[\to Y\rangle| &= |{}^{C}[\to Y\rangle| = |\operatorname{Hom}(C,Y)| \\ &= \sum_{i=1}^{t'} |\operatorname{Hom}(M_i, M)| < \sum_{i=1}^{t} |\operatorname{Hom}(M_i, M)| \\ &= |\operatorname{Hom}(M,Y)| = |{}^{M}[\to Y\rangle| \leq |[\to Y\rangle| \end{aligned}$$

a contradiction. This shows that $C = M$. □

Remark As we see, the indecomposable modules M_i which occur as direct summands of a minimal module C with $[\to Y\rangle = {}^{C}[\to Y\rangle$ are modules with $\operatorname{Hom}(M_i, Y) \neq 0$. But this is not surprising, since the minimal right determiner $C(f)$ of any morphism f has as indecomposable direct summands only modules C_i with $\operatorname{Hom}(C_i, Y) \neq 0$, see

3.9. Also the converse should be stressed, namely the following part of the assertion (b): any indecomposable module M_i with $\mathrm{Hom}(M_i, Y) \neq 0$ is needed as a direct summand of C.

Corollary 17.2 *Let Y be a module. The following conditions are equivalent:*

(i) *There are only finitely many isomorphism classes of indecomposable modules X with* $\mathrm{Hom}(X, Y) \neq 0$.
(ii) *The lattice $[\rightarrow Y\rangle$ has finite height.*
(iii) *There is a module C with $[\rightarrow Y\rangle = {}^{C}[\rightarrow Y\rangle$.*

Proof The implication (ii) $\Longrightarrow$ (i) has been shown in 17.1(a), the implication (i) $\Longrightarrow$ (iii) in 15.1(b). Any lattice of the form ${}^{C}[\rightarrow Y\rangle$ is of finite height, thus obviously (iii) implies (ii). □

If S is simple and $[\rightarrow Q(S)\rangle$ is of finite height, then we deal just with the hammock corresponding to S, as considered in [40].

Example 21 Let Λ be the 3-subspace quiver as considered in example 9 and let $Y = Q(a)$. As before, we denote the indecomposable modules which are neither projective nor injective by $N(i) = \tau Q(b_i)$ with $1 \leq i \leq 3$ and $M = \tau Q(a)$. Here is the lattice $[\rightarrow Y\rangle$

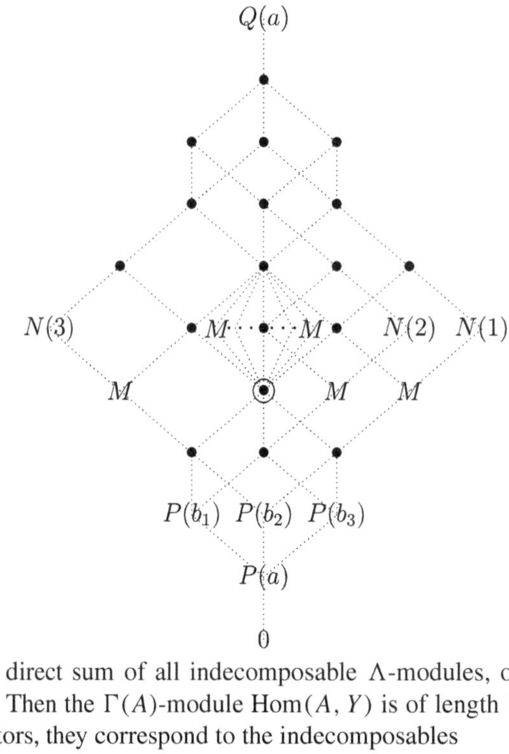

Let A be the direct sum of all indecomposable Λ-modules, one from each isomorphism class. Then the $\Gamma(A)$-module $\mathrm{Hom}(A, Y)$ is of length 10 with 9 different composition factors, they correspond to the indecomposables

$$P(a);\ P(b_1), P(b_2), P(b_3);\ M;\ N(1), N(2), N(3);\ Q(a),$$

and the composition factor corresponding to M occurs with multiplicity 2 in $\mathrm{Hom}(A, Y)$.

In the picture above, only join irreducible elements (as well as the zero element) have been labeled. Note that in the middle layer of the non-distributive interval of length 2, all but three elements are join irreducible, and have the label M; here we deal with the various epimorphisms $M \to Q(0)$. Altogether there are 18 non-zero elements which are not join-irreducible.

Note that the picture is obtained from the free modular lattice in 3 generators as presented by Dedekind [11] in 1900 by inserting in the non-distributive interval of length 2 further diagonals (one may call it the free k-modular lattice in 3 generators).

There is an obvious action of the symmetric group S_3 of order 6 on the 3-subspace quiver, and thus also on the lattice $[\to Y\rangle$. Six vertices of $[\to Y\rangle$ are invariant under this action, five of them correspond to important maps ending in Y:

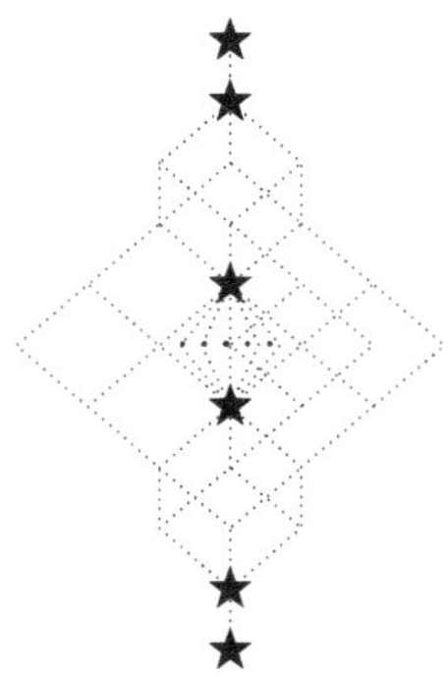

the remaining one is the universal map from add M to Y.

Instead of looking at representatives f in right equivalence classes, we may also draw the attention on the right equivalence classes $[f\rangle$ themselves. For example, $[1_Y\rangle$ is the class of all split epimorphisms ending in Y. If $g : X \to Y$ is minimal right almost split, then $[g\rangle$ is the class of all right almost split maps ending in Y and finally, $[0 \to Y\rangle$ is the class all zero maps ending in Y.

Remark An attempt to deal in a similar way with the n-subspace quivers for $n \geq 4$ was given by Gelfand and Ponomarev in a series of papers [16–19], see also [13] and [32]. In order to get information about the free modular lattice in n generators, they described part of the lattice $[\to Q(a)\rangle$ where Δ is the n-subspace quiver and $Q(a)$ is the indecomposable injective $k\Delta$-module with a the sink of Δ. Of course, for $n \geq 4$, this lattice $[\to Q(a)\rangle$ is of infinite height! It may be worthwhile to look for an interpretation of the results of Gelfand-Ponomarev in terms of the Auslander bijections.

18 Some serial modules Hom(C, Y)

The Auslander bijections are defined for any pair of Λ-modules C, Y and one of the posets involved is $\mathcal{S}\,\mathrm{Hom}(C, Y)$. Assume that J is an ideal of Λ which annihilates both modules C, Y so that we may consider C and Y as Λ'-modules, with $\Lambda' = \Lambda/J$. On the one hand, we have $\mathrm{Hom}_\Lambda(C, Y) = \mathrm{Hom}_{\Lambda'}(C, Y)$. On the other hand, we have to distinguish the set ${}^C[\to Y\rangle_{\Lambda'}$ of right equivalence classes of right C-determined Λ'-modules ending in Y from ${}^C[\to Y\rangle_\Lambda = {}^C[\to Y\rangle$. Using the Auslander bijections for Λ as well as for Λ', it is clear that the posets ${}^C[\to Y\rangle_{\Lambda'}$ and ${}^C[\to Y\rangle_\Lambda$ are isomorphic, however the modules present in ${}^C[\to Y\rangle_{\Lambda'}$ usually will be completely different from those present in ${}^C[\to Y\rangle_\Lambda$. The following examples will show such deviations.

Let Λ be a local uniserial ring. When dealing with a local uniserial ring, the indecomposable module of length n will be denoted just be n. Also, if C is a module, we will write $\mathrm{rad}^t(C, C)$ instead of $(\mathrm{rad}\,\mathrm{End}(C))^t$.

Example 22 **Let** $C = Y = 4$. First, let Λ be of length at least 8, so that $\mathrm{Hom}(4, \mathcal{P}, 4) = 0$.

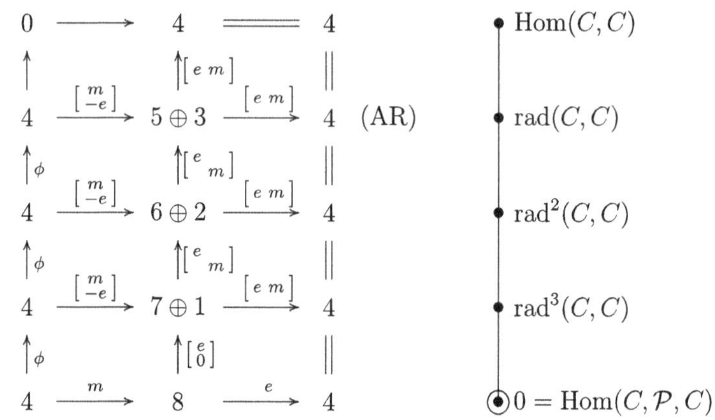

here, we have denoted by m the canonical inclusion maps, by e the canonical projections and ϕ is a radical generator $\mathrm{End}(C)$. The Auslander–Reiten sequence is marked as (AR).

Second, let Λ be of length 6, so that $\mathrm{Hom}(4, \mathcal{P}, 4) = \mathrm{rad}^2(C, C)$.

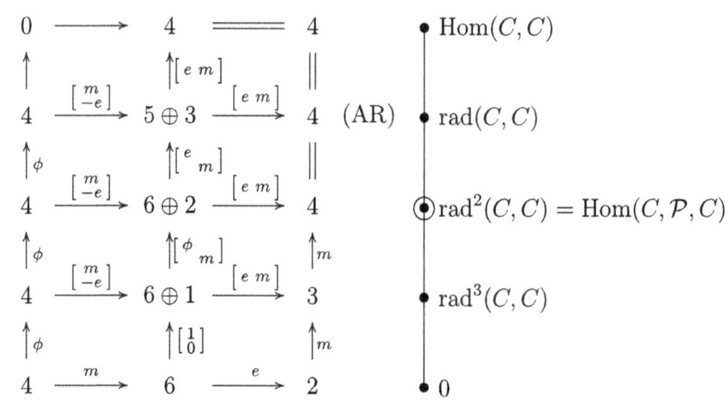

again, m stands for a canonical inclusion map, e for a canonical projection map and ϕ for a radical generator of the endomorphism ring of a uniserial module.

Finally, let Λ be of length 4, thus C is projective and therefore $\mathrm{Hom}(C, \mathcal{P}, C) = \mathrm{Hom}(C, C)$. Since C is projective, all the right minimal, right C-determined maps are inclusion maps:

$$\begin{array}{ccccccl}
0 & \longrightarrow & 4 & = & 4 & & \odot\ \mathrm{Hom}(C,C) = \mathrm{Hom}(C,\mathcal{P},C) \\
\uparrow & & \uparrow{\scriptstyle m} & & \uparrow{\scriptstyle m} & & \\
0 & \longrightarrow & 3 & = & 3 & & \bullet\ \mathrm{rad}(C,C) \\
\uparrow & & \uparrow{\scriptstyle m} & & \uparrow{\scriptstyle m} & & \\
0 & \longrightarrow & 2 & = & 2 & & \bullet\ \mathrm{rad}^2(C,C) \\
\uparrow & & \uparrow{\scriptstyle m} & & \uparrow{\scriptstyle m} & & \\
0 & \longrightarrow & 1 & = & 1 & & \bullet\ \mathrm{rad}^3(C,C) \\
\uparrow & & \uparrow & & \uparrow & & \\
0 & \longrightarrow & 0 & = & 0 & & \bullet\ 0
\end{array}$$

Example 23 **Let** $C = 1 \oplus 2$ **and** $Y = 3$. The ring $\Gamma(C) = \mathrm{End}(C)^{\mathrm{op}}$ is the Nakayama algebra with Kupisch series 2, 3. The $\Gamma(C)$-module $\mathrm{Hom}(C, Y)$ is the indecomposable projective module of length 3. If we work over a uniserial ring Λ of length at least 5, so that $\mathrm{Hom}(C, \mathcal{P}, Y) = 0$, then the situation is as follows:

$$\begin{array}{ccccccl}
0 & \longrightarrow & 3 & = & 3 & & \bullet\ \mathrm{Hom}(C,Y) \\
\uparrow & & \uparrow{\scriptstyle [e\ m]} & & \| & & \\
2 & \xrightarrow{\left[\begin{smallmatrix} m \\ -e \end{smallmatrix}\right]} & 4 \oplus 1 & \xrightarrow{[e\ m]} & 3 & & \bullet \\
\uparrow & & \uparrow{\scriptstyle \left[\begin{smallmatrix} 1 \\ 0 \end{smallmatrix}\right]} & & \| & & \\
1 & \xrightarrow{m} & 4 & \xrightarrow{e} & 3 & & \bullet \\
\uparrow & & \uparrow{\scriptstyle [e\ m]} & & \| & & \\
2 & \xrightarrow{m} & 5 \oplus 1 & \xrightarrow{[e\ m]} & 3 & & \odot\ 0 = \mathrm{Hom}(C,\mathcal{P},Y)
\end{array}$$

Next, assume that Λ is of length 4.

$$\begin{array}{ccccccl}
0 & \longrightarrow & 3 & = & 3 & & \bullet\ \mathrm{Hom}(C,Y) \\
\uparrow & & \uparrow{\scriptstyle [e\ m]} & & \| & & \\
2 & \xrightarrow{\left[\begin{smallmatrix} m \\ -e \end{smallmatrix}\right]} & 4 \oplus 1 & \xrightarrow{[e\ m]} & 3 & & \bullet \\
\uparrow & & \uparrow{\scriptstyle \left[\begin{smallmatrix} 1 \\ 0 \end{smallmatrix}\right]} & & \| & & \\
1 & \xrightarrow{m} & 4 & \xrightarrow{e} & 3 & & \odot\ \mathrm{Hom}(C,\mathcal{P},Y) \\
\uparrow & & \uparrow{\scriptstyle \phi} & & \uparrow{\scriptstyle m} & & \\
2 & \xrightarrow{m} & 4 & \xrightarrow{e} & 2 & & \bullet\ 0
\end{array}$$

Finally, we consider the case where Λ is of length 3.

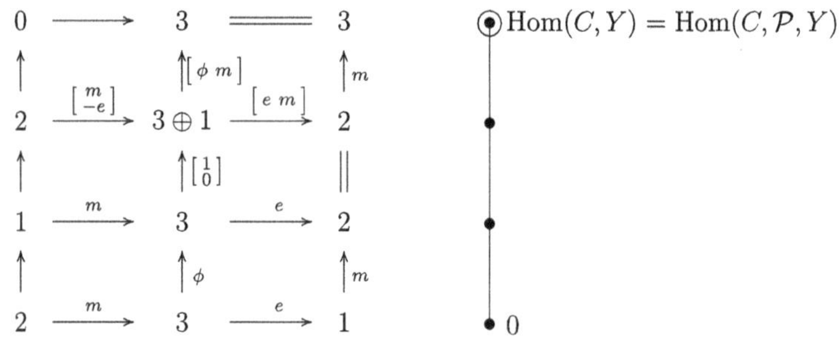

19 Final remarks

19.1 Duality

For all the results presented here, there is a dual version which has the same importance. Note that we deal with an artin algebra Λ and consider finitely generated modules, thus there is a duality functor. Namely, by assumption, Λ is a module-finite k-algebra, where k is a commutative artinian ring. A minimal injective cogenerator Q in mod k yields the duality functor $\mathrm{Hom}_k(-, Q)$.

We did not attempt to formulate the dual definitions and statements, but the reader is advised to do so. To start with, we need the notion of left equivalence in order to introduce $\langle Y \rightarrow]$ as the set of left equivalence classes of maps starting in Y (or of left minimal maps). Then we need the notion of left C-determination in order to define $\langle Y \rightarrow]^C$ as the set of the left equivalence classes maps starting in Y which are left C-determined.

19.2 Proofs of Auslander's two main theorems

Both results are presented in detail in the Philadelphia notes [2], see also [3]. There is a different treatment in the book [5] of Auslander–Reiten–Smalø: Auslander's Second Theorem is presented in Theorem XI.3.9., for the First Theorem, see Theorem XI.2.10 and Proposition XI.2.4 (this actually improves the assertion given in the Philadelphia Notes). A concise proof of the First Theorem can also be found in [38]. As one of the main ingredients for the proof of the Second Theorem, one may use Auslander's defect formula. For a direct approach to the defect formula, we recommend the paper [24] by Krause.

19.3 The universal character of the Auslander bijections

The Auslander bijections are Auslander's approach to describe say the module category for an artin algebra completely: not just to provide some invariants. The importance of using invariants usually relies on the fact that they will allow to distinguish certain objects, but they may not say much about other ones—the most effective invariants are often those which attach to objects just one of the values 1 or 0 (thus "yes" or "no"). Of course, the use of such an invariant is restricted to some specific problem. Now Auslander's approach is an attempt to describe a module category completely, thus we may ask whether it does not have to be tautological: if we don't forget some of the structure, we will not see the remaining structure in more detail. Indeed, the Auslander bijections focus on parts of the category, namely the sets ${}^{C}[\to Y\rangle$, but the decisive feature is the possibility to change the focus by enlarging C (adding direct summands). The universality of this approach is due to the fact that the category is covered completely by such subsets.

19.4 The irritation of the wording "morphisms determined by modules"

Let us insert a short discussion of our hesitation to say that a morphism f is right determined by a module C. One possible interpretation is to assert that even if the morphism f itself is not determined by C, there is a certain factorization property of f which is determined by C. But there is another way out:

When Auslander asserts that *every morphism in* mod Λ *is right determined by a* Λ-module, one expects a classification of the (right minimal) morphisms in mod Λ using as invariants just Λ-modules. One even may strengthen Auslander's assertion by saying *every morphism in* mod Λ *is right determined by the isomorphism class of a multiplicity-free* Λ*-module.* Clearly, such a formulation is irritating, since the set of isomorphism classes of multiplicity-free modules may outnumber the set of right equivalence classes of right minimal morphisms by far: Consider just the special case of a representation finite artin algebra of uncountable cardinality, then there are only finitely many isomorphism classes of multiplicity-free modules, but usually uncountably many right equivalence classes of right minimal morphisms. So how should it be possible that finitely many modules C determine uncountably many morphisms α? The solution is rather simple: it is not just the module C which is needed to recover a morphism $\alpha : X \to Y$ but one actually needs a submodule of $\mathrm{Hom}(C, Y)$, with $\mathrm{Hom}(C, Y)$ being considered as an $\mathrm{End}(C)^{\mathrm{op}}$-module. In the setting where Λ is representation-finite and uncountable, one should be aware that usually the modules $\mathrm{Hom}(C, Y)$ will have uncountably many submodules, thus we are no longer in trouble. So if we assert that the morphism $\alpha : X \to Y$ is determined by the module C, then one should keep in mind that C is only part of the data which are required to recover α; in addition to C one will need a submodule of $\mathrm{Hom}(C, Y)$.

19.5 Modules versus morphisms, again

The following feature seems to be of interest: The concept of the determination of morphisms by modules concerns the category of maps with fixed target Y, namely one wants to decide whether two elements in $[\to Y\rangle$ are comparable. The theory asserts that there is a test set of modules, namely the indecomposable direct summands of

$C(f)$. For the testing procedure, they are just modules, but any such object L comes equipped with a non-zero (thus right minimal) map $L \to Y$.

19.6 Logic and category theory

Let us stress that the setting of the Auslander bijections is well-accepted both in mathematical logic and in category theory. The lattice of Λ-pointed modules, as studied in model theory (see for example [28]) is precisely the lattice $\langle \Lambda \to]$. In this way, all the results concerning pp-formulae and pp-definable subgroups concern the Auslander setting.

In the terminology of abstract category theory, we deal with a comma category, namely the category of *objects over* Y: its objects are the maps $X \to Y$, a map from $f : X \to Y$ to $f' : X' \to Y$ being given by a map $h : X \to X'$ such that $f = f'h$.

Of course, the use of the representable functors $\mathrm{Hom}(X, -)$ and $\mathrm{Hom}(-, Y)$ is standard in representation theory. To provide here all relevant references would overload our presentation due to the abundance of such papers. So we restrict to mention only few names: of course Auslander himself, but also Gabriel [14] as well as Gelfand-Ponomarev.

19.7 Generalisations

In this survey we have restricted the attention on the module category of an artin algebra Λ, or, what is the same, a length k-category $\mathcal{C}$, where k is a commutative artinian ring, such that $\mathcal{C}$ is Hom- and Ext-finite, has only finitely many simple objects and all objects have bounded Loewy length. Auslander's investigations were devoted to larger classes of rings, and many of our considerations should be of interest in a broader context.

Krause [25] has considered the general setting of dualising varieties in the sense of Auslander and Reiten [4] and has obtained the precise analogues of the Auslander Theorems which form the basis of our survey. In particular, he was looking also at triangulated categories and showed the relationship to the existence of Serre duality. The interested reader may try to work out in which way our presentation has to be modified in this context.

Acknowledgments Some of the material presented here has been exhibited in 2012 in lectures at SJTU, Shanghai, at USTC, Hefei, and at the ICRA conference at Bielefeld, in 2013 in lectures at Woods Hole and at NEU, Boston. The author is grateful to the audience for questions and remarks. He also wants to thank Xiao-Wu Chen, Lutz Hille, Henning Krause and Idun Reiten for helpful comments.

Appendix: Modular lattices

Let L be a modular lattice of height h. The set of elements in L of height i will be denoted by L_i or also L^{h-i}. Here are some typical modular lattices:

The *chain* $\mathbb{I}(h)$ of height h, this is the poset of the integers $0, 1 \ldots, h$ with the usual order relation. Note that $\mathbb{I}(h)_i$ consists of a single element, for $0 \leq i \leq h$. Here are the lattices $\mathbb{I}(i)$ with $0 \leq i \leq 4$.

$\mathbb{I}(0)$ $\mathbb{I}(1)$ $\mathbb{I}(2)$ $\mathbb{I}(3)$ $\mathbb{I}(4)$

Let k be a field. A *projective geometry* over k is the lattice of all subspaces of a vector space over k. In accordance to the notation used throughout the paper, we denote the lattice of subspaces of the d-dimensional vector space k^d by $\mathcal{S}(k^d)$ or also just by $\mathbb{G}(d)$ (this is called the projective geometry over k of *dimension* $d-1$, it is a lattice of height d). Here are the lattices $\mathbb{G}(d)$ for $0 \leq d \leq 4$:

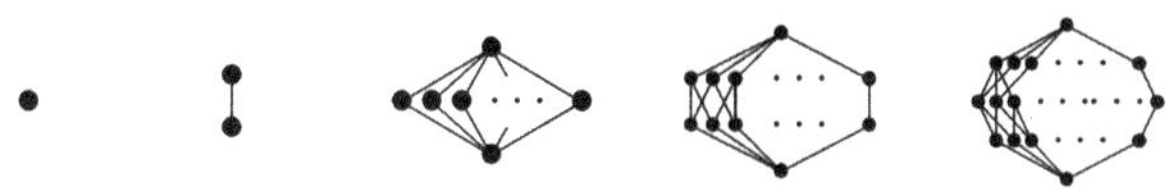

$\mathbb{G}(0)$ $\mathbb{G}(1)$ $\mathbb{G}(2)$ $\mathbb{G}(3)$ $\mathbb{G}(4)$

Of course, for $d \geq 2$, the number of elements of $\mathbb{G}(d)$ depends on the cardinality of k. If k is a field with q elements, the cardinality of $\mathbb{G}(2)_1 = \mathbb{G}(2)^1$ is $q+1$ (here, q is an arbitrary power of a prime number, thus if $\mathbb{G}(2)$ is finite, then $|\mathbb{G}(2)_1| = 3, 4, 5, 6, 8, 9, 10, 12, 14, \ldots$).

The subset $\mathbb{G}(d)_i$ is the set of subspaces of dimension i of a k-space of dimension d, it is often denoted by $\mathbb{G}(i, d-i)$ and called a *Grassmannian;* both sets $\mathbb{G}(d)_1$ and $\mathbb{G}(d)^1$ are also denoted by $\mathbb{P}^{d-1}$ and called the *projective space of dimensiond* -1:

$$\begin{aligned}
\mathbb{G}(d)_{d-1} &= \mathbb{G}(d)^1 &&= \mathbb{P}^{d-1} \\
\mathbb{G}(d)_i &= \mathbb{G}(d)^{d-i} &&= \mathbb{G}(i, d-i) \\
\mathbb{G}(d)_1 &= \mathbb{G}(d)^{d-1} &&= \mathbb{P}^{d-1}.
\end{aligned}$$

References

1. Auslander, M., Reiten, I., Smalø, S.: Representation theory of Artin algebras. In: Cambridge Studies in Advanced Mathematics (1994)
2. Auslander, M.: Functors and morphisms determined by objects. In: Representation Theory of Algebras. Lecture Notes in Pure Appl. Math., vol. 37, pp. 1–244. Marcel Dekker, New York (1978). Also in: Selected Works of Maurice Auslander. Amer. Math. Soc. (1999)
3. Auslander, M.: Applications of morphisms determined by objects. In: Representation Theory of Algebras. Lecture Notes in Pure Appl. Math., vol. 37, pp. 245–327. Marcel Dekker, New York (1978). Also in: Selected Works of Maurice Auslander, Amer. Math. Soc. (1999)
4. Auslander, M., Reiten, I.: Stable equivalences of dualizing R-varieties. Adv. Math. **12**, 306–366 (1974)
5. Auslander, M., Reiten, I., Smalø, S.: Representation Theory of Artin Algebras. Cambridge Studies in Advanced Mathematics, vol. 36. Cambridge University Press, Cambridge (1997)
6. Baez, J.: The n-category cafe: quivering with excitement. Blog May 4, 2012.

7. Bernstein, I.N., Gelfand, I.M., Ponomarev, V.A.: Coxeter functors and Gabriel's theorem. Russ. Math. Surv. **28**, 17–32 (1973)
8. Bongartz, K., Huisgen-Zimmermann, B.: Varieties of uniserial representations IV. Kinship to geometric quotients. Trans. Am. Math. Soc. **353**, 2091–2113 (2001)
9. Caldero, P., Chapoton, F.: Cluster algebras as Hall algebras of quiver representations. Comment. Math. Helv. **81**, 595–616 (2006)
10. Crawley-Boevey, W.: On homomorphisms from a fixed representation to a general representation of a quiver. Trans. Am. Math. Soc. **348**(5), 1909–1919 (1996)
11. Dedekind, R.: Über die von drei Moduln erzeugte Dualgruppe. Math. Ann. **53**, 371–403 (1900)
12. Derksen, H., Huisgen-Zimmermann, B., Weymann, J.: Top stable degenerations of finite dimensional, representations II. (2013, to appear)
13. Dlab, V., Ringel, C.M.: Perfect elements in the free modular lattices. Math. Ann. **247**, 95–100 (1980)
14. Gabriel, P.: Representations indecomposables. Sem. Bourbaki (1973–1974). LNM, vol. 431, pp. 143–160. Springer, Berlin (1975)
15. Gelfand, I.M., Ponomarev, V.A.: Indecomposable representations of the Lorentz group. Russ. Math. Surv. **23**, 1–58 (1968)
16. Gelfand, I.M., Ponomarev, V.A.: Free modular lattices and their representations. Russ. Math. Surv. **29**, 1–56 (1974)
17. Gelfand, I.M., Ponomarev, V.A.: Lattices, representations and algebras connected with them I. Russ. Math. Surv. **31**, 67–85 (1976)
18. Gelfand, I.M., Ponomarev, V.A.: Lattices, representations and algebras connected with them II. Russ. Math. Surv. **32**, 91–114 (1977)
19. Gelfand, I.M., Ponomarev, V.A.: Representations of graphs. Perfect subrepresentations. Funkc. Anal. i Pril 13, 14–31. Funct. Anal. Appl. **14**(1980), 177–190 (1980)
20. Grätzer, G.: Lattice Theory: Foundation. Birkhäuser, Basel (2011)
21. Hille, L.: Tilting line bundles and moduli of thin sincere representations of quivers. An. St. Univ. Ovidius Constantza **4**, 76–82 (1996)
22. Huisgen-Zimmermann, B.: The geometry of uniserial representations of finite dimensional algebras I. J. Pure Appl. Algebra **127**, 39–72 (1998)
23. Kraft, H.: Geometric methods in representation theory. In: Representations of Algebras, Lecture Notes in Math., vol. 944, pp 180–257. Springer, New York (1980)
24. Krause, H.: A short proof for Auslander's defect formula. Linear Algebra Appl. **365**, 267–270 (2003)
25. Krause, H.: Morphisms determined by objects in triangulated categories. arXiv:1110.5625
26. Le Bruyn, L.: Quiver Grassmannins can be anything. Blog, May 2, 2012.
27. MacLane, S.: Categories for the Working Mathematician. Springer, Berlin (1998)
28. Prest, M.: Purity, Spectra and Localisation, Encyclopedia of Mathematics and its Applications, vol. 121. Cambridge University Press, Cambridge (2009)
29. Reineke, M.: Every projective variety is a quiver Grassmannian. Algebras and Representation Theory (to appear). arXiv:1204.5730
30. Ringel, C.M.: Report on the Brauer–Thrall conjectures. In: Proceedings ICRA 2. LNM, vol. 831, pp. 104–136. Springer, Berlin (1980)
31. Ringel, C.M.: The rational invariants of tame quivers. Invent. Math. **58**, 217–239 (1980)
32. Ringel, C.M.: Four papers on problems in lniear algebra. In: Representation Theory. Selected Papers by Gelfand et al. London Math Soc. Lecture Note Series, vol. 69. Cambridge University Press, Cambridge (1982)
33. Ringel, C.M.: Tame algebras and integral quadratic forms. In: Lecture Notes in Math., vol. 1099. Springer, New York (1984)
34. Ringel, C.M.: Hall polynomials for the representation-finite hereditary algebras. Adv. Math. **84**, 137–178 (1990)
35. Ringel, C.M.: The Gabriel-Roiter measure. Bull. Sci. Math. **129**, 726–748 (2005)
36. Ringel, C.M.: Foundation of the representation theory of artin algebras, using the Gabriel-Roiter measure. In: de la Pena, Bautista (eds.) Trends in Representation Theory of Algebras and Related Topics. Contemporary Math. vol. 406, pp. 105–135. Amer. Math. Soc., New York (2006)
37. Ringel, C.M.: The first Brauer-Thrall conjecture. In: Models, Modules and Abelian Groups. In: Goldsmith, B., Göbel, R. (eds.) Memory of A. L. S. Corner, pp. 369–374. Walter de Gruyter, Berlin (2008)

38. Ringel, C.M.: Morphisms determined by objects: the case of modules over artin algebras. Illinois J. (2013, to appear)
39. Ringel, C.M.: Quiver Grassmannians and Auslander varieties for wild algebras (in preparation)
40. Ringel, C.M., Vossieck, D.: Hammocks. Proc. London Math. Soc. **54**(3), 216–246 (1987)
41. Roiter, A.V.: Unboundedness of the dimensions of the indecomposable representations of an algebra which has infinitely many indecomposable representations. Izv. Akad. Nauk SSSR Ser. Mat. **32**, 1275–1282 (1968)
42. Schofield, A.: General representations of quivers. Proc. Lond. Math. Soc. **65**(3) (1992)

4

On the determination of a developable spherical orthotomic ruled surface

Ö. Gökmen Yıldız · Sıddıka Ö. Karakuş · H. Hilmi Hacısalihoğlu

Abstract In this paper, a method for determination of developable spherical orthotomic ruled surfaces is given by using dual vector calculus. We show that dual vectorial expression of a developable spherical orthotomic ruled surface can be obtained from coordinates and the first derivatives of the base curve. The paper concludes with an example related to this method.

Keywords Ruled surface · Developable surface · Dual vector · Dual unit sphere · Orthotomic · Spherical orthotomic

Mathematics Subject Classification 53A04 · 53A05 · 53A25 · 53A40

1 Introduction

In geometry, a surface is called a ruled surface if it is swept out by a moving line in $\mathbb{R}^3$. The theory of ruled surfaces is a classical subject in differential geometry. It is well known that developable surfaces are special cases of ruled surfaces [18]. Developable

---

Communicated by S. K. Jain.

Ö. G. Yıldız · S. Ö. Karakuş (✉) · H. H. Hacısalihoğlu
Department of Mathematics, Faculty of Sciences and Arts,
Bilecik Şeyh Edebali University, 11210 Bilecik, Turkey
e-mail: siddika.karakus@bilecik.edu.tr

Ö. G. Yıldız
e-mail: ogokmen.yildiz@bilecik.edu.tr

H. H. Hacısalihoğlu
e-mail: h.hacisali@bilecik.edu.tr

surfaces have a very important place in mathematics and engineering such as motion analysis or designing cars and ships.

Let α be a regular curve and $\overrightarrow{T}$ be its tangent, and let u be a source. An orthotomic of α with respect to the source (u) is defined as a locus of reflection of u about tangents $\overrightarrow{T}$ [8]. Bruce and Giblin applied the unfolding theory to the study of the evolutes and orthotomics of plane and space curves [3–5]. Georgiou, Hasanis and Koutroufiotis investigated the orthotomics in the Euclidean (n + 1) space [7]. Alamo and Criado studied the antiorthotomics in the Euclidean (n + 1)-space [1]. Xiong defined the spherical orthotomic and the spherical antiorthotomic [20]. Also, an orthotomic concept can be applied to surface. For a given surface S and a fixed point (source) P, orthotomic surface of S relative to P is defined as a locus of reflection of P about all tangent planes of S [11].

Dual numbers were defined by Clifford (1873) with the aid of real numbers. After him E. Study has done fundamental research with dual numbers and dual vectors on the geometry of lines and kinematics [2] introducing what is now called the Study mapping. This mapping constitutes a one-to-one correspondence between the dual points of dual unit sphere S^2 and the directed lines of space of lines $\mathbb{R}^3$ [19]. Therefore, the motion locus of a straight line in $\mathbb{R}^3$ can be identified by a point on the source of dual unit sphere S^2 in $\mathbb{D}$ Module $\mathbb{D}^3$. Afterwards Guggenheimer studied relationship between a ruled surface in $\mathbb{R}^3$ and unique dual curve on the surface of S^2 [9].

Köse introduced a method for determination of developable ruled surfaces [13]. In this study, by using methods given [13], we construct a method for determination of developable orthotomic ruled surfaces and obtain a linear differential equation of first order.

2 Preliminaries

If a and a^* are real numbers and $\varepsilon^2 = 0$ with $\varepsilon \notin \mathbb{R}$, a dual number can be written as $\widehat{a} = a + \varepsilon a^*$, where $\varepsilon = (0, 1)$ is the dual unit.

The set $\{\widehat{a} = a + \varepsilon a^* \,|a, a^* \in \mathbb{R}\}$ of dual numbers forms a commutative ring over $\mathbb{R}$ and is denoted by $\mathbb{D}$. Then the set

$$\mathbb{D}^3 = \left\{\overrightarrow{\widehat{a}} = (\widehat{a}_1, \widehat{a}_2, \widehat{a}_3)\,|\widehat{a}_i \in \mathbb{D}, 1 \leq i \leq 3\right\}$$

becomes a module over the ring $\mathbb{D}$ which is called $\mathbb{D}$ Module, under addition and scalar multiplication on $\mathbb{D}$ [14]. The elements of $\mathbb{D}^3$ are called dual vectors. Thus, a dual vector has the form $\overrightarrow{\widehat{a}} = \overrightarrow{a} + \varepsilon \overrightarrow{a}^*$, where $\overrightarrow{a}$ and $\overrightarrow{a}^*$ are real vectors in $\mathbb{R}^3$. For any vectors $\overrightarrow{\widehat{a}}$ and $\overrightarrow{\widehat{b}}$ in $\mathbb{D}^3$, the inner product and the vector product of these vectors are defined as

$$\left\langle \overrightarrow{\widehat{a}}, \overrightarrow{\widehat{b}} \right\rangle = \left\langle \overrightarrow{a}, \overrightarrow{b} \right\rangle + \varepsilon \left(\left\langle \overrightarrow{a}, \overrightarrow{b}^* \right\rangle + \left\langle \overrightarrow{a}^*, \overrightarrow{b} \right\rangle \right)$$

and

$$\overrightarrow{\tilde{a}} \wedge \overrightarrow{\tilde{b}} = \vec{a} \wedge \vec{b} + \varepsilon \left(\vec{a} \wedge \vec{b}^* + \vec{a}^* \wedge \vec{b} \right),$$

respectively. The norm $\left\| \overrightarrow{\tilde{a}} \right\|$ of $\overrightarrow{\tilde{a}} = \vec{a} + \varepsilon \vec{a}^*$ is defined as

$$\left\| \overrightarrow{\tilde{a}} \right\| = \| \vec{a} \| + \varepsilon \frac{\langle \vec{a}, \vec{a}^* \rangle}{\| \vec{a} \|}, \ \vec{a} \neq 0.$$

The dual vector $\overrightarrow{\tilde{a}}$ with norm 1 is called a dual unit vector. The set of dual unit vectors

$$\widehat{S}^2 = \left\{ \overrightarrow{\tilde{a}} = \vec{a} + \varepsilon \vec{a}^* \in \mathbb{D}^3 \,\middle|\, \left\| \overrightarrow{\tilde{a}} \right\| = 1; \vec{a}, \vec{a}^* \in \mathbb{R}^3 \right\}$$

is called the dual unit sphere.

Theorem 1 *The Study Mapping is a one-to-one mapping between the dual points of a dual unit sphere, in $\mathbb{D}$ Module, and the oriented lines in the Euclidean line space $\mathbb{E}^3$ [10].*

According to Theorem 1, the unit dual vector $\overrightarrow{\tilde{a}} = \vec{a} + \varepsilon \vec{a}^*$ corresponds to only oriented line; where the real vector $\vec{a}$ shows the direction of this line and the real vector $\vec{a}^*$ shows the vectorial moment of the unit vector $\vec{a}$ with respect to the origin point.

Let $\widehat{S}^2$, O and $\left\{ O; \overrightarrow{\tilde{e}}_1, \overrightarrow{\tilde{e}}_2, \overrightarrow{\tilde{e}}_3 \right\}$ denote the unit dual sphere, the center of $\widehat{S}^2$ and dual orthonormal system at O, respectively, where

$$\overrightarrow{\tilde{e}}_i = \vec{e}_i + \varepsilon \vec{e}_i^*; \quad 1 \leq i \leq 3.$$

Let S_3 be the group of permutations of $\{1, 2, 3\}$, then it can be written as

$$\left. \begin{array}{r} \overrightarrow{\tilde{e}}_{\sigma(1)} = \text{sgn}(\sigma) \overrightarrow{\tilde{e}}_{\sigma(2)} \wedge \overrightarrow{\tilde{e}}_{\sigma(3)}, \text{sgn}(\sigma) = \pm 1, \\ \sigma = \begin{pmatrix} 1 & 2 & 3 \\ \sigma(1) & \sigma(2) & \sigma(3) \end{pmatrix} \end{array} \right\}.$$

In the case that the orthonormal system

$$\{O; \vec{e}_1, \vec{e}_2, \vec{e}_3\}$$

is the system of $\mathbb{E}^3$.

By using the Study Mapping, we can conclude that there exists a one-to-one correpondence between the dual orthonormal system and the real orthonormal system.

Now, define of spherical normal, spherical tangent and spherical orthotomic of a spherical curve α. Let $\left\{ \vec{T}, \vec{N}, \vec{B} \right\}$ be the Frenet frame of α. The spherical normal of α is the great circle, passing through $\alpha(s)$, that is normal to α at $\alpha(s)$ and given by

$$\begin{cases} \left\langle \vec{x}, \vec{x} \right\rangle = 1, \\ \left\langle \vec{x}, \vec{T} \right\rangle = 0, \end{cases}$$

where x is an arbitrary point of the spherical normal. The spherical tangent of α is the great circle which tangent to α at $\alpha(s)$ and given by

$$\begin{cases} \left\langle \vec{y}, \vec{y} \right\rangle = 1, \\ \left\langle \vec{y}, (\vec{\alpha} \wedge \vec{T}) \right\rangle = 0, \end{cases} \tag{2.1}$$

where y is an arbitrary point of the spherical tangent.

Let u be a source on a sphere. Then, Xiong defined the spherical orthotomic of α relative to u as to be the set of reflections of u about the planes, lying on the above great circles (2.1) for all $s \in I$ and given by

$$\vec{\tilde{u}} = 2\left\langle (\vec{\alpha} - \vec{u}), \vec{v} \right\rangle \vec{v} + \vec{u} \tag{2.2}$$

where $\vec{v} = \dfrac{\vec{B} - \left\langle \vec{B}, \vec{\alpha} \right\rangle \vec{\alpha}}{\left\| \vec{B} - \left\langle \vec{B}, \vec{\alpha} \right\rangle \vec{\alpha} \right\|}$ [21].

3 The dual vector formulation

Let L be a line and x denotes the direction and p be the position vector of any point on L. Dual vector representation allows us the Plucker vectors x and $p \wedge x$. Thus, dual vector $\widehat{x}(t)$ can be written as

$$\widehat{x}(t) = x + \varepsilon(p \wedge x) = x + \varepsilon x^*$$

where ε is called the dual unit and $\varepsilon^2 = 0$.

By using the dual vector function $\widehat{x}(t) = x(t) + \varepsilon(p(t) \wedge x(t)) = x(t) + \varepsilon x^*(t)$, a ruled surface can be given as

$$m(u, t) = p(t) + u x(t).$$

It is known that the dual unit vector $\widehat{x}(t)$ is a differentiable curve on a dual unit sphere and also having unit magnitude [15]:

$$\begin{aligned} \langle \widehat{x}, \widehat{x} \rangle &= \langle x + \varepsilon p \wedge x, x + \varepsilon p \wedge x \rangle \\ &= \langle x, x \rangle + \langle 2\varepsilon x, p \wedge x \rangle + \varepsilon^2 \langle p \wedge x, p \wedge x \rangle \\ &= \langle x, x \rangle \\ &= 1. \end{aligned}$$

The dual arc-length of the ruled surface $\widehat{x}(t)$ is defined by

$$\hat{s}\,(t) = \int_{0}^{t} \left\| \frac{d\hat{x}}{dt} \right\| dt. \tag{3.1}$$

The integrant of the Eq. (3.1) is the dual speed $\widehat{\delta}$ of $\widehat{x}(t)$ and is

$$\widehat{\delta} = \left\| \frac{d\widehat{x}}{dt} \right\| = \left\| \frac{dx}{dt} \right\| \left(1 + \varepsilon \frac{\left\langle \frac{dx}{dt}, \frac{dp}{dt} \wedge x \right\rangle}{\left\| \frac{dx}{dt} \right\|^2} \right) = \left\| \frac{dx}{dt} \right\| (1 + \varepsilon \Delta).$$

The curvature function

$$\Delta = \frac{\left\langle \frac{dx}{dt}, \frac{dp}{dt} \wedge x \right\rangle}{\left\| \frac{dx}{dt} \right\|^2} = \frac{\left\langle \frac{dx}{dt}, \frac{dx^*}{dt} \right\rangle}{\left\| \frac{dx}{dt} \right\|^2} \tag{3.2}$$

is the well-known distribution parameter (drall) of the ruled surface.

4 The determination of a developable ruled surface

Let $\widehat{x}(t)$ be a point on the dual unit sphere. The dual coordinates of $\widehat{x}(t) = x_i + \varepsilon x_i^*$ can be expressed as

$$\begin{aligned} \widehat{x_1} &= x_1 + \varepsilon x_1^* = \cos\widehat{\varphi}\sin\widehat{\theta} \\ \widehat{x_2} &= x_2 + \varepsilon x_2^* = \sin\widehat{\varphi}\sin\widehat{\theta} \\ \widehat{x_3} &= x_3 + \varepsilon x_3^* = \cos\widehat{\theta} \end{aligned} \tag{4.1}$$

where $\widehat{\varphi} = \varphi + \varepsilon\varphi^*$, $\widehat{\theta} = \theta + \varepsilon\theta^*$ are dual angles.

Since $\varepsilon^2 = \varepsilon^3 = \cdots = 0$ according to the Taylor series expansion from (4.1), we obtain the real parts as

$$\begin{aligned} x_1 &= \cos\varphi\sin\theta, \\ x_2 &= \sin\varphi\sin\theta, \\ x_3 &= \cos\theta, \end{aligned}$$

and the dual parts as

$$\begin{aligned} x_1^* &= \theta^*\cos\varphi\cos\theta - \varphi^*\sin\varphi\sin\theta \\ x_2^* &= \theta^*\sin\varphi\cos\theta + \varphi^*\cos\varphi\sin\theta \\ x_3^* &= -\theta^*\sin\theta. \end{aligned}$$

Hence, we conclude that a dual curve $\widehat{x}(t) = x(t) + \varepsilon x^*(t)$ can be written as

$$\begin{aligned}\widehat{x}(t) &= (\cos\varphi(t)\sin\theta(t), \sin\varphi(t)\sin\theta(t), \cos\theta(t)) \\ &+\varepsilon\begin{pmatrix}\theta^*(t)\cos\varphi(t)\cos\theta(t) - \varphi^*(t)\sin\varphi(t)\sin\theta(t), \\ \theta^*(t)\sin\varphi(t)\cos\theta(t) + \varphi^*(t)\cos\varphi(t)\sin\theta(t), -\theta^*(t)\sin\theta(t)\end{pmatrix}.\end{aligned}$$

Let $\widehat{\sigma}(t) = \sigma(t) + \varepsilon\sigma^*(t)$ be spherical orthotomic of the great circle, which lies on the $\overrightarrow{e}_2\overrightarrow{e}_3$ plane, relative to the dual curve $\widehat{x}(t)$. By (2.2), we get $\widehat{\sigma}(t) = (-\widehat{x}_1, \widehat{x}_2, \widehat{x}_3)$ where $\widehat{x}_i$'s are the coordinates of $\widehat{x}(t)$ for $i = 1, 2, 3$. By considering the spherical orthotomic dual curve, we have:

$$\begin{aligned}\widehat{\sigma}(t) &= (-\cos\varphi(t)\sin\theta(t), \sin\varphi(t)\sin\theta(t), \cos\theta(t)) \\ &+\varepsilon\begin{pmatrix}-\theta^*(t)\cos\varphi(t)\cos\theta(t) + \varphi^*(t)\sin\varphi(t)\sin\theta(t), \\ \theta^*(t)\sin\varphi(t)\cos\theta(t) + \varphi^*(t)\cos\varphi(t)\sin\theta(t), -\theta^*(t)\sin\theta(t)\end{pmatrix}\end{aligned}$$

on the dual unit sphere corresponding to ruled surface

$$m(u,t) = q(t) + u\sigma(t),$$

where $q(t)$ is a base curve of the ruled surface. Since $\sigma^*(t) = q(t) \wedge \sigma(t)$, we have the following system of linear equations in variables Q_1, Q_2, Q_3;

$$\begin{aligned}Q_2\cos\theta - Q_3\sin\varphi\sin\theta &= -\theta^*\cos\varphi\cos\theta + \varphi^*\sin\varphi\sin\theta, \\ -Q_1\cos\theta - Q_3\cos\varphi\sin\theta &= \theta^*\sin\varphi\cos\theta + \varphi^*\cos\varphi\sin\theta, \\ Q_1\sin\varphi\sin\theta + Q_2\cos\varphi\sin\theta &= -\theta^*\cos\theta\end{aligned}$$

where Q_i's are the coordinates of $q(t)$ for $i = 1, 2, 3$. The matrix of coefficients of unknows Q_1, Q_2, and Q_3 is skew a symmetric matrix

$$\begin{bmatrix}0 & \cos\theta & -\sin\varphi\sin\theta \\ -\cos\theta & 0 & -\cos\varphi\sin\theta \\ \sin\varphi\sin\theta & \cos\varphi\sin\theta & 0\end{bmatrix}$$

and therefore its rank is 2 with $\theta(t) \neq 2k\pi$, $k \in \mathbb{Z}$. The rank of the augmented matrix

$$\begin{bmatrix}0 & \cos\theta & -\sin\varphi\sin\theta & -\theta^*\cos\varphi\cos\theta + \varphi^*\sin\varphi\sin\theta \\ -\cos\theta & 0 & -\cos\varphi\sin\theta & \theta^*\sin\varphi\cos\theta + \varphi^*\cos\varphi\sin\theta \\ \sin\varphi\sin\theta & \cos\varphi\sin\theta & 0 & -\theta^*\cos\theta\end{bmatrix}$$

is 2. Thus this system of linear equations have infinite solutions given by

$$\begin{aligned}Q_1 &= -\left(Q_3 + \varphi^*\right)\cos\varphi\tan\theta - \theta^*\sin\varphi, \\ Q_2 &= \left(Q_3 + \varphi^*\right)\sin\varphi\tan\theta - \theta^*\cos\varphi, \\ Q_3 &= Q_3.\end{aligned} \tag{4.2}$$

Since $Q_3(t)$ can be chosen arbitrarily, we may take that $Q_3(t) = -\varphi^*(t)$. In this case, (4.2) reduces to

$$\begin{aligned} Q_1 &= -\theta^* \sin\varphi, \\ Q_2 &= -\theta^* \cos\varphi, \\ Q_3 &= -\varphi^*. \end{aligned} \tag{4.3}$$

The distribution parameter of the ruled surface given (3.2) is obtained as follows

$$\Delta = \frac{\left\langle \frac{d\sigma}{dt}, \frac{d\sigma^*}{dt} \right\rangle}{\left\| \frac{d\sigma}{dt} \right\|^2} = \frac{\frac{d\varphi^*}{dt}\frac{d\varphi}{dt}\sin^2\theta + \theta^*\left(\frac{d\varphi}{dt}\right)^2 \sin\theta\cos\theta + \frac{d\theta^*}{dt}\frac{d\theta}{dt}}{\left(\frac{d\varphi}{dt}\right)^2 \sin^2\theta + \left(\frac{d\theta}{dt}\right)^2}. \tag{4.4}$$

If this spherical orthotomic ruled surface is developable, then $\Delta = 0$ and by (4.4), we get

$$\frac{d\varphi^*}{dt}\frac{d\varphi}{dt}\sin^2\theta + \theta^*\left(\frac{d\varphi}{dt}\right)^2 \sin\theta\cos\theta + \frac{d\theta^*}{dt}\frac{d\theta}{dt} = 0.$$

Setting

$$y(t) = \cot\theta, \quad A(t) = -\frac{\theta^*\left(\frac{d\varphi}{dt}\right)^2}{\frac{d\theta^*}{dt}}, \quad B(t) = -\frac{\frac{d\varphi^*}{dt}\frac{d\varphi}{dt}}{\frac{d\theta^*}{dt}}$$

we are lead to a linear differential equation of first order

$$\frac{dy}{dt} + A(t)y + B(t) = 0. \tag{4.5}$$

If we assume that $\varphi(t)$ and $\theta(t)$ are both constants, then (4.5) is identically zero, that is, the spherical orthotomic ruled surface $\widehat{\sigma}(t)$ is a cylinder.

Let $q(t)$ be a curve. Then we can find a developable spherical orthotomic ruled surface such that, its base curve is the curve $q(t)$ and by (4.3), we have

$$\tan\varphi = \frac{Q_1}{Q_2}, \quad \theta^* = \pm\sqrt{Q_1^2 + Q_2^2} \quad \text{and} \quad \varphi^* = -Q_3. \tag{4.6}$$

Now we need to determine $\theta(t)$. The solution of (4.5) gives $\cot\theta$. This solution includes an integral constant, thus we have infinitely many developable spherical orthotomic ruled surfaces such that each of them has a base curve $q(t)$.

Moreover, note that $\theta^*(t)$, given by (4.6), has two values; by using the minus sign, then we obtain the reciprocal of the spherical orthotomic ruled surfaces obtained by using the plus sign for a given integral constant.

Example 1 Consider the curve $q(t) = (t^2, t^2, 2t + 1)$
Then

$$\tan\varphi = 1,\ \theta^* = t\sqrt{2}, \quad \text{and} \quad \varphi^* = -2t - 1$$

and

$$\frac{d\varphi^*}{dt} = 2,\ \frac{d\varphi}{dt} = 0 \quad \text{and} \quad \frac{d\theta^*}{dt} = \sqrt{2}$$

we substitute these values into (4.5) and solve this differential equation

$$y = \cot\theta = C, \quad C \in \mathbb{R}.$$

Hence, the family of the developable ruled surface is given by

$$m(u, t) = q(t) + u\sigma(t),$$

where

$$\sigma(t) = \left(-\frac{Q_1}{\theta^*}\sin\theta, -\frac{Q_2}{\theta^*}\sin\theta, \cos\theta\right).$$

The graph of the developable spherical orthotomic ruled surface given by this equation for $C = 0$ in domain

$$D : \begin{cases} -2 \le t \le 2 \\ -4 \le u \le 4 \end{cases}$$

is given in Figure 1.

For the spherical orthotomic of the great circle which lies on the $\vec{e}_1\vec{e}_3$ plane, relative to the dual curve $\widehat{x}$ is

$$\begin{aligned}\widehat{\sigma}(t) = {} & (\cos\varphi(t)\sin\theta(t), -\sin\varphi(t)\sin\theta(t), \cos\theta(t)) \\ & +\varepsilon\begin{pmatrix} \theta^*(t)\cos\varphi(t)\cos\theta(t) - \varphi^*(t)\sin\varphi(t)\sin\theta(t), \\ -\theta^*(t)\sin\varphi(t)\cos\theta(t) - \varphi^*(t)\cos\varphi(t)\sin\theta(t), -\theta^*(t)\sin\theta(t) \end{pmatrix}.\end{aligned}$$

We obtain the distribution parameter same as the first choosing. But in this case unknowns Q_1, Q_2, and Q_3 are

$$\begin{aligned} Q_1 &= \theta^*\sin\varphi, \\ Q_2 &= \theta^*\cos\varphi, \\ Q_3 &= -\varphi^*. \end{aligned} \tag{4.7}$$

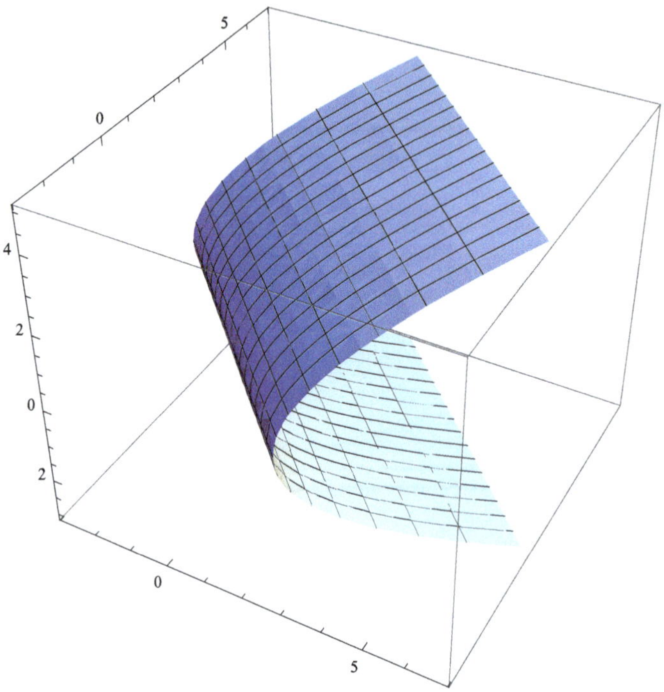

Fig. 1 Spherical orthotomic ruled surface

Consider the same curve $q(t) = (t^2, t^2, 2t+1)$, the spherical orthotomic ruled surface is obtained as

$$m(u,t) = \left(t^2 + \frac{ut}{\sqrt{2}}, t^2 - \frac{ut}{\sqrt{2}}, 2t+1\right).$$

The graph is the same as the previous.

For the plane spanned by $\left\{\overrightarrow{e}_1, \overrightarrow{e}_2\right\}$, we obtain the same distribution parameter and the same differential equation with

$$Q_1 = \theta^* \sin\varphi,$$
$$Q_2 = -\theta^* \cos\varphi,$$
$$Q_3 = -\varphi^*.$$

For $q(t) = (t^2, t^2, 2t+1)$, the spherical orthotomic ruled surface is

$$m(u,t) = \left(t^2 - \frac{ut}{\sqrt{2}}, t^2 + \frac{ut}{\sqrt{2}}, 2t+1\right).$$

and the graph is the same as the other ones.

Acknowledgments The authors would like to thank the anonymous referees for their helpful suggestions and comments which improved significantly the presentation of the paper.

References

1. Alamo, N., Criado, C.: Generalized antiorthotomics and their singularities. Inverse Problems **18**(3), 881–889 (2002)
2. Blaschke, W.: Vorlesungen Über Differential geometry I. Verlag von Julieus Springer in Berlin, pp. 89 (1930)
3. Bruce, J.W.: On singularities, envelopes and elementary differential geometry. Math. Proc. Cambridge Philos. Soc. **89**(1), 43–48 (1981)
4. Bruce, J.W., Giblin, P.J.: Curves and singularities. A geometrical introduction to singularity theory. University Press, Cambridge (1992)
5. Bruce, J.W., Giblin, P. J.: One-parameter families of caustics by reflexion in the plane. Quart. J. Math. Oxford Ser. (2), 35(139), 243–251 (1984).
6. Ekici, C., Özüsağlam, E.: On the method of determination of a developable timelike ruled surface. Kuwait J. Sci. Eng. **39**(1A), 19–41 (2012)
7. Georgiou, C., Hasanis, T., Koutroufiotis, D.: On the caustic of a convex mirror. Geom. Dedicata **28**(2), 153–169 (1988)
8. Elementary geometry of differentiable curves. Cambridge University Press, Cambridge (2011)
9. Differential geometry. Dover, New York (1977)
10. Hacısalihoğlu, H.H.: Acceleration axes in spatial kinematics I. Communications de la Faculte des Sciences de L'Universite d'Ankara. Serie A, Tome 20 A, Annee pp. L-15 (1971).
11. Hoschek, J.: Smoothing of curves and surfaces. Computer aided geometric design, Vol. 2, No. 1–3, special issue, pp. 97–105 (1985).
12. Karadağ, H.B., Kılıç, E., Karadağ, M.: On the developable ruled surfaces kinematically generated in Minkowski 3-space. Kuwait J. Sci. Eng. **41**(1), 19–41 (2012)
13. Köse, Ö.: A method of the determination of a developable ruled surface. Mech. Mach. Theory. **34**, 1187–1193 (1999)
14. Köse, Ö.: Contributions to the theory of integral invariants of a closed ruled surface. Mechanism and Machine Theory **32**(2), 261–277 (1997)
15. McCarthy, J.M.: On the scalar and dual formulations of the curvature theory of line trajectories. ASME Journal of Mechanisms, Transmissions and Automation in Design 109/101 (1987).
16. Müller, H.R.: Kinematik dersleri, Ankara University Press, pp. 247-267-271, (1963).
17. Nomizu, K.: Fundamentals of linear algebra, Mc Graw-Hill, Book Company. London Lib. Cong. Cat. Card Numb. **65–28732**, 52–67 (1966)
18. O'Neill, B.: Semi-Riemannian geometry. Academic Press, New York (1983)
19. Study, E.: Geometrie der Dynamen, Leibzig (1903).
20. Xiong, J.F.: Geometry and singularities of spatial and spherical curves. University of Hawai, The degree of Doctor of Philosophy, Hawai (2004)
21. Xiong, J.F.: Spherical orthotomic and spherical antiorthotomic. Acta Mathematica Sinica **23**(9), 1673–1682 (2007)

5

On Pogorelov estimates in optimal transportation and geometric optics

Feida Jiang · Neil S. Trudinger

Abstract In this paper, we prove global and interior second derivative estimates of Pogorelov type for certain Monge–Ampère type equations, arising in optimal transportation and geometric optics, under sharp hypotheses. Specifically, for the case of generated prescribed Jacobian equations, as developed recently by the second author, we remove barrier or subsolution hypotheses assumed in previous work by Trudinger and Wang (Arch Ration Mech Anal 192:403–418, 2009), Liu and Trudinger (Discret Contin Dyn Syst Ser A 28:1121–1363, 2010), Jiang et al. (Calc Var Partial Differ Equ 49:1223–1236, 2014), resulting in new applications to optimal transportation and near field geometric optics.

Keywords Monge–Ampère equations · Optimal transportation · Geometric optics · Second derivative estimates

Mathematics Subject Classification 35J96 · 90B06 · 78A05

Communicated by S.K. Jain.

Research supported by Australian Research Council Grant, National Natural Science Foundation of China (No. 11401306) and the Jiangsu Natural Science Foundation of China (BK20140126).

F. Jiang · N. S. Trudinger (✉)
Centre for Mathematics and Its Applications, The Australian National University,
Canberra, ACT 0200, Australia
e-mail: neil.trudinger@anu.edu.au

F. Jiang
e-mail: Feida.Jiang@anu.edu.au

1 Introduction

This paper is concerned with global and interior second derivative estimates of solutions to partial differential equations of Monge–Ampère type, (MATEs) that arise in optimal transportation and geometric optics and are embraced by the notion of "generated prescribed Jacobian equation" (GPJE) introduced in [16]. Such equations can be written in the general form

$$\det\{D^2u - A(\cdot, u, Du)\} = B(\cdot, u, Du), \tag{1.1}$$

where A and B are respectively $n \times n$ symmetric matrix valued and scalar functions on a domain $\mathcal{U} \subset \mathbb{R}^n \times \mathbb{R} \times \mathbb{R}^n$, Du and D^2u denote respectively the gradient vector and Hessian matrix of the scalar function u, which is defined on a bounded domain $\Omega \subset \mathbb{R}^n$ for which the sets

$$\mathcal{U}_x = \{(u, p) \in \mathbb{R} \times \mathbb{R}^n \mid (x, u, p) \in \mathcal{U}\}$$

are non-empty for all $x \in \Omega$. A solution $u \in C^2(\Omega)$ of (1.1) is called elliptic, (degenerate elliptic), whenever $D^2u - A(\cdot, u, Du) > 0$, (≥ 0), which implies $B > 0$, (≥ 0).

We shall establish the second derivative bounds for generated prescribed Jacobian equations under the hypotheses G1, G2, G1*, G3w and G4w introduced in [16], which extend the conditions A1, A2 and A3w for optimal transportation equations introduced in [20]. These are degenerate versions of the conditions A1, A2 and A3 for regularity originating in [11]. In fact we will relabel them here replacing G in [16] by A to make this extension clearer. First we need to suppose that there exists a smooth generating function g defined on a domain $\Gamma \subset \mathbb{R}^n \times \mathbb{R}^n \times \mathbb{R}$ for which the sets

$$\Gamma_{x,y} = \{z \in \mathbb{R} \mid (x, y, z) \in \Gamma\}$$

are convex, (and hence open intervals). Setting

$$\mathcal{U} = \{(x, g(x, y, z), g_x(x, y, z)) \mid (x, y, z) \in \Gamma\},$$

we then assume

A1: For each $(x, u, p) \in \mathcal{U}$, there exists a unique point $(x, y, z) \in \Gamma$ satisfying

$$g(x, y, z) = u, \ g_x(x, y, z) = p.$$

A2: $g_z < 0$, $\det E \neq 0$, in Γ, where E is the $n \times n$ matrix given by

$$E = [E_{x,y}] = [g_{x,y} - (g_z)^{-1} g_{x,z} \otimes g_y].$$

Defining $Y(x, u, p) = y$, $Z(x, u, p) = z$ in A1, we thus obtain mappings $Y : \mathcal{U} \to \mathbb{R}^n$, $Z : \mathcal{U} \to \mathbb{R}$ with $Y_p = E^{-1}$. The matrix function A in the corresponding generated prescribed Jacobian equation is then given by

$$A(x, u, p) = -Y_p^{-1}(Y_x + Y_u \otimes p) = g_{xx}[x, Y, Z], \tag{1.2}$$

where $Y = Y(x, u, p)$ and $Z = Z(x, u, p)$. Note that the Jacobian determinant of the mapping $(y, z) \to (g_x, g)(x, y, z)$ is $g_z \det E$, $\neq 0$ by A2, so that Y and Z are accordingly smooth. In particular, it suffices to have $g \in C^2(\Gamma)$ to define $Y, Z \in C^1(\mathcal{U})$, $A \in C^0(\mathcal{U})$. The reader is referred to [16] for more details. We also mention though that the special case of optimal transportation is given by taking

$$g(x, y, z) = c(x, y) - z, \tag{1.3}$$

where c is a cost function defined on a domain $\mathcal{D}$ in $\mathbb{R}^n \times \mathbb{R}^n$ so that $\Gamma = \mathcal{D} \times \mathbb{R}$, $g_x = c_x$, $g_z = -1$, $E_{x,y} = c_{x,y}$ and conditions A1 and A2 are equivalent to those in [11,19]. Note that here we follow the same sign convention as in [15,20] so that our cost functions are the negatives of those usually considered. We remark also that the case when Y is independent of u is equivalent to the optimal transportation case.

Crucial to us in this paper is the dual condition to A1, namely

A1*: The mapping $Q := -g_y/g_z$ is one-to-one in x, for all $(x, y, z) \in \Gamma$.

We will treat the implications of condition A1* as needed in Sect. 2. Meanwhile we note that the Jacobian matrix of the mapping $x \to Q(x, y, z)$ is $-E^t/g_z$, where E^t is the transpose of E, so its determinant is automatically non-zero when condition A2 holds. Note that in the optimal transportation case (1.3), we simply have $Q = c_y$ but the duality is more complicated in the general situation and will be amplified further in Sect. 2.

The next conditions are expressed in terms of the matrix function A and extend condition A3w in the optimal transportation case [20]. For these we assume that A is twice differentiable in p and once in u.

A3w: The matrix function A is regular in $\mathcal{U}$, that is A is codimension one convex with respect to p in the sense that,

$$A_{ij,kl}\xi_i\xi_j\eta_k\eta_l := (D_{p_k p_l} A_{ij})\xi_i\xi_j\eta_k\eta_l \geq 0,$$

in $\mathcal{U}$, for all $\xi, \eta \in \mathbb{R}^n$ such that $\xi \cdot \eta = 0$.

A4w: The matrix A is monotone increasing with respect to u in $\mathcal{U}$, that is

$$D_u A_{ij}\xi_i\xi_j \geq 0,$$

in $\mathcal{U}$, for all $\xi \in \mathbb{R}^n$.

An explicit formula for $D_p^2 A$ in terms of the generating function g is given in [16]. In order to use conditions A3w and A4w for our barrier constructions in Sect. 2, we also assume that $\mathcal{U}$ is sufficiently large in that the sets $\mathcal{U}_x$ are convex in p, for each $(x, u) \in \mathbb{R}^n \times \mathbb{R}$, (for A3w), and $\mathcal{U}_x = \mathcal{J}_x \times \mathcal{P}_x$, for each $x \in \mathbb{R}^n$, where $\mathcal{J}_x$ is an open interval in $\mathbb{R}$, (possibly empty), and $\mathcal{P}_x \subset \mathbb{R}^n$, (for A4w). When we use both A3w and A4w, we would also assume $\mathcal{P}_x$ to be convex. We note that in [4], $\mathcal{U} = \mathbb{R}^n \times \mathbb{R} \times \mathbb{R}^n$

so that these conditions are automatically satisfied and in the optimal transportation case $J_x = \mathbb{R}$. Similar conditions are also needed for the convexity theory in [16].

To formulate our second derivative estimates, we first denote the one-jet of a function $u \in C^1(\Omega)$ by

$$J_1[u](\Omega) = \{x, u(x), Du(x) \mid x \in \Omega\},$$

and recall from [16] that a g-affine function in Ω is a function of the form

$$g_0 = g(\cdot, y_0, z_0),$$

where y_0 and z_0 are fixed so that $(x, y_0, z_0) \in \Gamma$ for all $x \in \Omega$.

Theorem 1.1 *Let $u \in C^4(\Omega) \cap C^2(\bar{\Omega})$ be an elliptic solution of Eq. (1.1) in Ω, where $B \in C^2(\mathcal{U})$, $\inf B > 0$ and $A \in C^2(\mathcal{U})$ is given by (1.2) with generating function g satisfying conditions A1, A2, A1*, A3w and A4w. Suppose also $u \le g_0$ in Ω for some g-affine function g_0 with the one jets $J_1[u](\Omega)$, $J_1[g_0](\Omega) \Subset \mathcal{U}$. Then we have the estimate*

$$\sup_{\Omega} |D^2 u| \le C \left(1 + \sup_{\partial\Omega} |D^2 u|\right), \tag{1.4}$$

where the constant C depends on $n, \mathcal{U}, g, B, \Omega, g_0$ and $J_1[u]$.

Note that the inequality $u \le g_0$ is trivially satisfied if there exists a g-affine function on $\bar{\Omega}$ which is greater than $\sup u$ and moreover can be dispensed with in the optimal transportation case, where Theorem 1.1 improves Theorems 3.1 and 3.2 in [20] and Theorem 1.1 in [4]. Also we may assume in place of A4w the reversed monotonicity condition:

A4*w: The matrix A is monotone decreasing with respect to u in $\mathcal{U}$, that is

$$D_u A_{ij}\xi_i\xi_j \le 0, \quad \text{for all } \xi \in \mathbb{R}^n,$$

in which case Theorem 1.1 continues to hold provided the inequality $u \le g_0$ is replaced by $u \ge g_0$.

Next we formulate the corresponding extension of the Pogorelov interior estimate [2] for generated prescribed Jacobian equations.

Theorem 1.2 *Let $u \in C^4(\Omega) \cap C^{0,1}(\bar{\Omega})$ be an elliptic solution of (1.1) satisfying $u = g_0$ on $\partial\Omega$, for some g-affine function g_0, where $B \in C^2(\mathcal{U})$, $\inf B > 0$ and $A \in C^2(\mathcal{U})$ is given by (1.2) with generating function g satisfying conditions A1, A2, A1*, A3w, A4w and $J_1[u](\Omega)$, $J_1[g_0](\Omega) \Subset \mathcal{U}$. Then we have for any $\Omega' \Subset \Omega$,*

$$\sup_{\Omega'} |D^2 u| \le C, \tag{1.5}$$

where the constant C depends on $n, \mathcal{U}, g, B, \Omega, \Omega', g_0$ and $J_1[u]$.

In the optimal transportation case, Theorem 1.2 improves Theorem 1.1 in [9] and Theorem 2.1 in [4] by removing the barrier and subsolution hypotheses assumed there. As in [9], we may also replace g_0 by any strict supersolution. We also remark that the dependence on $J_1[u]$ in Theorems 1.1 and 1.2 is more specifically determined by $\sup_\Omega(|u| + |Du|)$ and $\mathrm{dist}(J_1[u](\Omega), \partial\mathcal{U})$.

This paper is organised as follows. In the next section we will construct a function in Lemma 2.1 for which the Eq. (1.1) is uniformly elliptic and use it to prove an appropriate version, Lemma 2.2, of the key Lemma 2.1 in [4]. From here we obtain Theorems 1.1 and 1.2 by following the same arguments as in [4]. In Sect. 3, we will also discuss the application of Theorem 1.1 to global second derivative estimates for the Dirichlet and second boundary value problems for optimal transportation and generated prescribed Jacobian equations, yielding improvements of the corresponding results in [4,13,14,20]. In Sect. 4 we focus on applications to optimal transportation and near field geometric optics. In the optimal transportation case, we obtain a different and more direct proof of the improved global regularity result in [15]. Then we conclude by treating some examples of generating functions in geometric optics satisfying our hypotheses, which arise from the reflection and refraction of parallel beams.

To conclude this introduction we should also emphasise that while we have considered the more general case of generated prescribed Jacobian equations our results are also new for optimal transportation. Further we note that when we strengthen condition A3w to the strict inequality A3, the corresponding second derivative estimates as treated in [11], (see also [20] and [18]), are much simpler to prove although even here the constructions in Sect. 2 are still new.

2 Barrier constructions

Our first lemma shows the existence of a uniformly elliptic function under conditions A1, A2 and A1*.

Lemma 2.1 *Let $g \in C^4(\Gamma)$ be a generating function satisfying conditions A1, A2 and A1*, and suppose g_0 is a g-affine function on $\bar\Omega$, where Ω is a bounded domain in $\mathbb{R}^n$. Then there exists a function $\bar u \in C^2(\bar\Omega)$ such that $J_1[\bar u] \Subset \mathcal{U}$, $\bar u > g_0$ in Ω and*

$$D^2\bar u - A(\cdot, \bar u, D\bar u) \geq a_0 I, \quad in\ \Omega, \tag{2.1}$$

for some positive constant a_0 depending on g_0 and Γ.

Proof Suppose $g_0 = g(\cdot, y_0, z_0)$ for some $(y_0, z_0) \in \mathbb{R}^n \times \mathbb{R}$, let $B_\rho = B_\rho(y_0)$ be a ball of radius ρ and centre y_0 in $\mathbb{R}^n$, and consider the function $v = v_\rho$ in B_ρ, given by

$$v_\rho(y) = z_0 - \sqrt{\rho^2 - |y - y_0|^2}. \tag{2.2}$$

Our desired function $\bar u$ is now defined as the g^*-transform of v, in accordance with the notion of duality introduced in [16], namely

$$\bar u(x) = v^*_\rho(x) = \sup_{y \in B_\rho} g(x, y, v_\rho(y)). \tag{2.3}$$

Note that $\bar{u}$ is well defined for $x \in \bar{\Omega}$ for ρ sufficiently small to ensure the set

$$\Gamma_0 = \bar{\Omega} \times \bar{B}_\rho(y_0) \times [z_0 - \rho, z_0] \subset \Gamma.$$

Furthermore, the supremum in (2.3) will be taken on at a unique point $y = Tx \in B_\rho$. To prove this assertion, we fix a point $y \in \partial B_\rho$ and set $y_\lambda = (1-\lambda)y + \lambda y_0$ for some $\lambda \in (0,1)$. Since $g_z < 0$ in Γ, we then have, by the mean value theorem,

$$\begin{aligned} &g(x, y, v_\rho(y)) - g(x, y_\lambda, v_\rho(y_\lambda)) \\ &\leq \left(\sup_{\Gamma_0} |g_y|\right)|y - y_\lambda| + \left(\sup_{\Gamma_0} g_z\right)(v_\rho(y) - v_\rho(y_\lambda)) \\ &\leq C\lambda\rho - \delta\rho\sqrt{2\lambda - \lambda^2}, \end{aligned}$$

for positive constants C and δ. Consequently, choosing $0 < \lambda < 2/1 + (\frac{C}{\delta})^2$, we have

$$g(x, y, v_\rho(y)) < g(x, y_\lambda, v_\rho(y_\lambda))$$

and hence the supremum in (2.3) cannot occur on ∂B_ρ. By differentiation, we then have at an interior maximum point y,

$$g_y + g_z Dv_\rho(y) = 0 \tag{2.4}$$

and hence by condition A1*, there exists a unique $x = T^*y$ corresponding to y. From our construction (2.3) the function $\bar{u}$ is g-convex in Ω and its g-normal mapping $\chi[\bar{u}](x)$ consists of those y maximizing (2.3). Here we recall from [16] that a function $u \in C^0(\Omega)$ is g-convex in Ω if for each point $x_0 \in \Omega$, there exists $y_0 \in \mathbb{R}^n$ and $z_0 \in I(\Omega, y_0) := \cap_\Omega \Gamma_{\cdot, y_0}$, such that

$$u(x) \geq g(x, y_0, z_0), \quad u(x_0) = g(x_0, y_0, z_0),$$

for all $x \in \Omega$ and the g-normal mapping $\chi[u](x_0)$ of u at x_0 is given by

$$\chi[u](x_0) = \{y_0 \in \mathbb{R}^n \mid u(x) \geq g(x, y_0, z_0) \text{ for all } x \in \Omega\},$$

and moreover, $\chi[u](\Omega) = \cup_\Omega \chi[u](\cdot)$. From (2.4), we also have

$$|Dv_\rho(y)| \leq \sup_{\Gamma_0} \frac{|g_y|}{-g_z} \leq C \tag{2.5}$$

whence $\chi[\bar{u}](\Omega) \subset B_{(1-\lambda)\rho}(y_0)$ for some constant $\lambda > 0$. To show $y = Tx$ is uniquely determined by x, we need to prove $\chi[\bar{u}](x)$ is a single point. For this we invoke the duality considerations in [16]. Letting

$$\Gamma^* = \{(x, y, u) \in \mathbb{R}^n \times \mathbb{R}^n \times \mathbb{R} \mid u = g(x, y, z) \text{ for } (x, y, z) \in \Gamma\},$$

we let $g^* \in C^4(\Gamma^*)$ denote the dual generating function of g, given by

$$g(x, y, g^*(x, y, u)) = u.$$

Also denoting

$$\mathcal{U}^* = \{(y, z, q) \in \mathbb{R}^n \times \mathbb{R} \times \mathbb{R}^n \mid q = Q(x, y, z) \text{ for } (x, y, z) \in \Gamma\},$$

the dual matrix, corresponding to g^*, is given by

$$\begin{aligned} A^*(y, z, q) &= g^*_{yy}[X, y, g(X, y, z)] \\ &= -\left[\left(\frac{g_y}{g_z}\right)_y (X, y, z) + \left(\frac{g_y}{g_z}\right)_z (X, y, z) \otimes q\right], \end{aligned} \tag{2.6}$$

where $X = X(y, z, q)$ is the unique mapping, given by A1*, satisfying

$$Q(x, y, z) = g^*_y(X, y, g(X, y, z)) = q. \tag{2.7}$$

Note that the mapping T^* constructed above is given by

$$x = T^*y = X(y, v_\rho(y), Dv_\rho(y)). \tag{2.8}$$

Since the graph of v_ρ is a hemisphere of radius ρ and hence has constant curvature $1/\rho$, we have, on $\chi[\bar{u}](\Omega)$

$$D^2 v_\rho - A^*(\cdot, v_\rho, Dv_\rho) \geq a_0^* I \tag{2.9}$$

for some fixed constant $a_0^* > 0$, provided ρ is taken sufficiently small, which implies in particular that v_ρ is locally strictly g^*-convex, that is for any $y_1 \in \chi[\bar{u}](x_0)$, $x_0 \in \Omega$, we have

$$v_\rho(y) > g^*(x_0, y, \bar{u}(x_0)) \tag{2.10}$$

for all $y \neq y_1$, near y_1. Clearly, by again taking ρ sufficiently small, we infer that the function v_ρ is strictly globally g^*-convex, that is inequality (2.10) holds for all $y \neq y_1, \in B_{(1-\lambda)\rho}(y_1)$. Consequently $y_1 = Tx_0$ is uniquely determined by $x_0 \in \Omega$, whence the mapping T^* is invertible with inverse T given by

$$Tx = Y(x, \bar{u}(x), D\bar{u}(x))$$

and $\bar{u}$ is differentiable with $J_1[\bar{u}] \Subset \mathcal{U}$. Since T^* is C^1 smooth so also is T with Jacobian matrix

$$DT(x) = [D_y T^*(y)]^{-1}, \quad y = Tx.$$

Using the relation

$$D\bar{u}(x) = g_x(x, Tx, g^*(x, Tx, \bar{u}(x))$$

we then have $\bar{u} \in C^2(\bar{\Omega})$. Since

$$\begin{aligned} DT(x) &= Y_p\left[D^2\bar{u} - A(\cdot, \bar{u}, D\bar{u})\right], \\ D_yT^*(y) &= X_q\left[D^2 v_\rho - A^*(\cdot, v_\rho, Dv_\rho)\right], \end{aligned}$$

and $Y_p = E^{-1}$, $X_q = -g_z(E^t)^{-1}$, we thus obtain (2.1) from condition A2. Finally we observe that

$$\begin{aligned} \bar{u}(x) &= \sup_{y\in B_\rho(y_0)} g(x, y, v_\rho(y)) \\ &\geq g(x, y_0, v_\rho(y_0)) \\ &= g(x, y_0, z_0 - \rho) \\ &> g(x, y_0, z_0), \end{aligned}$$

by using $g_z < 0$ in the last inequality. We then obtain $\bar{u} > g_0$ as asserted and the proof is complete. □

Remark 2.1 For a matrix A, not arising from optimal transportation or generated Jacobian mappings, Lemma 2.1 may not be true even if A satisfies the regularity condition A3w in a strong way. To see this, suppose $A = A(p) \geq a_1|p|^2 I$ for some constant $a_1 > 0$. Then the uniform ellipticity condition, $D^2u - A(Du) \geq a_0 I, a_0 > 0$, implies u is convex in a convex domain Ω with

$$\det D^2u \geq (a_0 + a_1|Du|^2)^n.$$

By integration, we obtain

$$|\Omega| \leq \int_{\mathbb{R}^n} (a_0 + a_1|p|^2)^{-n} dp,$$

which can only be satisfied for $|\Omega|$ sufficiently small.

We now use Lemma 2.1 to construct a barrier for the linearized operator of Eq. (1.1), in accordance with Remark 2.4 in [5]. Letting $u \in C^2(\Omega)$ be elliptic for Eq. (1.1) with $J_1[u] \subset \mathcal{U}$ we set $F[u] = \log(\det(D^2u - A(x, u, Du)))$ and define an associated linear operator L by

$$L = L[u] = F^{ij}(D_{ij} - D_{p_k}A_{ij}(x, u, Du)D_k), \tag{2.11}$$

where $F^{ij} = \frac{\partial F}{\partial w_{ij}}$ and $\{F^{ij}\} = \{w^{ij}\}$ denotes the inverse matrix of $\{w_{ij}\} \triangleq \{u_{ij} - A_{ij}(x, u, Du)\}$. It is known that F is a concave operator with respect to w_{ij} for elliptic

u. We then have the following improvement and extension of the fundamental lemma, Lemma 2.1 in [4] in the optimal transportation case.

Lemma 2.2 *Let $g \in C^4(\Gamma)$ be a generating function satisfying conditions A1, A2, A1*, A3w, A4w and let $u \in C^2(\bar{\Omega})$ be elliptic with respect to F, $J_1[u] \Subset \mathcal{U}$ and $u \le g_0$ in Ω for some g-affine function, g_0, on $\bar{\Omega}$. Then*

$$L\left(e^{K(\bar{u}-u)}\right) \ge \epsilon_1 \sum_i F^{ii} - C, \tag{2.12}$$

holds in Ω, for sufficiently large positive constant K, uniform positive constants ϵ_1, C, depending on $n, \mathcal{U}, g, \Omega, g_0, J_1[u]$ and $\sup_\Omega F[u]$.

Proof By Lemma 2.1, $\bar{u} \in C^2(\bar{\Omega})$ is a uniformly elliptic function satisfying $J_1[\bar{u}] \Subset \mathcal{U}$, $\bar{u} > g_0$ in Ω and

$$F[\bar{u}] = \log(\det(D^2\bar{u} - A(x, \bar{u}, D\bar{u}))) \ge n \log a_0,$$

where a_0 is the positive constant in (2.1). For any $x_0 \in \Omega$, the perturbation $\bar{u}_\epsilon = \bar{u} - \frac{\epsilon}{2}|x - x_0|^2$ is still a uniformly elliptic function with $J_1[\bar{u}_\epsilon] \Subset \mathcal{U}$, for sufficiently small $\epsilon > 0$, satisfying

$$F[\bar{u}_\epsilon] = \log(\det(D^2\bar{u}_\epsilon - A(x, \bar{u}_\epsilon, D\bar{u}_\epsilon))) \ge a_1,$$

for some constant $a_1 = n \log \frac{a_0}{2}$. From our construction of $\bar{u}$, we can also choose ϵ sufficiently small such that $\bar{u}_\epsilon \ge g_0 \ge u$ in Ω.

Setting $v = \bar{u} - u$, $v_\epsilon = \bar{u}_\epsilon - u$ and $\eta = Kv = K(\bar{u} - u)$, we have $v = v_\epsilon + \frac{\epsilon}{2}|x - x_0|^2$. By detailed calculation, we have

$$\begin{aligned}
Le^\eta &= e^\eta(L\eta + F^{ij} D_i\eta D_j\eta) \\
&= Ke^\eta(Lv + KF^{ij} D_i v D_j v) \\
&= Ke^\eta\left(Lv_\epsilon + L\left(\frac{\epsilon}{2}|x - x_0|^2\right) + KF^{ij} D_i v D_j v\right) \\
&= Ke^\eta \Bigg\{ \epsilon \sum_i F^{ii} - \epsilon F^{ij} D_{p_k} A_{ij}(x, u, Du)(x - x_0)_k \\
&\quad + F^{ij}(D_{ij} v_\epsilon - D_{p_k} A_{ij}(x, u, Du) D_k v_\epsilon) + KF^{ij} D_i v D_j v \Bigg\} \\
&= Ke^\eta \Bigg\{ \epsilon \sum_i F^{ii} - \epsilon F^{ij} D_{p_k} A_{ij}(x, u, Du)(x - x_0)_k \\
&\quad + F^{ij}[D_{ij}(\bar{u}_\epsilon - u) - (A_{ij}(x, \bar{u}_\epsilon, D\bar{u}_\epsilon) - A_{ij}(x, u, Du))] \\
&\quad + F^{ij}[A_{ij}(x, u, D\bar{u}_\epsilon) - A_{ij}(x, u, Du) - D_{p_k} A_{ij}(x, u, Du) D_k v_\epsilon] \\
&\quad + F^{ij}[A_{ij}(x, \bar{u}_\epsilon, D\bar{u}_\epsilon) - A_{ij}(x, u, D\bar{u}_\epsilon)] + KF^{ij} D_i v D_j v \Bigg\}.
\end{aligned} \tag{2.13}$$

By the concavity of F with respect to w_{ij}, we have

$$\begin{aligned} &F^{ij}[D_{ij}(\bar{u}_\epsilon - u) - (A_{ij}(x,\bar{u}_\epsilon, D\bar{u}_\epsilon) - A_{ij}(x,u,Du))] \\ &\geq F[\bar{u}_\epsilon] - F[u] \geq a_1 - F[u]. \end{aligned} \tag{2.14}$$

Since $\bar{u}_\epsilon \geq u$ in Ω, we then have, by the mean value theorem and condition A4w

$$F^{ij}\left[A_{ij}(x,\bar{u}_\epsilon, D\bar{u}_\epsilon) - A_{ij}(x,u,D\bar{u}_\epsilon)\right] = F^{ij}D_u A_{ij}(x,\hat{u}, D\bar{u}_\epsilon)(\bar{u}_\epsilon - u) \geq 0, \tag{2.15}$$

where $\hat{u} = \lambda\bar{u}_\epsilon + (1-\lambda)u$ for some $\lambda \in (0,1)$. Note that from our conditions on $\mathcal{U}$, we have $u(x)$,$\hat{u}(x) \in J_x$ for all $x \in \Omega$ so that $A_{ij}(x,u,D\bar{u}_\epsilon)$ is well defined in (2.13), and $A_{ij}(x,\hat{u},D\bar{u}_\epsilon)$ is well defined in (2.15). Using the Taylor expansion, we have

$$\begin{aligned} &A_{ij}(x,u,D\bar{u}_\epsilon) - A_{ij}(x,u,Du) - D_{p_k}A_{ij}(x,u,Du)D_k v_\epsilon \\ &= \frac{1}{2}A_{ij,kl}(x,u,\hat{p})D_k v_\epsilon D_l v_\epsilon = \frac{1}{2}A_{ij,kl}(x,u,\hat{p})D_k v D_l v \\ &-\epsilon A_{ij,kl}(x,u,\hat{p})(x-x_0)_k\left[D_l v - \frac{\epsilon}{2}(x-x_0)_l\right], \end{aligned} \tag{2.16}$$

where $\hat{p} = (1-\theta)Du + \theta D\bar{u}_\epsilon$ for some $\theta \in (0,1)$. Here we are also using the convexity of the set $\mathcal{P}_x$ for each $x \in \Omega$ which ensures the point $(x,u,\hat{p}) \in \mathcal{U}$ so that the term $A_{ij,kl}(x,u,\hat{p})$ in (2.16) is also well defined.

By (2.14), (2.15), (2.16), we have from (2.13)

$$\begin{aligned} Le^\eta \geq Ke^\eta \Bigg\{ &\epsilon\sum_i F^{ii} - \epsilon F^{ij}[D_{p_k}A_{ij}(x,u,Du) \\ &+A_{ij,kl}(x,u,\hat{p})(D_l v - \frac{\epsilon}{2}(x-x_0)_l)](x-x_0)_k \\ &+\frac{1}{2}F^{ij}A_{ij,kl}(x,u,\hat{p})D_k v D_l v + KF^{ij}D_i v D_j v + a_1 - F[u]\Bigg\}. \end{aligned} \tag{2.17}$$

Next we choose a finite family of balls $B_\rho(x^i)$, with centres at $x^i, i = 1\cdots N$, covering $\bar{\Omega}$ and with fixed radii $\rho < \min\{1/2n\Theta, 1\}$, where $\Theta = \max_{\bar{\Omega}}\left[|D_{p_k}A_{ij}(x,u,Du)| + |A_{ij,kl}(x,u,\hat{p})|(|D_l v| + \frac{1}{2})\right]$. Then we select $x_0 = x^i$ for some i such that $x \in B_\rho(x^i)$. Accordingly, we have for a fixed small positive ϵ,

$$\begin{aligned} Le^\eta \geq Ke^\eta \Bigg\{ &\epsilon\sum_i F^{ii} - \frac{\epsilon}{2n}\sum_{i,j}|F^{ij}| + \frac{1}{2}F^{ij}A_{ij,kl}(x,u,\hat{p})D_k v D_l v \\ &+ KF^{ij}D_i v D_j v + a_1 - F[u]\Bigg\}, \end{aligned} \tag{2.18}$$

holds for $x \in B_\rho(x^i)$. We see that (2.18) holds in all balls $B_\rho(x^i)$, $i = 1 \cdots N$, with a fixed positive constant ϵ. Then by the finite covering, (2.18) holds in Ω with a uniform positive constant ϵ.

Without loss of generality, assume that $Dv = (D_1 v, 0, \dots, 0)$ at a given point in Ω, we get

$$\begin{aligned}
Le^\eta \geq K e^\eta & \left\{ \epsilon \sum_i F^{ii} - \frac{\epsilon}{2n} \sum_{i,j} |F^{ij}| + \frac{1}{2} F^{ij} A_{ij,11}(x, u, \hat{p})(D_1 v)^2 \right. \\
& \left. + K F^{11}(D_1 v)^2 + a_1 - F[u] \right\} \\
\geq K e^\eta & \left\{ \epsilon \sum_i F^{ii} - \frac{\epsilon}{2n} \sum_{i,j} |F^{ij}| + \frac{1}{2} \sum_{i \text{ or } j=1} F^{ij} A_{ij,11}(x, u, \hat{p})(D_1 v)^2 \right. \\
& \left. + K F^{11}(D_1 v)^2 + a_1 - F[u] \right\} \\
= K e^\eta & \left\{ \epsilon \sum_i F^{ii} - \frac{\epsilon}{2n} \sum_{i,j} |F^{ij}| + \sum_i F^{1i} A_{1i,11}(x, u, \hat{p})(D_1 v)^2 \right. \\
& \left. - \frac{1}{2} F^{11} A_{11,11}(x, u, \hat{p})(D_1 v)^2 + K F^{11}(D_1 v)^2 + a_1 - F[u] \right\},
\end{aligned}$$

where condition A3w is used in the second inequality.

Since the matrix $\{F^{ij}\}$ is positive definite, any 2×2 diagonal minor has positive determinant. By the Cauchy's inequality, we have

$$|F^{ij}| \leq \sqrt{F^{ii} F^{jj}} \leq \frac{1}{2}(F^{ii} + F^{jj}), \quad \left(\text{which leads to} \sum_{i,j} |F^{ij}| \leq n \sum_i F^{ii} \right),$$

and

$$|F^{1i}| \leq \sqrt{F^{11} F^{ii}} \leq \alpha F^{ii} + \frac{1}{4\alpha} F^{11},$$

for any positive constant α.

Thus, we have

$$\begin{aligned}
Le^\eta \geq K e^\eta & \left\{ \sum_i F^{ii} \left[\frac{\epsilon}{2} - \alpha |A_{1i,11}(x, u, \hat{p})| (D_1 v)^2 \right] \right. \\
& \left. + F^{11} \left[K - \frac{1}{4\alpha} \sum_i |A_{1i,11}(x, u, \hat{p})| - \frac{1}{2} |A_{11,11}(x, u, \hat{p})| \right] (D_1 v)^2 + a_1 - F[u] \right\}.
\end{aligned}$$

By first choosing α small such that $\alpha \le \frac{\epsilon}{4\max\limits_i \max\limits_{\bar\Omega}\{|A_{1i,11}(x,u,\hat p)|(D_1 v)^2\}}$, and then choosing K large such that $K \ge \max_{\bar\Omega}\left(\frac{1}{4\alpha}\sum_i |A_{1i,11}(x,u,\hat p)| + \frac{1}{2}|A_{11,11}(x,u,\hat p)|\right)$, we obtain

$$Le^{\eta} \ge Ke^{\eta}\left\{\frac{\epsilon}{4}\sum_i F^{ii} + a_1 - F[u]\right\}.$$

By choosing $\epsilon_1 = \min_{\bar\Omega}\{\frac{\epsilon}{4}Ke^{\eta}\}$ and $C = \max_{\bar\Omega}\{Ke^{\eta}|F[u] - a_1|\}$, the conclusion of this lemma is proved. □

Remark 2.2 When we apply the operator L to the function $e^{K(\bar u - u)}$, the second derivatives of u appear in the coefficients of F^{ij} which are controlled by using the concavity property of F. Accordingly ϵ_1 in (2.12) depends on the one jet $J_1[u]$ but not on D^2u. We note that the constant C in (2.12) has the additional dependence on $\sup_\Omega F[u]$, which does involve second derivatives of u, but when we apply Lemma 2.2 to elliptic solutions u and uniformly elliptic function $\bar u$, such a dependence is reduced. For example, when we consider the Eq. (1.1), namely $F[u] = \log B(x, u, Du)$, the additional dependence of C in (2.12) is just $\sup_\Omega(\log B(x, u, Du))$.

Remark 2.3 When we assume the reverse monotonicity A4*w and $u \ge g_0$, then by modifying $\bar u$ appropriately and using (2.15), we still obtain the key barrier inequality (2.12).

3 Applications to second derivative estimates

Theorems 1.1 and 1.2 follow from Lemma 2.2 and the proofs of Theorem 3.1 in [20] and Theorem 1.1 in [9]. These latter results are proved under an additional hypothesis that Ω is A-bounded on $J_1[u](\Omega)$, that is there exists a function $\varphi \in C^2(\Omega)\cap C^{0,1}(\bar\Omega)$ satisfying

$$D^2\varphi - D_pA(\cdot, u, Du)\cdot D\varphi \ge I, \tag{3.1}$$

in Ω. However, as pointed out in [4], we need only assume the weaker inequality

$$L\varphi \ge w^{ii} - C_0, \tag{3.2}$$

for some constant C_0 to carry out the same proofs. By virtue of Lemma 2.2, the barrier inequality (3.2) is satisfied by the function

$$\varphi = \frac{1}{\epsilon_1}e^{K(\bar u - u)}. \tag{3.3}$$

We will first treat the global estimates for boundary value problems, and subsequently the interior estimate, Theorem 1.2, as we will need to modify slightly the proof in [9] to accommodate the dependance of A on u.

Taking account of the above remarks we may extend the statement of Theorem 3.1 in [20] as follows.

Lemma 3.1 *Let $u \in C^4(\Omega) \cap C^{1,1}(\bar{\Omega})$ be an elliptic solution of Eq.* (1.1) *in Ω with $A, B \in C^2(\mathcal{U})$. Suppose that A is regular on $J_1 = J_1[u](\Omega)$, that is*

$$D_{p_k p_l} A_{ij}(\cdot, u, Du)\xi_i \xi_j \eta_k \eta_l \geq 0 \tag{3.4}$$

in Ω, for all $\xi, \eta \in \mathbb{R}^n$ such that $\xi \cdot \eta = 0$, and that A satisfies (3.2) *for some $\varphi \in C^2(\Omega) \cap C^{0,1}(\bar{\Omega})$, with $\inf_{J_1} B > 0$. Then we have the estimate*

$$\sup_{\Omega} |D^2 u| \leq C\left(1 + \sup_{\partial\Omega} |D^2 u|\right), \tag{3.5}$$

where the constant C depends on A, B, Ω, $|u|_{0,1;\Omega}$, $|\varphi|_{0,1;\Omega}$ and C_0.

Clearly, Theorem 1.1 follows immediately from Lemmas 2.2 and 3.1. As an immediate consequence of Theorem 1.1, the hypothesis of the existence of a subsolution $\underline{u}$ in Theorem 1.1 of [4] can be removed altogether in the optimal transportation case, while for the Dirichlet problem in Theorem 1.2 of [4], we need only assume the existence of a subsolution $\underline{u}$ in a neighbourhood of the boundary, whose boundary trace is the prescribed boundary function. However, the natural boundary condition for optimal transportation and the more general prescribed Jacobian equations is the prescription of the image of the mapping, $Tu := Y(\cdot, u, Du)$, that is

$$Tu(\Omega) = \Omega^* \tag{3.6}$$

for some given target domain $\Omega^* \in Y(\mathcal{U})$. Global estimates for solutions of Eq. (1.1) subject to (3.6) now follow from Theorem 1.1 under appropriate convexity hypotheses on Ω and Ω^*. As with (3.1) and (3.4), we will express these initially in terms of $J_1[u]$. Accordingly, we assume the domain Ω is uniformly A-convex with respect to $J_1[u]$ in the sense that there exists a defining function $\varphi \in C^2(\bar{\Omega})$ satisfying $\varphi = 0$, $D\varphi \neq 0$ on $\partial\Omega$ together with the "uniform convexity" condition (3.1) in a neighbourhood $\mathcal{N}$ of $\partial\Omega$, while for the domain Ω^* we assume the dual condition, that Ω^* is uniformly Y^*-convex with respect to $J_1[u]$, namely that there exists a defining function $\varphi^* \in C^2(\bar{\Omega}^*)$ satisfying $\varphi^* = 0$, $D\varphi^* \neq 0$ on $\partial\Omega^*$ together with the "dual uniform convexity" condition:

$$D^2\varphi^* - A^*_{ij,k}(\cdot, u, Du)D_k\varphi^* \geq I \tag{3.7}$$

in a neighbourhood $\mathcal{N}^*$ of $\partial\Omega^*$, where $A^*_{ij,k}$ is given by

$$[A^*_{ij,k}] = Y_p^{-1} D_{pp} Y^k (Y_p^{-1})^t \tag{3.8}$$

for each $k = 1, \ldots, n$. Taking account of (1.2) we see that these conditions can be formulated for general prescribed Jacobian equations where $Y \in C^2(\mathcal{U})$ satisfies $\det Y_p \neq 0$, and are not necessarily restricted to those determined through generating functions. Furthermore, writing

$$G(x, u, p) = \varphi^* \circ Y(x, u, p), \tag{3.9}$$

condition (3.7) may be expressed in the form

$$G_{pp}(x, u, Du) \geq \delta_0^* I \tag{3.10}$$

for a positive constant δ_0^*, whenever $Y(x, u, Du) \in \mathcal{N}^*$ and moreover the boundary condition (3.6) implies a nonlinear oblique boundary condition,

$$G(\cdot, u, Du) = 0, \tag{3.11}$$

on $\partial\Omega$, for an elliptic solution $u \in C^2(\overline{\Omega})$ of Eq. (1.1) [14]. Building on the optimal transportation case in [20], an obliqueness estimate

$$G_p(\cdot, u, Du) \cdot \gamma \geq \kappa_0 \tag{3.12}$$

on $\partial\Omega$, for a positive constant κ_0, where γ denotes the unit outer normal to $\partial\Omega$, is derived in [10,14] for elliptic solutions u of (1.1) (3.6), in $C^3(\overline{\Omega})$, under the above hypotheses that Ω and Ω^* are respectively uniformly A-convex and uniformly Y^*-convex with respect to $J_1[u]$.

To proceed from (3.12) to global second derivative estimates for the second boundary value problem, we need a boundary estimate for second derivatives, which we can extract from [20], Sect. 4, as a refinement of the Monge–Ampère case in [21].

Lemma 3.2 *Let $u \in C^4(\bar{\Omega})$ be an elliptic solution of Eq.* (1.1) *in Ω subject to a nonlinear oblique boundary condition (3.11), (3.12) on $\partial\Omega \in C^4$ with $A, B, G \in C^2(\mathcal{U})$. Suppose that A is regular on $J_1 = J_1[u](\Omega) \Subset \mathcal{U}$, Ω is uniformly A-convex with respect to J_1, $\inf_{J_1} B > 0$ and G satisfies the uniform convexity condition (3.10). Then we have the estimate,*

$$\sup_{\partial\Omega} |D^2 u| \leq C \left(1 + \sup_{\Omega} |D^2 u|\right)^{\frac{2n-3}{2(n-1)}}, \tag{3.13}$$

where the constant C depends on A, B, G, Ω and $|u|_{0,1;\Omega}$.

Using the estimates, (3.12) and (3.13), we can thus conclude from Theorem 1.1, the following global second derivative estimate for generated prescribed Jacobian equations.

Theorem 3.1 *Let $u \in C^4(\bar{\Omega})$ be an elliptic solution of the second boundary value problem (1.1), (3.6) in Ω, where $A \in C^2(\mathcal{U})$ is given by (1.2) with generating function $g \in C^4(\Gamma)$, satisfying conditions A1, A2, A1*, A3w and A4w (or A4*w), $B > 0, \in C^2(\mathcal{U})$ and the domains $\Omega, \Omega^* \in C^4$ are respectively uniformly A-convex and uniformly Y^*-convex with respect to $J_1 = J_1[u](\Omega) \Subset \mathcal{U}$. Suppose also $u \leq g_0$ (or $u \geq g_0$) in Ω for some g-affine function g_0 on $\bar{\Omega}$. Then we have the estimate,*

$$\sup_{\Omega} |D^2 u| \leq C, \tag{3.14}$$

where the constant C depends on $n, \mathcal{U}, g, B, \Omega, \Omega^*, g_0$ and $J_1[u]$.

We may express the domain convexity hypotheses in Theorem 3.1 in terms of boundary data depending on the solution u or more specifically, as in [10], intervals containing the range of u. For this purpose we will say that Ω is uniformly Y-convex with respect to (Ω^*, u) if

$$(D_i\gamma_j - D_pA_{ij}(x, u(x), p)\cdot D\gamma)\tau_i\tau_j \geq \delta_0, \tag{3.15}$$

for all $x \in \partial\Omega$, $Y(x, u(x), p) \in \Omega^*$, unit outer normal γ and unit tangent vector τ, for some constant $\delta_0 > 0$. Similarly the target domain Ω^* is uniformly Y^*-convex with respect to (Ω, u) if

$$\left[D_i\gamma_j^*(y) - A_{ij,k}^*(x, u(x), p)\gamma_k^*(y)\right]\tau_i\tau_j \geq \delta_0^*, \tag{3.16}$$

for all $x \in \Omega$, $y = Y(x, u(x), p) \in \partial\Omega^*$, unit outer normal γ^* and unit tangent vector τ, for some constant $\delta_0^* > 0$. It then follows that the uniform convexity assumptions in Theorem 3.1 may be replaced by Ω, Ω^* being respectively uniformly Y-convex, Y^*-convex with respect to (Ω^*, u), (Ω, u). Moreover these assumptions are equivalent to the convexity notions defined in terms of the generating function g in [16]. Assuming Ω and Ω^* are connected and Tu is a diffeomorphism, then Ω is uniformly Y-convex with respect to (Ω^*, u) if and only if Ω is uniformly g-convex with respect to v, that is the images $Q(\cdot, y, v(y))(\Omega)$ are uniformly convex for all $y \in \bar{\Omega}^*$, where v is the g-transform of u on Ω^*, given by

$$v(y) = g^*(x, y, u(x)) = Z(x, u(x), Du(x)), \tag{3.17}$$

for $y = Tu(x)$, while Ω^* is uniformly Y^*-convex with respect to (Ω, u) if and only if Ω^* is uniformly g^*-convex with respect to u, that is the images $P(x, \cdot, u(x))(\Omega^*)$ are uniformly convex for all $x \in \bar{\Omega}$, where $P(x, y, u) = g_x(x, y, g^*(x, y, u))$ for $(x, y, u) \in \Gamma^*$; see [16] for more details.

Returning to the Dirichlet problem, we state the following analogue of Theorem 3.1, which follows from Theorem 1.1 and our boundary estimates in [4].

Theorem 3.2 *Let $u \in C^4(\bar{\Omega})$ be an elliptic solution of (1.1) in Ω, satisfying $u = u_0$ on $\partial\Omega$, where $A \in C^2(\mathcal{U})$ is given by (1.2) with generating function $g \in C^4(\Gamma)$, satisfying conditions A1, A2, A1*, A3w and A4w (or A4*w), $B > 0, \in C^2(\mathcal{U})$, $u_0 \in C^4(\bar{\Omega})$ and the domain $\Omega \in C^4$ is uniformly A-convex with respect to $J_1 = J_1[u](\Omega) \Subset \mathcal{U}$. Suppose also $u \leq g_0$ (or $u \geq g_0$) in Ω for some g-affine function g_0 on $\bar{\Omega}$. Then we have the estimate,*

$$\sup_{\Omega} |D^2u| \leq C, \tag{3.18}$$

where the constant C depends on $n, \mathcal{U}, g, B, \Omega, \varphi, g_0$ and $J_1[u]$.

To reduce the proof of Theorem 3.2 to our boundary estimates in Section 3 of [4], we replace $\underline{u}$ by u_0 in inequality (3.5) there and use the barrier $v = -C\varphi$ in place of $v = 1 - \phi$, for sufficiently large constant C, in the rest of the proof, where φ is a defining function in the definition of A-convexity. Note that under the assumption of uniform A-convexity we do not need Lemma 2.2 to obtain the second derivative boundary estimate, which is already asserted in [13]. Note also that in these arguments the dependence of the coefficients on u can be removed by replacing u by $u(x)$.

To conclude this section we indicate how Theorem 1.2 follows from Lemma 2.2 and [9]. In the optimal transportation case (1.3), when A is independent of u, we simply use (3.3) in the proof of Theorem 1.1 in [9] instead of (3.1), with u_0 replaced by g_0. In the general case we proceed the same way, first noting that the monotonicity condition A4w ensures $\eta = g_0 - u > 0$ in Ω, by virtue of the strong maximum principle [2] and the fact that g_0 is a degenerate elliptic solution of the homogeneous equation, ($B = 0$). We then consider as in [9], the corresponding auxiliary function

$$v = \tau \log \eta + \log(w_{ij}\xi_i\xi_j) + \frac{1}{2}\beta|Du|^2 + e^{\kappa\varphi}, \tag{3.19}$$

where β, κ, τ are positive constants to be determined and estimate $\tilde{L}v$ from below at an interior maximum point $x = \bar{x}$, where

$$\tilde{L}v = Lv - D_p(\log B)\cdot Dv.$$

As in Section 3 of [20], the additional terms arising from the the u dependence, in the differentiations of (1.1), are absorbed in the lower bounds (18) and (19) in [9] so it remains to extend the lower bound for $\tilde{L}\eta$ in inequalities (20) and (21) in [9], where now $\eta = g_0 - u$. This is done similarly to the argument in the proof of Lemma 2.2, as follows. By calculation, we have

$$\begin{aligned}
\tilde{L}\eta &= w^{ij}[D_{ij}\eta - D_{p_k}A_{ij}(x, u, Du)D_k\eta] - D_{p_k}(\log B)D_k\eta \\
&= w^{ij}[-w_{ij} + A_{ij}(x, g_0, Dg_0) - A_{ij}(x, u, Du) - D_{p_k}A_{ij}(x, u, Du)D_k\eta] \\
&\quad - D_{p_k}(\log B)D_k\eta \\
&\geq -C + w^{ij}[A_{ij}(x, g_0, Dg_0) - A_{ij}(x, u, Du) - D_{p_k}A_{ij}(x, u, Du)D_k\eta] \\
&\geq -C + w^{ij}[A_{ij}(x, u, Dg_0) - A_{ij}(x, u, Du) - D_{p_k}A_{ij}(x, u, Du)D_k\eta] \\
&= -C + \frac{1}{2}w^{ij}A_{ij,kl}(x, u, \bar{p})D_k\eta D_l\eta,
\end{aligned} \tag{3.20}$$

where $\bar{p} = (1-\theta)Dg_0 + \theta Du$ for some $\theta \in (0, 1)$, condition A4w is used in the second inequality and Taylor's formula is used in the last equality. By assuming $\{w_{ij}\}$ is diagonal, we have from (3.20)

$$\begin{aligned}
\tilde{L}\eta &\geq -Cn + \frac{1}{2}w^{ii}A_{ii,kl}(x, u, \bar{p})D_k\eta D_l\eta, \\
&\geq -C + \sum_{k\neq i} w^{ii}A_{ii,ik}(x, u, \bar{p})D_i\eta D_k\eta + \frac{1}{2}w^{ii}A_{ii,ii}(x, u, \bar{p})(D_i\eta)^2 \\
&\geq -C(1 + w^{ii}|D_i\eta|),
\end{aligned} \tag{3.21}$$

where condition A3w is used in the second inequality and the constant C depends on Ω, A, B, g_0 and $J_1[u]$. We remark that the particular form of the estimate (3.21) is actually crucial in the ensuing estimations in [9].

Accordingly we obtain from [9], an estimate,

$$(g_0 - u)^\tau |D^2 u| \le C \tag{3.22}$$

for positive constants τ and C, corresponding to the estimate (9) in [9]. Our desired estimate (1.5) follows, noting also that the estimate (28) in [9] for $u_0 - u$ from below extends automatically when we substitute $u(x)$ for u in A and B.

4 Optimal transportation and geometric optics

4.1 Optimal transportation

For our applications to optimal transportation when g reduces to a cost function c through (1.3), it will suffice to assume $c \in C^4(\bar{\Omega} \times \bar{\Omega}^*)$ where the domains Ω and Ω^* are respectively uniformly c and c^*-convex with respect to each other, that is the images $c_y(\cdot, y)(\Omega)$ and $c_x(x, \cdot)(\Omega^*)$ are uniformly convex in $\mathbb{R}^n$, for each $y \in \bar{\Omega}^*$, $x \in \bar{\Omega}$.

Conditions A1, A1*, A2 and A3w can then be written as:

A1: The mappings $c_x(x, \cdot)$, $c_y(\cdot, y)$ are one-to-one for each $x \in \Omega$, $y \in \Omega^*$;
A2: $\det c_{x,y} \neq 0$ in $\bar{\Omega} \times \bar{\Omega}^*$;
A3w: $(D_{p_k p_l} A_{ij}(x, p))\xi_i \xi_j \eta_k \eta_l \ge 0$, for all $(x, p) \in \Omega \times \mathbb{R}^n$ such that $Y(x, p) \in \Omega^*$ and $\xi, \eta \in \mathbb{R}^n$ such that $\xi \cdot \eta = 0$.

Here we also have the simpler formulae,

$$Y(x, u, p) = Y(x, p) = c_x^{-1}(x, \cdot)(p), \quad A(x, u, p) = A(x, p) = c_{xx}(x, Y). \tag{4.1}$$

We then conclude from Theorem 3.2 the following global second derivative estimate which improves Theorem 1.1 in [20] and Theorem 2.1 in [15].

Corollary 4.1 *Let $u \in C^4(\bar{\Omega})$ be an elliptic solution of the second boundary value problem (1.1), (3.6) in Ω, where A is given by (4.1) with cost function c satisfying conditions A1, A2 and A3w and $B > 0, \in C^2(\bar{J}_1[u])$ and the domains $\Omega, \Omega^* \in C^4$ are respectively uniformly c-convex and uniformly c^*-convex with respect to each other. Then we have the estimate,*

$$\sup_\Omega |D^2 u| \le C, \tag{4.2}$$

where the constant C depends on n, c, B, Ω, Ω^* and $J_1[u]$.

For application to optimal transportation, the function B has the form

$$B(\cdot, u, p) = \frac{|\det c_{x,y}|\rho}{\rho^* \circ Y(\cdot, p)} \tag{4.3}$$

where ρ and ρ^* are positive densities defined on Ω and Ω^* respectively, satisfying the mass balance condition

$$\int_\Omega \rho = \int_{\Omega^*} \rho^*, \tag{4.4}$$

which is a necessary condition for a solution u for which the mapping $T_u = Y(\cdot, Du)$ is a diffeomorphism. From Corollary 4.1, we obtain following [20] the existence of an elliptic solution $u \in C^3(\bar{\Omega})$ which is unique up to additive constants, as formulated in Theorem 1.1 in [15]. However our proof of Corollary 4.1 in this case is more direct and more general than the approach in [15] which depends on a strict convexity regularity result of Figalli et al. [1]. From the existence we obtain the existence of a smooth optimal mapping solving the transportation problem as formulated in Corollary 1.2 in [15].

Examples of cost functions which satisfy condition A3w and not A3 are given in [20] and [9]. However for these examples the global second derivative bounds in Theorems 1.1 and 4.1 are proved in [20] from other structures such as c-boundedness or duality so that Corollary 4.1 is not new in these cases. This is not the case though for the interior estimate, Theorem 1.2, which for example is new for the cost function $c(x, y) = |x - y|^{-2}$. For explicit calculations in this case the reader is referred to [8], as well as [20].

4.2 Geometric optics

Keeping to domains in Euclidean space, we will just consider here parallel beams in $\mathbb{R}^{n+1}$, directed in the e_{n+1} direction, through a domain $\Omega \subset \mathbb{R}^n \times \{0\}$, illuminating targets which are graphs over domains Ω^* in $\mathbb{R}^n \times \{0\}$. The formulae are easily adjusted to cover more general situations and we may also consider point sources of light as in [6]. The special cases where the targets are themselves domains in $\mathbb{R}^n \times \{0\}$ yield strictly regular matrix functions A and global estimates for the second boundary value problem are already proved in [10,14]. Consequently our main interest here is with situations where the resultant matrices A satisfy A3w and not necessarily A3, although as we indicated in Sect. 1 our lemmas in Sect. 2 are still new in this case.

(i) *Reflection.* We modify slightly the example in [16] to allow for non-flat targets; see also [7] for a thorough description of the geometric picture.
Let $\mathcal{D}$ be a domain in $\mathbb{R}^n \times \mathbb{R}^n$, containing $\Omega \times \Omega^*$, and consider the generating function:
$$g(x, y, z) = \Phi(y) + \frac{1}{2z} - \frac{z}{2}|x - y|^2, \tag{4.5}$$
defined for $(x, y) \in \mathcal{D}$ and $z > 0$ where Φ is a smooth function on $\mathbb{R}^n$. By differentiation we have
$$g_x(x, y, z) = z(y - x) \tag{4.6}$$
so that if there are mappings $y = Y(x, u, p)$, $z = Z(x, u, p)$ satisfying A1, we must have $y = x + p/z$ and from (4.5),
$$u = h(z) = \Phi\left(x + \frac{p}{z}\right) + \frac{1}{2z}\left(1 - |p|^2\right). \tag{4.7}$$

By differentiation we then obtain,

$$h'(z) = -\frac{1}{2z^2}(2p\cdot D\Phi + 1 - |p|^2) < 0, \tag{4.8}$$

provided

$$2p\cdot D\Phi + 1 - |p|^2 > 0,$$

that is,

$$z^2|y-x|^2 + 2z(x-y)\cdot D\Phi - 1 < 0. \tag{4.9}$$

Accordingly conditions A1 and A2 will be satisfied for $0 < z < z_0(x, y)$, where

$$z_0 = \frac{1}{\sqrt{|x-y|^2 + |(x-y)\cdot D\Phi|^2} + (x-y)\cdot D\Phi}, \tag{4.10}$$

$Y(x, u, p) = x + p/Z(x, u, p)$ and Z, given implicitly by (4.7), satisfies $Z_u < 0$. Now from (4.6), we see that the matrix A is given by

$$A(x, u, p) = -Z(x, u, p)I \tag{4.11}$$

and hence conditions A3w and A4w are both satisfied whenever the dual function Z is concave in the gradient variables. It also follows that condition A1* is also satisfied when (4.9) holds, that is the mapping Q given by

$$Q(x, y, z) = 2z^2\frac{D\Phi + z(x-y)}{1 + z^2|x-y|^2}$$

is one to one in x, for $0 < z < z_0(x, y)$, which can be proved for example from the quadratic structure of the equation, $Q(x, y, z) = q$. Furthermore by taking z sufficiently small, we see that there exist arbitrarily large g-affine functions and hence all the hypotheses of Theorems 1.1 and 1.2 are satisfied by the generating function g if Z is concave in p. Note that when Φ is constant, then $Z = (1 - |p|^2)/2(u - \Phi)$, $u > \Phi$, $|p| < 1$ and the strict condition A3 holds, [16]. By direct calculation, we have an explicit formula for the matrix E,

$$E = z\left\{I - \frac{2z(x-y)\otimes[z(x-y) + D\Phi]}{1 + z^2|x-y|^2}\right\},$$

and its determinant

$$\det E = -z^n\frac{z^2|x-y|^2 + 2z(x-y)\cdot D\Phi - 1}{1 + z^2|x-y|^2} > 0,$$

when (4.9) holds. By the Sherman–Morrison formula, we also have a formula for the inverse of E,

$$E^{-1} = \frac{1}{z}\left\{ I - \frac{2z(x-y)\otimes[z(x-y)+D\Phi]}{z^2|x-y|^2 + 2z(x-y)\cdot D\Phi - 1} \right\}.$$

(ii) *Refraction.* We consider refraction from media I to media II with respective refraction indices $n_1, n_2 > 0$ and set $\kappa = n_1/n_2$. For $\kappa \neq 1$, we consider now generating functions,

$$g(x,y,z) = \Phi(y) - \frac{1}{|\kappa^2 - 1|}\left(\kappa z + \sqrt{z^2 + (\kappa^2-1)|x-y|^2}\right), \tag{4.12}$$

where again $(x,y) \in \mathcal{D}$, $z > \sqrt{1-\kappa^2}|x-y|$ for $0 < \kappa < 1$, > 0 for $\kappa > 1$ and Φ is a smooth function on $\mathbb{R}^n$. For more details about the geometric and physical aspects of this model, see for example [12]. We will now restrict attention to the case $0 < \kappa < 1$, in which case by setting $\kappa' = \sqrt{1-\kappa^2}$ and rescaling $z \to z/\kappa'$, $\Phi \to \kappa'\Phi$, $g \to \kappa' g$, we can write

$$g(x,y,z) = \Phi(y) - \kappa z - \sqrt{z^2 - |x-y|^2}. \tag{4.13}$$

By differentiation we now have

$$g_x(x,y,z) = \frac{x-y}{\sqrt{z^2 - |x-y|^2}} \tag{4.14}$$

so that if there are mappings $y = Y(x,u,p)$, $z = Z(x,u,p)$ satisfying A1, we must have

$$y = x - \frac{zp}{\sqrt{1+|p|^2}},$$

so from (4.13),

$$u = h(z) = \Phi\left(x - \frac{zp}{\sqrt{1+|p|^2}}\right) - \kappa z - \frac{z}{\sqrt{1+|p|^2}}. \tag{4.15}$$

By differentiation we then obtain,

$$h'(z) = -\frac{1}{\sqrt{1+|p|^2}}\left(p\cdot D\Phi + 1 + \kappa\sqrt{1+|p|^2}\right) < 0, \tag{4.16}$$

provided

$$p\cdot D\Phi + 1 + \kappa\sqrt{1+|p|^2} > 0,$$

that is,

$$\sqrt{z^2 - |x-y|^2} + \kappa z + (x-y)\cdot D\Phi > 0. \tag{4.17}$$

Accordingly conditions A1 and A2 will be satisfied for $z > z_0 = z_0(x, y)$, where $z_0 = |x-y|$, if $\kappa|x-y| \le (y-x)\cdot D\Phi$ and z_0 is given by

$$z_0 = \frac{1}{1-\kappa^2}\left(\kappa(x-y)\cdot D\Phi + \sqrt{(1-\kappa^2)|x-y|^2 + |(x-y)\cdot D\Phi(y)|^2}\right), \tag{4.18}$$

if $\kappa|x-y| \ge (y-x)\cdot D\Phi$. Furthermore the mapping Y is given by $Y(x, u, p) = x + Z(x, u, p)p/\sqrt{1+|p|^2}$ and Z, given implicitly by (4.15), satisfies $Z_u < 0$. From (4.14), we now have

$$A(x, u, p) = \frac{\sqrt{1+|p|^2}}{Z(x, u, p)}(I + p\otimes p). \tag{4.19}$$

It follows then that conditions A3w and A4w are both satisfied whenever the function $p \to \sqrt{1+|p|^2}/Z(x, u, p)$ is convex. Again the condition A1* is also satisfied when (4.17) holds so that the hypotheses of Theorems 1.1 and 1.2 are fulfilled provided in the case of Theorem 1.1 the solution u has a g-affine majorant, which is satisfied automatically from the boundary condition, $u = g_0$ on $\partial\Omega$ in the case of Theorem 1.2. Note that when Φ is constant, we obtain

$$Z = \frac{(\Phi - u)\sqrt{1+|p|^2}}{1+\kappa\sqrt{1+|p|^2}}, \quad u < \Phi,$$

so again the strict condition A3 holds. By direct calculation, we now have an explicit formula for the matrix E,

$$E = -\frac{1}{\sqrt{z^2-|x-y|^2}}\left\{I + \frac{(x-y)\otimes[\kappa(x-y)+zD\Phi]}{\sqrt{z^2-|x-y|^2}(z+\kappa\sqrt{z^2-|x-y|^2})}\right\},$$

and its determinant

$$\det E = \frac{(-1)^n z\left[\sqrt{z^2-|x-y|^2}+\kappa z+(x-y)\cdot D\Phi\right]}{(z^2-|x-y|^2)^{\frac{n+1}{2}}(z+\kappa\sqrt{z^2-|x-y|^2})} \ne 0,$$

when (4.17) holds. We also obtain the following formula for the inverse of E,

$$E^{-1} = -\sqrt{z^2-|x-y|^2}\left\{I - \frac{(x-y)\otimes[\kappa(x-y)+zD\Phi]}{z[\sqrt{z^2-|x-y|^2}+\kappa z+(x-y)\cdot D\Phi]}\right\}.$$

In the case $\kappa > 1$, the monotonicity is reversed in the case Φ = constant. For general Φ and taking now $\kappa' = \sqrt{\kappa^2-1}$, we have after rescaling $z \to z/\kappa'$,

$\Phi \to \kappa'\Phi$, $g \to \kappa' g$,

$$g(x, y, z) = \Phi(y) - \kappa z - \sqrt{z^2 + |x - y|^2}, \tag{4.20}$$

and hence we obtain in place of (4.14) to (4.16),

$$\begin{aligned} g_x(x, y, z) &= \frac{y - x}{\sqrt{z^2 + |x - y|^2}}, \quad y = x + \frac{zp}{\sqrt{1 - |p|^2}}, \quad |p| < 1, \\ u = h(z) &= \Phi\left(x + \frac{zp}{\sqrt{1 - |p|^2}}\right) - \kappa z - \frac{z}{\sqrt{1 - |p|^2}}, \\ h'(z) &= \frac{1}{\sqrt{1 - |p|^2}}\left(p \cdot D\Phi - 1 - \kappa\sqrt{1 - |p|^2}\right) < 0, \end{aligned} \tag{4.21}$$

provided

$$(y - x) \cdot D\Phi < \kappa z + \sqrt{z^2 + |x - y|^2}.$$

However now we have in place of (4.19),

$$A(x, u, p) = -\frac{\sqrt{1 - |p|^2}}{Z(x, u, p)}(I - p \otimes p) \tag{4.22}$$

so that we obtain A4*w instead of A4w. Hence for A4w we need to assume the reverse inequality, $h'(z) > 0$, that is $0 < z < z_0$, where now z_0 is given by

$$z_0 = \frac{1}{\kappa^2 - 1}\left(\kappa(y - x) \cdot D\Phi - \sqrt{(\kappa^2 - 1)|x - y|^2 + |(x - y) \cdot D\Phi(y)|^2}\right), \tag{4.23}$$

which implies at least that Ω and Ω^* are disjoint and excludes the case $\Phi =$ constant. However with $z > \max\{z_0, 0\}$, Theorems 1.1, 3.1 and 3.2 are still applicable provided the function $p \to -\sqrt{1 - |p|^2}/Z(x, u, p)$ is convex as there exist arbitrarily small g-affine functions. Also then the case of constant Φ is embraced with

$$Z = \frac{(\Phi - u)\sqrt{1 - |p|^2}}{1 + \kappa\sqrt{1 - |p|^2}}, \quad u < \Phi, \ |p| < 1.$$

By direct calculation, we again have an explicit formula for the matrix E,

$$E = \frac{1}{\sqrt{z^2 + |x - y|^2}}\left\{I + \frac{(x - y) \otimes [-\kappa(x - y) + zD\Phi]}{\sqrt{z^2 + |x - y|^2}(z + \kappa\sqrt{z^2 + |x - y|^2})}\right\},$$

and its determinant

$$\det E = \frac{z[\sqrt{z^2 + |x-y|^2} + \kappa z + (x-y)\cdot D\Phi]}{(z^2 + |x-y|^2)^{\frac{n+1}{2}}(z + \kappa\sqrt{z^2 + |x-y|^2})} \neq 0,$$

when $h'(z) \neq 0$. By the Sherman-Morrison formula, we also obtain the inverse of E,

$$E^{-1} = \sqrt{z^2 + |x-y|^2}\left\{I - \frac{(x-y)\otimes[-\kappa(x-y) + zD\Phi]}{z[\sqrt{z^2 + |x-y|^2} + \kappa z + (x-y)\cdot D\Phi]}\right\}.$$

As in the previous cases the condition A1* will hold if either $h'(z) < 0$, (that is $\det E > 0$), or $h'(z) > 0$, (that is $\det E < 0$), which again follows from the equivalent quadratic structure of the equation $Q(x, y, z) = q$.

(iii) *Further remarks.* It is interesting to note that when we pass to the dual functions in the above examples the monotonicity of A is reversed but, as shown in general in [16], the condition A3w is preserved. In particular for the dual in the reflection case (4.5) we have from [16],

$$g^*(x, y, u) = \frac{1}{u - \Phi(y) + \sqrt{(u - \Phi(y))^2 + |x-y|^2}}, \tag{4.24}$$

while in the refraction cases (4.12), a simple calculation gives

$$g^*(x, y, u) = +, (-)\left(\kappa(u - \Phi(y)) + \sqrt{(u - \Phi(y))^2 + |x-y|^2}\right), \tag{4.25}$$

for $\kappa < 1$, ($\kappa > 1$) respectively. Note that by replacing z by $-1/z$ in (4.5), the dual function in (4.24) is the limit as $\kappa \to 1$ of the case $\kappa > 1$ in (4.25).

Also we note that when we consider instead the complementary ellipticity condition, $D^2u - A(\cdot, u, Du) < 0$, in Eq. (1.1), condition A3w is replaced by substituting $-A$ for A but conditions A4w and A4*w are maintained. Thus in the above examples we need to replace Z by $-Z$ in the convexity conditions, (and interchange the inequalities $u < g_0, u > g_0$), to fulfil our hypotheses.

As well as the monotonicity properties, another interesting ingredient of the above examples is that we cannot infer, as in the optimal transportation examples mentioned in the previous section, that bounded domains are A-bounded so second derivative estimates do not follow from the relevant arguments in [14,20].

4.3 Final remarks

The existence of locally smooth solutions to the second boundary value problem for generated prescribed Jacobian equations is treated in [16] under conditions A1, A2, A1*, A3 and A4w. We remark that there that the montonicity condition A4w may also be replaced by condition A4* for local regularity. In [17] and ensuing work the monotonicity condition A4w is relaxed and local boundary regularity is also considered, yielding an alternative approach to that in [10] which deals with estimates and existence of globally smooth solutions. We remark that the strict monotonicity of Z, in our examples in Sect. 4.2, also follows from a general formula in [17],

$$Z_u = \frac{\det g_{x,y}}{g_z \det E}.$$

In a follow up work [3] we will use our estimates here to extend the global theory when A3 is weakened to A3w, without A-boundedness conditions, as well as further develop the optics examples. We also develop an alternative duality approach to second derivative estimates when the matrix A depends only on u and p. As already noted in [20], this situation is much simpler in two dimensions.

Finally we wish to thank the anonymous referee of this paper for their careful checking and useful comments.

References

1. Figalli, A., Kim, Y.-H., McCann, R.J.: Hölder continuity and injectivity of optimal maps. Arch. Ration. Mech. Anal. **209**, 747–795 (2013)
2. Gilbarg, D., Trudinger, N.S.: Elliptic Partial Differential Equation of Second Order. Springer, Berlin (2001)
3. Jiang, F., Trudinger, N.S.: On the second boundary value problem for Monge–Ampère type equations and geometric optics (in preparation)
4. Jiang, F., Trudinger, N.S., Yang, X.-P.: On the Dirichlet problem for Monge–Ampère type equations. Calc. Var. Partial Differ. Equ. **49**, 1223–1236 (2014)
5. Jiang, F., Trudinger, N.S., Yang, X.-P.: On the Dirichlet problem for a class of augmented Hessian equations. Preprint (2014)
6. Karakhanyan, A., Wang, X.-J.: On the reflector shape design. J. Differ. Geom. **84**, 561–610 (2010)
7. Karakhanyan, A.: Existence and regularity of the reflector surfaces in $\mathbb{R}^{n+1}$. Arch. Ration. Mech. Anal. **213**, 833–885 (2014)
8. Kim, Y.-H., Kitagawa, J.: On the degeneracy of optimal transportation. Commun. Partial Differ. Equ. **39**, 1329–1363 (2014)
9. Liu, J., Trudinger, N.S.: On Pogorelov estimates for Monge–Ampère type equations. Discret. Contin. Dyn. Syst. Ser. A. **28**, 1121–1135 (2010)
10. Liu, J., Trudinger, N.S.: On classical solutions of near field reflection problems. Preprint (2014)
11. Ma, X.-N., Trudinger, N.S., Wang, X.-J.: Regularity of potential functions of the optimal transportation problem. Arch. Ration. Mech. Anal. **177**, 151–183 (2005)
12. Oliker, V.: Differential equations for design of a freeform single lens with prescribed irradiance properties. Opt. Eng. **53**, 031302-1-10 (2014)
13. Trudinger, N.S.: Recent developments in elliptic partial differential equations of Monge–Ampère type. ICM. Madr. **3**, 291–302 (2006)
14. Trudinger, N.S.: On the prescribed Jacobian equation. In: Proceedings of International Conference for the 25th Anniversary of Viscosity Solutions, Gakuto International Series. Math. Sci. Appl. **20**, 243–255 (2008)

15. Trudinger, N.S.: A note on global regularity in optimal transportation. Bull. Math. Sci. **3**, 551–557 (2013)
16. Trudinger, N.S.: On the local theory of prescribed Jacobian equations. Discret. Contin. Dyn. Syst. **34**, 1663–1681 (2014)
17. Trudinger, N.S.: On the local theory of prescribed Jacobian equations revisited (in preparation) (2014)
18. Trudinger, N.S., Wang, X.-J.: The Monge–Ampère equation and its geometric applications. In: Handbook of geometric analysis. No. 1, Adv. Lect. Math. (ALM), vol. 7, pp. 467–524. International Press, Somerville (2008)
19. Trudinger, N.S., Wang, X.-J.: On strict convexity and continuous differentiability of potential functions in optimal transportation. Arch. Ration. Mech. Anal. **192**, 403–418 (2009)
20. Trudinger, N.S., Wang, X.-J.: On the second boundary value problem for Monge–Ampère type equations and optimal transportation. Ann. Scuola Norm. Sup. Pisa Cl. Sci. **VIII**, 143–174 (2009)
21. Urbas, J.: Oblique boundary value problems for equations of Monge–Ampère type. Calc. Var. Partial Differ. Equ. **7**, 19–39 (1998)

6

On global inversion of homogeneous maps

Michael Ruzhansky · **Mitsuru Sugimoto**

Abstract In this note we prove a global inverse function theorem for homogeneous mappings on $\mathbb{R}^n$. The proof is based on an adaptation of the Hadamard's global inverse theorem which provides conditions for a function to be globally invertible on $\mathbb{R}^n$. For the latter adaptation, we give a short elementary proof assuming a topological result.

Keywords Inverse function theorem · Homogeneous mappings · Global inverse

Mathematics Subject Classification Primary 26B10; Secondary 26B05 · 26-01

1 Introduction

The purpose of this note is to prove a global inverse function for homogeneous mappings on $\mathbb{R}^n$. The main difficulty is in the fact that such mappings are not smooth at the origin and thus, the known global inverse function theorems on $\mathbb{R}^n$ are not readily applicable. The main motivation for studying such inverses are applications to the global invertibility of Hamiltonian flows, and further applications to construction of

---

Communicated by Ari Laptev.

M. Ruzhansky was supported by the EPSRC Leadership Fellowship EP/G007233/1 and by EPSRC Grant EP/K039407/1.

M. Ruzhansky (✉)
Department of Mathematics, Imperial College London, 180 Queen's Gate, London SW7 2AZ, UK
e-mail: m.ruzhansky@imperial.ac.uk

M. Sugimoto
Graduate School of Mathematics, Nagoya University, Furocho Chikusa-ku, Nagoya 464-8602, Japan
e-mail: sugimoto@math.nagoya-u.ac.jp

suitable phase functions of Fourier integral operators, a topic that will be addressed elsewhere.

Thus, our aim here is to establish the existence and the following properties for the global inverses of homogeneous mappings.

Theorem 1.1 *Let $n \geq 3$ and $1 \leq k \leq \infty$. Let $f : \mathbb{R}^n \backslash 0 \to \mathbb{R}^n$ be a positively homogeneous mapping of order $\varkappa > 0$, i.e.*

$$f(\tau\xi) = \tau^{\varkappa} f(\xi) \quad \textit{for all } \tau > 0,\ \xi \neq 0.$$

Assume that $f \in C^k(\mathbb{R}^n \backslash 0)$ and that its Jacobian never vanishes on $\mathbb{R}^n \backslash 0$. Then f is bijective from $\mathbb{R}^n \backslash 0$ to $\mathbb{R}^n \backslash 0$, its global inverse satisfies $f^{-1} \in C^k(\mathbb{R}^n \backslash 0)$ and is positively homogeneous of order $1/\varkappa$. Moreover, if we extend f to $\mathbb{R}^n$ by setting $f(0) = 0$, the extension is a global homeomorphism on $\mathbb{R}^n$.

We remark that for $n = 2$ the conclusion is not always true because, for example, the Jacobian of $f(x, y) = (x^2 - y^2, 2xy)$ never vanishes on $\mathbb{R}^2 \backslash 0$ but f is not globally invertible since $f(x, y) = f(-x, -y)$.

Our proof is based on the application of an adaptation of the Hadamard global inverse theorem. Thus, the second aim of this note is to give a short elementary proof for it assuming a well-known topological result (very likely also known to Hadamard).

This kind of global inverse function theorem is a classical subject and of independent interest, but sometimes it also plays an important role when we discuss the global L^2 boundedness of oscillatory integrals. For example, Asada-Fujiwara [1] established this boundedness based on the global invertibility of maps defined by phase functions which are smooth everywhere. Generalising such results to the case of homogeneous phase as in Theorem 1.1 is also important but it is not straightforward because of the singularity at the origin. The global L^2 boundedness of oscillatory integrals with phases as in Theorem 1.1 has been analysed by the authors in [7] and applied to questions of global smoothing for partial differential equations in [8]. The application of Theorem 1.1 to the global analysis of hyperbolic partial differential equations and the corresponding Hamiltonian flows will appear elsewhere.

2 Global inverses

Let us start with a topological result we assume. A differentiable map between manifolds is called a C^1-diffeomorphism if it is one-to-one and its inverse is also differentiable. A mapping f is called proper if $f^{-1}(K)$ is compact whenever K is compact.

Theorem 2.1 *Let M and N be connected, oriented, d-dimensional C^1-manifolds, without boundary. Let $f : M \to N$ be a proper C^1-map such that the Jacobian $J(f)$ never vanishes. Then f is surjective. If N is simply connected in addition, then f is also injective.*

This fact was known to Hadamard, but a rigorous proof for surjectivity can be found in [6]. As for the injectivity, a precise proof can be found in [2, Sect. 3]. We remark that it follows in principle from the general fact that a proper submersion between smooth manifolds without boundary is a fibre bundle, meaning in our setting that it is a covering map. Since a simply connected manifold is its own universal covering space, it implies that f is a diffeomorphism.

We will mostly discuss the mappings $f : \mathbb{R}^n \to \mathbb{R}^n$, in which case we denote its Jacobian by $J(f) := \det Df := \det(\partial f_i/\partial x_j)$. The Hadamard global inverse function theorem states:

Theorem 2.2 *A C^1-map $f : \mathbb{R}^n \to \mathbb{R}^n$ is a C^1-diffeomorphism if and only if the Jacobian* $\det Df(w)$ *never vanishes and $|f(y)| \to \infty$ whenever $|y| \to \infty$.*

This theorem goes back to Hadamard [3–5]. In fact, in 1972 W. B. Gordon wrote "*This theorem goes back at least to Hadamard, but it does not appear to be 'well-known'. Indeed, I have found that most people do not believe it when they see it and that the skepticism of some persists until they see two proofs.*" The reason behind this is that while we know that the function is locally a C^1-diffeomorphism by the usual local inverse function theorem, the condition that $|f(y)| \to \infty$ as $|y| \to \infty$, guarantees that the function is both *injective* and, more importantly, *surjective* on the whole of $\mathbb{R}^n$. And indeed, W. B. Gordon proceeds in [2] by giving two different proofs for it, for C^2 and for C^1 mappings.

It turns out that for dimensions $n \geq 3$ the conclusion remains still valid even if we relax the assumption, in some sense rather substantially, almost removing it at a finite number of points. While often this is not the case (the famous one being the 'hairy ball theorem', which fails completely if we assume that the vector field may be not differentiable at one point), the theory of covering spaces assures that it is the case here. In fact, here we have the following:

Theorem 2.3 *Let $n \geq 3$. Let $a \in \mathbb{R}^n$. Let $f : \mathbb{R}^n \to \mathbb{R}^n$ be such that f is C^1 on $\mathbb{R}^n \backslash \{a\}$, with* $\det Df \neq 0$ *on $\mathbb{R}^n \backslash \{a\}$, and that f is continuous at a. Let $b := f(a)$, and assume that $f(\mathbb{R}^n \backslash \{a\}) \subset \mathbb{R}^n \backslash \{b\}$ and that $|y| \to \infty$ implies $|f(y)| \to \infty$. Then the mapping $f : \mathbb{R}^n \to \mathbb{R}^n$ is a global homeomorphism and its restriction $f : \mathbb{R}^n \backslash \{a\} \to \mathbb{R}^n \backslash \{b\}$ is a global C^1-diffeomorphism.*

While general topological considerations as outlined above are possible here, we prefer to also give an elementary proof in Sect. 3 of the fact that Theorem 2.1 implies Theorem 2.3, also noting that exactly the same proof yields the following further version:

Theorem 2.4 *Let $n \geq 3$. Let $A \subset \mathbb{R}^n$ be a closed set. Let $f : \mathbb{R}^n \to \mathbb{R}^n$ be such that f is C^1 on $\mathbb{R}^n \backslash A$, with* $\det Df \neq 0$ *on $\mathbb{R}^n \backslash A$, that f is continuous and injective on A, and that $\mathbb{R}^n \backslash f(A)$ is simply connected. Assume that $f(\mathbb{R}^n \backslash A) \subset \mathbb{R}^n \backslash f(A)$ and that $|y| \to \infty$ implies $|f(y)| \to \infty$. Then the mapping $f : \mathbb{R}^n \to \mathbb{R}^n$ is a global homeomorphism and its restriction $f : \mathbb{R}^n \backslash A \to \mathbb{R}^n \backslash f(A)$ is a global C^1-diffeomorphism.*

Given Theorem 2.3, we can apply it to derive Theorem 1.1:

Proof of Theorem 1.1 First, let us extend f to $\mathbb{R}^n$ by setting $f(0) = 0$ and show that f is continuous at 0. We observe that $f(\mathbb{S}^{n-1})$ has a finite maximum. Let $\xi_j \to 0$, and $\xi_j \neq 0$ for all j. Then

$$|f(\xi_j)| = |\xi_j|^{\varkappa} \left| f\left(\frac{\xi_j}{|\xi_j|}\right) \right| \leq C|\xi_j|^{\varkappa} \to 0,$$

so that f is continuous at 0.

Let us now check that other conditions of Theorem 2.3 are satisfied. We observe that $f(\mathbb{S}^{n-1})$ has a positive minimum $\min_{|\xi|=1} |f(\xi)| = c_0 > 0$. Indeed, if $f(\omega) = 0$ for some $\omega \in \mathbb{S}^{n-1}$, then $f(t\omega)(= t^{\kappa} f(\omega)) = 0$ for any $t > 0$. Differentiating it in t, we have $\omega = 0$ since the Jacobian of f never vanishes on $\mathbb{R}^n \backslash 0$, which is a contradiction. Then we have

$$|f(\xi)| = |\xi|^{\varkappa} \left| f\left(\frac{\xi}{|\xi|}\right) \right| \geq c_0 |\xi|^{\varkappa}, \quad \xi \neq 0,$$

which induces that $f(\mathbb{R}^n \backslash 0) \subset \mathbb{R}^n \backslash 0$ and that $|y| \to \infty$ implies $|f(y)| \to \infty$. Therefore, by Theorem 2.3, $f : \mathbb{R}^n \to \mathbb{R}^n$ is a homeomorphism, and f^{-1} is C^k on $\mathbb{R}^n \backslash 0$ by the usual local inverse function theorem. Let us finally show that f^{-1} is positively homogeneous of order $1/\varkappa$. Indeed, for every $\tau > 0$ and $\xi \neq 0$ we have $f^{-1}(\tau^{\varkappa} f(\xi)) = f^{-1}(f(\tau\xi)) = \tau\xi$. Since f is invertible, $\eta = f(\xi) \neq 0$, and we have $f^{-1}(\tau^{\varkappa}\eta) = \tau f^{-1}(\eta)$, or $f^{-1}(\tau\eta) = \tau^{1/\varkappa} f^{-1}(\eta)$. □

3 Proofs

First we observe that in the setting of Theorem 2.3, by translation (by a) in x and by subtracting b from f, we may assume without loss of generality that $a = b = 0$. To prove Theorem 2.3, we start with preliminary statements.

Lemma 3.1 *Let $F \subset \mathbb{R}^n \backslash 0$. Then F is compact in $\mathbb{R}^n \backslash 0$ if and only if it is compact in $\mathbb{R}^n$.*

Proof Assume that $F \subset \mathbb{R}^n \backslash 0$ is compact in $\mathbb{R}^n \backslash 0$. Let $F \subset \bigcup_\alpha V_\alpha$ for a family of sets V_α which are open in $\mathbb{R}^n$. Then

$$F \subset \left(\bigcup_\alpha V_\alpha \right) \cap (\mathbb{R}^n \backslash 0) = \bigcup_\alpha \left(V_\alpha \cap (\mathbb{R}^n \backslash 0) \right),$$

so that F is covered by a family of sets $V_\alpha \cap (\mathbb{R}^n \backslash 0)$ which are open in $\mathbb{R}^n \backslash 0$. Since F is compact in $\mathbb{R}^n \backslash 0$, there is a finite subfamily V_j, $j = 1, \ldots, m$, such that

$$F \subset \bigcup_{j=1}^{m} (V_j \cap (\mathbb{R}^n \backslash 0)) = \left(\bigcup_{j=1}^{m} V_j \right) \cap (\mathbb{R}^n \backslash 0).$$

Hence $F \subset \bigcup_{j=1}^m V_j$, so that F is compact in $\mathbb{R}^n$.

Conversely, assume that $F \subset \mathbb{R}^n \backslash 0$ is compact in $\mathbb{R}^n$, and let $F \subset \bigcup_\alpha U_\alpha$, for a family of sets $U_\alpha \subset \mathbb{R}^n \backslash 0$ which are open in $\mathbb{R}^n \backslash 0$. Then $U_\alpha = V_\alpha \cap (\mathbb{R}^n \backslash 0)$, for some V_α open in $\mathbb{R}^n$. Hence $F \subset \bigcup_\alpha V_\alpha$, and by compactness of F in $\mathbb{R}^n$, there is a finite subcovering $F \subset \bigcup_{j=1}^m V_j$. Since $F \subset \mathbb{R}^n \backslash 0$, we have

$$F \subset \left(\bigcup_{j=1}^m V_j \right) \cap (\mathbb{R}^n \backslash 0) = \bigcup_{j=1}^m \big(V_j \cap (\mathbb{R}^n \backslash 0)\big) = \bigcup_{j=1}^m U_j,$$

which proves that F is compact in $\mathbb{R}^n \backslash 0$. □

We recall that a mapping f is called proper if $f^{-1}(K)$ is compact whenever K is compact.

Lemma 3.2 *Let $f : \mathbb{R}^n \to \mathbb{R}^n$ be proper and such that $f(0) = 0$ and $f(\mathbb{R}^n \backslash 0) \subset \mathbb{R}^n \backslash 0$. Then the restriction $f : \mathbb{R}^n \backslash 0 \to \mathbb{R}^n \backslash 0$ is proper.*

Proof Let $K \subset \mathbb{R}^n \backslash 0$ be compact in $\mathbb{R}^n \backslash 0$. By Lemma 3.1 it is compact in $\mathbb{R}^n$, and, since f is proper, the set $f^{-1}(K)$ is compact in $\mathbb{R}^n$. We notice that if $0 \in f^{-1}(K)$ then we would have $0 = f(0) \in K$, which is impossible since $K \subset \mathbb{R}^n \backslash 0$. Hence $f^{-1}(K) \subset \mathbb{R}^n \backslash 0$, and by Lemma 3.1 again, the set $f^{-1}(K)$ is compact in $\mathbb{R}^n \backslash 0$. Hence the restriction $f : \mathbb{R}^n \backslash 0 \to \mathbb{R}^n \backslash 0$ is proper. □

Lemma 3.3 *Let $f : \mathbb{R}^n \to \mathbb{R}^n$ be continuous everywhere. Then f is proper if and only if $|y| \to \infty$ implies $|f(y)| \to \infty$.*

Proof We show the if part. Let $K \subset \mathbb{R}^n$ be compact. Then it is closed and hence $f^{-1}(K) \subset \mathbb{R}^n$ is closed. Suppose $f^{-1}(K)$ is not bounded. Then there is a sequence $y_j \in f^{-1}(K)$ such that $|y_j| \to \infty$. Hence $f(y_j) \in K$ and also $|f(y_j)| \to \infty$ by the assumption on f, which yields a contradiction with the boundedness of K. The converse implication is clearly also true. □

Lemma 3.4 *Let $f : \mathbb{R}^n \to \mathbb{R}^n$ be proper, injective, and continuous. Then f is an open map.*

Proof Let us assume that $f(U)$ is not open for an open subset $U \subset \mathbb{R}^n$. Then there is a point $a \in U$ such that $f(a)$ is on the boundary of $f(U)$, and we can construct a sequence $y_j \in \mathbb{R}^n \backslash U$ such that $f(y_j) \to f(a)$. Since f is proper, there exists a subsequence y'_j which converges to some point $b \in \mathbb{R}^n \backslash U$. Note that $b \neq a$. Since f is continuous, $f(y'_j) \to f(b)$, but we also have $f(y'_j) \to f(a)$ which contradicts to the fact that f is injective. □

The following result is a straight forward consequence of Theorem 2.1.

Corollary 3.5 *Let $n \geq 2$. Let $f : \mathbb{R}^n \to \mathbb{R}^n$ be proper and such that and $f(0) = 0$ and $f(\mathbb{R}^n \backslash 0) \subset \mathbb{R}^n \backslash 0$. Moreover, assume that f is C^1 on $\mathbb{R}^n \backslash 0$, with $\det Df \neq 0$ on $\mathbb{R}^n \backslash 0$. Then the restriction $f : \mathbb{R}^n \backslash 0 \to \mathbb{R}^n \backslash 0$ is surjective. If $n \geq 3$ in addition, $f : \mathbb{R}^n \backslash 0 \to \mathbb{R}^n \backslash 0$ is also injective.*

Proof By Lemma 3.2, the restriction of f is a proper map from $M = \mathbb{R}^n \backslash 0$ to $N = \mathbb{R}^n \backslash 0$. Note that N is simply connected if $n \geq 3$. Then Theorem 2.1 implies the statement. □

With all these facts, Theorem 2.3 is immediate:

Proof of Theorem 2.3 By Corollary 3.5 and Lemma 3.3, the map $f : \mathbb{R}^n \backslash 0 \to \mathbb{R}^n \backslash 0$ is bijective. Hence it is a global C^1 diffeomorphism by the usual local inverse function theorem. Furthermore, the map $f : \mathbb{R}^n \to \mathbb{R}^n$ is also bijective since $f(0) = 0$, hence the global inverse $f^{-1} : \mathbb{R}^n \to \mathbb{R}^n$ exists and is continuous by Lemma 3.4. Since $f : \mathbb{R}^n \to \mathbb{R}^n$ is also continuous, it is a homeomorphism. □

Acknowledgments The authors would like to thank Professor Adi Adimurthi for valuable remarks on the first version of our manuscript, leading to its considerable improvement.

References

1. Asada, K., Fujiwara, D.: On some oscillatory integral transformations in $L^2(R^n)$. Jpn. J. Math. **4**(2), 299–361 (1978)
2. Gordon, W.B.: On the diffeomorphisms of Euclidean space. Am. Math. Mon. **79**, 755–759 (1972)
3. Hadamard, J.: Sur les correspondances ponctuelles. Paris: Oeuvres, Editions du Centre Nationale de la Researche Scientifique, pp. 383–384 (1998)
4. Hadamard, J.: Sur les transformations planes. C. R. Math. Acad. Sci. Paris **142**, 74 (1906)
5. Hadamard, J.: Sur les transformations ponctuelles. Bull. Soc. Math. Fr. **34**, 71–84 (1906)
6. Nijenhuis, A., Richardson Jr, R.W.: A theorem on maps with non-negative Jacobians. Mich. Math. J. **9**, 173–176 (1962)
7. Ruzhansky, M., Sugimoto, M.: Global L^2-boundedness theorems for a class of Fourier integral operators. Commun. Partial Differ. Equ. **31**(4–6), 547–569 (2006)
8. Ruzhansky, M., Sugimoto, M.: A smoothing property of Schrödinger equations in the critical case. Math. Ann. **335**(3), 645–673 (2006)

7

Indecomposable decomposition and couniserial dimension

A. Ghorbani · S. K. Jain · Z. Nazemian

Abstract Dimensions like Gelfand, Krull, Goldie have an intrinsic role in the study of theory of rings and modules. They provide useful technical tools for studying their structure. We define and study couniserial dimension for modules. Couniserial dimension is a measure of how far a module deviates from being uniform. Despite their different objectives, it turns out that there are certain common properties between the couniserial dimension and Krull dimension. Among others, each module having such a dimension contains a uniform submodule and has finite uniform dimension. Like all dimensions, this is an ordinal valued invariant. Every module of finite length has couniserial dimension and its value lies between the uniform dimension and the length of the module. Modules with countable couniserial dimension are shown to possess indecomposable decomposition. In particular, a von Neumann regular ring with countable couniserial dimension is semisimple artinian. If the maximal right

---

Communicated by Efim Zelmanov.

This paper was presented at the OSU-Denison conference, May 10, 2014 and at the University of California, San Diego, June 21, 2014.

A. Ghorbani (✉) · Z. Nazemian
Department of Mathematical Sciences, Isfahan University of Technology,
P.O.Box: 84156-83111, Isfahan, Iran
e-mail: a_ghorbani@cc.iut.ac.ir

Z. Nazemian
e-mail: z.nazemian@math.iut.ac.ir

S. K. Jain
Department of Mathematics, Ohio University, Athens, OH 45701, USA
e-mail: jain@ohio.edu

S. K. Jain
King Abdulaziz University, Jeddah, Saudi Arabia

quotient ring of a semiprime right non-singular ring R has a couniserial dimension as an R-module, then R is a semiprime right Goldie ring. As one of the applications, it follows that all right R-modules have couniserial dimension if and only if R is a semisimple artinian ring.

Keywords Uniform module · Maximal right quotient ring · Indecomposable decomposition · Uniserial dimension · Couniserial dimension · Von Neumann regular ring · Semisimple module

Mathematics Subject Classification Primary 16D70 · 16D90 · 16P70 ; Secondary 03E10 · 13E10

1 Introduction

In this article we introduce a notion of dimension of a module, to be called couniserial dimension. It is an ordinal valued invariant that is in some sense a measure of how far a module is from being uniform. In order to define couniserial dimension for modules over a ring R, we first define, by transfinite induction, classes ζ_α of R -modules for all ordinals $\alpha \geq 1$. First we remark that if a module M is isomorphic to all its non-zero submodules, then M must be uniform. To start with, let ζ_1 be the class of all uniform modules. Next, consider an ordinal $\alpha > 1$; if ζ_β has been defined for all ordinals $\beta < \alpha$, let ζ_α be the class of those R-modules M such that for every non-zero submodule N of M, where $N \ncong M$, we have $N \in \bigcup_{\beta<\alpha} \zeta_\beta$. If an R-module M belongs to some ζ_α, then the least such α is called the *couniserial dimension* of M, denoted by c.u.dim(M). For $M = 0$, we define c.u.dim$(M) = 0$. If a non-zero module M does not belong to any ζ_α, then we say that c.u.dim(M) is not defined, or that M has no couniserial dimension. Equivalently, Proposition 3.3 shows that an R-module M has couniserial dimension if and only if for each descending chain of submodules of M, $M_1 \geq M_2 \geq \cdots$, there exists $n \geq 1$, either M_n is uniform or $M_n \cong M_k$ for all $k \geq n$. It is clear by the definition that every submodule and so every summand of a module with couniserial dimension has couniserial dimension. Also note that, for the integer number n, couniserial dimension of $\mathbb{Z}^n$ is n. An example is given to show that the direct sum of two modules each with couniserial dimension (even copies of a module) need not have couniserial dimension. In Sect. 3, we prove some basic properties of the couniserial dimension. In Sect. 4, we prove our main results. It is shown in Theorem 4.3 that a module of countable (finite or infinite) couniserial dimension can be decomposed into indecomposable modules. Theorem 4.5 shows that a Dedekind finite module with couniserial dimension is a finite direct sum of indecomposable modules. Theorem 4.10 in Sect. 4 shows that for a semiprime right non-singular ring R with maximal right quotient ring Q, if Q_R has couniserial dimension, then R is a semiprime right Goldie ring which is a finite product of piecewise domains. The reader may compare this with the well known result that a prime ring with Krull dimension is a right Goldie ring but need not be a piecewise domain. Furthermore, a prime right Goldie ring need not have couniserial dimension as is also the case for Krull dimension.

In Sect. 5, we give some applications of couniserial dimension. It is shown in Proposition 5.2 that a module M with finite length is semisimple if and only if for every submodule N of M the right R-module $\oplus_{i=1}^{\infty} M/N$ has couniserial dimension. As a consequence a commutative noetherian ring R is semisimple if and only if for every finite length module M the module $\oplus_{i=1}^{\infty} M$ has couniserial dimension. It is shown in Proposition 5.4 that if P is an anti-coHopfian projective right R-module and $\oplus_{i=1}^{\infty} E(P)$ has couniserial dimension, then P is injective. As another application we show that all right (left) R-module have couniserial dimension if and only if R is semisimple artinian (see Theorem 5.8). Several examples are included in the paper that demonstrates as to why the conditions imposed are necessary.

2 Definitions and notations

Recall that a semisimple module M is said to be *homogeneous* if M is a direct sum of pairwise isomorphic simple submodules. A module M has *finite uniform dimension* (or *finite Goldie rank*) if M contains no infinite direct sum of non-zero submodules, or equivalently, there exist independent uniform submodules $U_1, \ldots, U_n$ in M such that $\oplus_{i=1}^{n} U_i$ is an essential submodule of M. Note that n is uniquely determined by M. In this case, it is written $\text{u.dim}(M) = n$.

For any module M, we define $Z(M) = \{x \in M : \text{ann}(x)$ is an essential right ideal of $R\}$. It can be easily checked that $Z(M)$ is a submodule of M. If $Z(M) = 0$, then M is called a *non-singular* module. In particular, if we take $M = R_R$, then R is called right non-singular if $Z(R_R) = 0$.

A ring R is called a *right Goldie* ring if it satisfies the following two conditions: (i) R has ascending chain condition on right annihilator ideals and, (ii) $\text{u.dim}(R_R)$ is finite. Recall that a ring R is *right V-ring* if all right simple R-modules are injective. A ring R is called *fully right idempotent* if $I = I^2$, for every right ideal I. We recall that a right V-ring is fully right idempotent (see [19, Corollary 2.2]) and a prime fully right idempotent ring is right non-singular (see [2, Lemma 4.3]). So a prime right V-ring is right non-singular. Recall that a module M is called Σ*-injective* if every direct sum of copies of M is injective. A ring R is called *right* Σ*-V-ring* if each simple right module is Σ-injective.

In this paper, for a ring R, $Q = Q_{max}(R)$ stands by the maximal right quotient ring R. It is well known that if R is a right non-singular, then the injective hull of R_R, $E(R_R)$, is a ring and is equal to the maximal right quotient ring of R [9, Corollary 2.31].

A module M is called *Hopfian* if M is not isomorphic to any of its proper factor modules (equivalently, every onto endomorphism of M is 1-1). *Anti-Hopfian* modules are introduced by Hirano and Mogami [13]. Such modules are isomorphic to all their non-zero factor modules. A module M is called uniserial if the lattice of submodules are linearly ordered. Anti-Hopfian modules are uniserial artinian.

Recall that a module M is called *coHopfian* if it is not isomorphic to a proper submodule (equivalently, every 1-1 endomorphism of M is onto). Varadarjan [22] dualized the concept of anti-Hopfian module and called it anti-coHopfian module. With slight modification we will call a non-zero module to be *anti-coHopfian* if it is isomorphic to all its non-zero submodules. A non-zero module M is called *uniform* if the intersection of any two non-zero submodules is non-zero. We see an anti-coHopfian module is noetherian and uniform (see Lemma 3.1 below).

An R-module M has cancellation property if for every R-modules N and T, $M \oplus N \cong M \oplus T$ implies $N \cong T$. Every module with semilocal endomorphism ring has cancellation property [16]. Since the endomorphism ring of a simple module is a division ring, it has cancellation property.

Unless otherwise specified, throughout this paper R denotes an arbitrary ring and all modules are assumed to be unitary right modules. If N is a submodule (resp. proper submodule) of M we write $N \leq M$ (resp. $N < M$). Also, for a module M, $\oplus_{i=1}^{\infty} M$ stands for countably infinite direct sum of copies of M. If N is a submodule of M and $k > 1$, $\oplus_{i=k}^{\infty} N = \oplus_{i=1}^{\infty} N_i$ is a submodule of $\oplus_{i=1}^{\infty} M$ with $N_1 = N_2 = \cdots = N_{k-1} = 0$ and for $i \geq k$ $N_i = N$.

3 Basic and preliminary results

As defined in the introduction, couniserial dimension is an ordinal valued number. The reader may refer to [21] regarding ordinal numbers. We begin this section with a lemma and a remark on the definition of couniserial dimension.

Lemma 3.1 *An anti-coHopfian module is uniform noetherian.*

Proof Since M is isomorphic to each cyclic submodules, M is cyclic and every submodule of M is cyclic and so M is noetherian. Thus M has a uniform submodule, say U. Since $U \cong M$, M is uniform. □

Remark 3.2 We make the convention that a statement "c.u.dim$(M) = \alpha$" will mean that the couniserial dimension of M exists and equals α. By the definition of couniserial dimension, if M has couniserial dimension and N is a submodule of M, then N has couniserial dimension and c.u.dim$(N) \leq$ c.u.dim(M). Moreover, if M is not uniform and c.u.dim$(M) =$ c.u.dim(N), where N is a submodule of M, then $M \cong N$. On the other hand, since every set of ordinal numbers has supremum, it follows immediately from the definition that M has couniserial dimension if and only if for all submodules N of M with $N \ncong M$, c.u.dim(N) is defined. In the latter case, if $\alpha = \sup\{$ c.u.dim$(N) \mid N \leq M, N \ncong M\}$, then c.u.dim$(M) \leq \alpha + 1$.

The next proposition provides a working criterion for a module to have couniserial dimension.

Proposition 3.3 *An R-module M has couniserial dimension if and only if for every descending chain of submodules $M_1 \geq M_2 \geq \cdots$, there exists $n \geq 1$ such that M_n is uniform or $M_n \cong M_k$ for all $k \geq n$.*

Proof (⇒) Let $M_1 \geq M_2 \geq \cdots$ be a descending chain of submodules of M. Put $\gamma = \inf\{$c.u.dim$(M_n) \mid n \geq 1\}$. So $\gamma =$ c.u.dim (M_n) for some $n \geq 1$. If M_n is not uniform, then $M_n \cong M_k$ for all $k \geq n$, because γ is infimum.

(⇐) If M does not have couniserial dimension, then M is not uniform and so there exists a submodule M_1 of M such that $M_1 \ncong M$ and M_1 does not have couniserial dimension, by the above remark. So there exists a submodule M_2 of M_1 such that $M_2 \not\cong M_1$ and M_2 does not have couniserial dimension. Continuing in this manner, we obtain a descending chain of submodules $M_1 \geq M_2 \geq \cdots$, such that for every $i \geq 1$, M_i does not have couniserial dimension and $M_i \not\cong M_{i+1}$, a contradiction. This completes the proof. □

As a consequence, we have the following corollary.

Corollary 3.4 *Every artinian module has couniserial dimension.*

Lemma 3.5 *If M is an R-module and* c.u.dim$(M) = \alpha$, *then for any* $0 \leq \beta \leq \alpha$, *there exists a submodule N of M such that* c.u.dim$(N) = \beta$.

Proof The proof is by transfinite induction on c.u.dim$(M) = \alpha$. The case $\alpha = 1$ is clear. Let $\alpha > 1$ and $0 \leq \beta < \alpha$, then, using Remark 3.2, there exists a submodule K of M such that $K \not\cong M$ and $\beta \leq$ c.u.dim(K). Now since $\beta \leq$ c.u.dim$(K) < \alpha$, by induction hypothesis, there exists a submodule N of K such that c.u.dim$(N) = \beta$. □

As a consequence we have the following.

Lemma 3.6 *Every module with couniserial dimension has a uniform submodule.*

In the next proposition we observe that every module of finite couniserial dimension has finite uniform dimension.

Lemma 3.7 *Let M be an R-module of finite couniserial dimension. Then M has finite uniform dimension and* u.dim$(M) \leq$ c.u.dim(M).

Proof The proof is by induction on c.u.dim$(M) = n$. The case $n = 1$ is clear. Let $n > 1$ and N be a submodule of M such that c.u.dim$(N) = n - 1$. Thus by the inductive hypothesis, N has finite uniform dimension. Put $m =$ u.dim(N). If N is not essential in M, then there exists a uniform submodule U of M such that $N \cap U = 0$. Thus $N \oplus U$ is a submodule of M of uniform dimension $m + 1$. Then $(N \oplus U) \not\cong N$ and so $n - 1 <$ c.u.dim$(N \oplus U) \leq n$. Thus $(N \oplus U) \cong M$, by Remark 3.2. This proves the lemma. □

Example 3.8 There exist modules of infinite couniserial dimension but of finite uniform dimension. Take $M = \mathbb{Z}_{p^\infty} \oplus \mathbb{Z}_{p^\infty}$. Then M is an artinian $\mathbb{Z}$-module of infinite couniserial dimension but of finite uniform dimension 2.

In the following, we show that, in a special case, equality holds in the previous lemma.

Lemma 3.9 *Let M be an injective non-uniform R-module of finite couniserial dimension. Then* c.u.dim$(M) =$ u.dim(M) *if and only if M is a finitely generated semisimple module.*

Proof ($\Leftarrow$) is clear. ($\Rightarrow$). Let c.u.dim$(M) =$ u.dim$(M) = m > 1$. Then $M = E_1 \oplus \cdots \oplus E_m$, where E_i are uniform injective modules. If E_1 is not simple then there exists a non-injective submodule K of E_1. Thus $K \oplus E_2 \oplus \cdots \oplus E_m$ is not isomorphic to M. But clearly c.u.dim$(K \oplus E_2 \oplus \cdots \oplus E_m) \geq m$, a contradiction. This completes the proof. □

Note that the condition of being injective is necessary in the above proposition.

Example 3.10 We can see easily that for $M = \mathbb{Z} \oplus \mathbb{Z}$, c.u.dim$(M)$ = u.dim$(M) = 2$ but M is not semisimple. Also, the next lemma shows that there exists a module of finite uniform dimension without couniserial dimension.

The following lemma shows that the direct sum of two uniform modules may not have couniserial dimension.

Lemma 3.11 *Let D be a domain and S be a simple D-module. If $S \oplus D$ as D-module has couniserial dimension, then D is a principal right ideal domain.*

Proof Let I be a non-cyclic right ideal of D. Choose a non-zero element $x \in I$. Set $J_1 = xR$ which is isomorphic to D. Thus there exists a right ideal J_2 of D such that $J_2 \cong I$ and $J_2 \leq J_1$. Now let J_3 be a cyclic right ideal contained in J_2 and by continuing this manner we have a descending chain $J_1 \geq J_2 \geq \cdots$ of right ideals of D where for each odd integer i, J_i is cyclic and for each even integer i, J_i is not cyclic. Now consider the descending chain $S \oplus J_1 \geq S \oplus J_2 \geq \cdots$ of submodules of $S \oplus D$. Since S has cancellation property and for each i, $S \oplus J_i$ is not uniform, by using Proposition 3.3, we see that, for some n, $J_n \cong J_{n+1}$, a contradiction. Thus D is a principal right ideal domain. □

Remark 3.12 (1) The simple module S in the statement of the Lemma 3.11 can be replaced by any cancellable module. Indeed it follows from Theorem 4.10, proved later that, if the maximal right quotient ring Q of a domain D has couniserial dimension as D-module, then Q_D has cancellation property and so if $Q \oplus D$ as D-module has couniserial dimension, D must be right principal ideal domain.

(2) Also, since a Dedekind domain has cancellation property, similar proof shows that if D is a Dedekind domain which is not right principal ideal domain, then $D \oplus D$ does not have couniserial dimension. This example shows that even direct sum of a uniform module with itself may not have couniserial dimension.

The definition of addition two ordinal numbers can be given inductively. If α and β are two ordinal numbers then $\alpha + 0 = \alpha$, $\alpha + (\beta + 1) = (\alpha + \beta) + 1$ and if γ is a limit ordinal then $\alpha + \gamma$ is the limit of $\alpha + \beta$ for all $\beta < \gamma$ (see [21]).

Lemma 3.13 (See [21, Theorem 7.10]) *For ordinal numbers α, β and γ, we have the following:*

(1) *If $\alpha < \beta$, then $\gamma + \alpha < \gamma + \beta$.*
(2) *If $\alpha < \beta$, then $\alpha + \gamma \leq \beta + \gamma$.*

We call an R-module M *fully coHopfian* if every submodule of M is coHopfian. Note that artinian modules are fully coHopfian. If I is the set of prime numbers, then $\oplus_{p \in I} \mathbb{Z}_p$ is an example of fully coHopfian $\mathbb{Z}$-module that it is not artinian.

Proposition 3.14 *Let $M = M_1 \oplus M_2$ be a fully coHopfian R-module with couniserial dimension. Then c.u.dim$(M) \geq$ c.u.dim(M_1)+ c.u.dim(M_2).*

Proof We may assume $M_1, M_2 \neq 0$. We use transfinite induction on c.u.dim$(M_2) = \alpha$. Since $M_1 \ncong M$, c.u.dim$(M) \geq$ c.u.dim$(M_1) + 1$. So the case $\alpha = 1$ is clear. Thus, suppose $\alpha > 1$ and for every right R-module L of couniserial dimension less than α, c.u.dim$(M_1 \oplus L) \geq$ c.u.dim$(M_1)+$ c.u.dim(L). There are two possibilities: α is a successor ordinal, or a limit ordinal. Suppose first that α is a successor ordinal, then there exists an ordinal number γ such that c.u.dim$(M_2) = \gamma + 1$. Using Lemma 3.5, there exists a non-zero submodule K of M_2 such that c.u.dim$(K) = \gamma < \alpha$. So by the induction hypothesis

$$\text{c.u.dim}(M_1) + \gamma = \text{c.u.dim}(M_1) + \text{c.u.dim}(K) \leq \text{c.u.dim}(M_1 \oplus K).$$

Using our assumption and Remark 3.2, we have c.u.dim$(M_1 \oplus K) <$ c.u.dim(M) and hence c.u.dim$(M_1)+$ c.u.dim$(M_2) \leq$ c.u.dim(M).

If α is a limit ordinal and $1 \leq \beta < \alpha$, then by Remark 3.2, there exists a non-zero submodule K of M_2 such that $\beta \leq$ c.u.dim(K). Then by induction hypothesis c.u.dim$(M_1)+\beta \leq$ c.u.dim$(M_1)+$c.u.dim$(K) \leq$ c.u.dim$(M_1 \oplus K) <$ c.u.dim(M). Therefore c.u.dim$(M_1) + \alpha = \sup\{\text{c.u.dim}(M_1) + \beta \mid \beta < \alpha\} \leq$ c.u.dim(M). □

The condition fully coHopfian of Proposition 3.14 is necessary.

Example 3.15 For the $\mathbb{Z}$-modules $M = \oplus_{i=1}^{\infty}\mathbb{Z}_p$, and $L = \mathbb{Z}_p$, we have $M \cong M \oplus L$. One can see c.u.dim$(M) = \omega$ and so c.u.dim$(M) \ngeq$ c.u.dim$(M)+$ c.u.dim(L). Also, in general, we don't have the equality in Proposition 3.14. Consider the $\mathbb{Z}$-module $M = \mathbb{Z}_2 \oplus \mathbb{Z}_4$. Then, M is fully coHopfian and $3 =$ c.u.dim$(M) >$ c.u.dim$(\mathbb{Z}_2)+$ c.u.dim$(\mathbb{Z}_4)$.

Here we prove another result on fully coHopfian module:

Proposition 3.16 *Let M be an R-module and N be a cancellable module (for example a simple module) such that $N \oplus M$ has couniserial dimension. If M is fully coHopfian, then M is artinian.*

Proof Let M be fully coHopfian and let $M_1 \geq M_2 \geq \cdots$ be a descending chain of submodules of M. Then $N \oplus M_1 \geq N \oplus M_2 \geq \cdots$ is a descending chain of submodules of $N \oplus M$ and so for some n, $M_i \cong M_n$ for each $i \geq n$. Now since M is fully coHopfian, we have $M_i = M_n$ for each $i \geq n$. □

Let us recall the definition of uniserial dimension [20].

Definition 3.17 In order to define uniserial dimension for modules over a ring R, we first define, by transfinite induction, classes ζ_α of R-modules for all ordinals $\alpha \geq 1$. To start with, let ζ_1 be the class of non-zero uniserial modules. Next, consider an ordinal $\alpha > 1$; if ζ_β has been defined for all ordinals $\beta < \alpha$, let ζ_α be the class of those R-modules M such that, for every submodule $N < M$, where $M/N \ncong M$, we have $M/N \in \bigcup_{\beta<\alpha} \zeta_\beta$. If an R-module M belongs to some ζ_α, then the least such α is the *uniserial dimension of* M, denoted u.s.dim(M). For $M = 0$, we define u.s.dim$(M) = 0$. If M is non-zero and M does not belong to any ζ_α, then we say that "u.s.dim(M) is not defined," or that "M has no uniserial dimension."

Remark 3.18 Note that, in general, there is no relation between the existence of the uniserial dimension and the existence of the couniserial dimension of a module. For example, the polynomial ring in infinite number of commutative indeterminates over a field k, $R = k[x_1, x_2, \dots]$ has this property that $\text{c.u.dim}(R_R) = 1$ but R_R does not have uniserial dimension (see [6, Remark 2.3]). It follows by the definition that a semisimple module M has uniserial dimension if and only if M has couniserial dimension, in which case $\text{u.s.dim}(M) = \alpha$ if and only if $\text{c.u.dim}(M) = \alpha$. Furthermore a semisimple module M has couniserial dimension if and only if M is a finite direct sum of homogeneous semisimple modules (see [20, Proposition 1.18]).

Using the above remark we have the following interesting results.

Corollary 3.19 *All right semisimple modules over a ring R have couniserial dimension if and only if there exist only finitely many non-isomorphic simple right R-modules.*

Lemma 3.20 *Let M be a finitely generated cancellable module. Then:*

(1) *If $\oplus_{i=1}^{\infty} E(M)$ has couniserial dimension, then M is injective.*
(2) *Suppose that M is simple. Then $\oplus_{i=1}^{\infty} E(M)$ has couniserial dimension if and only if M is injective.*

Proof (1) Consider the descending chain

$$M \oplus (\oplus_{i=2}^{\infty} E(M)) \geq M^{(2)} \oplus (\oplus_{i=3}^{\infty} E(M)) \geq \cdots$$

of submodules of $\oplus_{i=1}^{\infty} E(M)$ where $M^{(n)} = \oplus_{i=1}^{\infty} M_i$ with $M_1 = \cdots = M_n = M$ and for each $i > n$, $M_i = 0$. Then, by Proposition 3.3, there exists $n \geq 1$ such that

$$M^{(n)} \oplus (\oplus_{i=n+1}^{\infty} E(M)) \cong M^{(n+1)} \oplus (\oplus_{i=n+2}^{\infty} E(M))$$

and so $\oplus_{i=n+1}^{\infty} E(M) \cong M \oplus (\oplus_{i=n+2}^{\infty} E(M))$, because M is cancellable . Since M is finitely generated, there exists a right module L such that for some k, $E(M)^k \cong M \oplus L$. This shows M is injective.

(2) It is clear by part 1 and the statement in Remark 3.18.

□

4 Main results

In this section we use our basic results to prove the main results.

Proposition 4.1 *Let M_R be an injective module and N_R be a cancellable module over a commmutative ring R (for example a simple module) such that $N \oplus M$ has couniserial dimension. Then M is Σ-injective.*

Proof According to [15, Theorem 6.17], it is enough to show that R satisfies the ascending chain condition on ideals of R that are annihilators of subsets of M. Let

$I_1 \leq I_2 \leq \cdots$ be a chain of such annihilator ideals. Then for each i, $M_i = \mathrm{ann}_M(I_i)$ is a submodule of M and so we have descending chain $N \oplus M_1 \geq N \oplus M_2 \geq \cdots$ of submodules of $N \oplus M$. Then there exists a positive integer n such that $M_n \cong M_i$ for all $i \geq n$. Thus $\mathrm{ann}(M_i) = \mathrm{ann}(M_n)$. Therefore for each $i \geq n$, $I_i = I_n$. □

Remark 4.2 If S is a simple module, then $S \oplus S$ has couniserial dimension and S is also a cancellable module. Thus, the above result provides another proof for the fact that commutative V-rings (i.e, von Neumann regular rings) are Σ-V-ring. Now if R is a right V-ring that is not right Σ-V-ring, see [15, Example, page 60], then there exists a simple right R-module S which is not Σ-injective. This shows that the commutativity condition cannot be dropped in the above proposition.

The next result shows that if a module has countable couniserial dimension then it can be decomposed into indecomposable modules.

Theorem 4.3 *For an R-module M, if* $\mathrm{c.u.dim}(M) \leq \omega$, *then M has indecomposable decomposition.*

Proof The proof is by induction on $\mathrm{c.u.dim}(M) = \alpha$. The case $\alpha = 1$ is clear. If $\alpha > 1$ and M is not indecomposable, then $M = N_1 \oplus N_2$, where N_1 and N_2 are non-zero submodules of M. If $\mathrm{c.u.dim}(N_i) < \mathrm{c.u.dim}(M)$, $i = 1, 2$, then by induction hypothesis M has indecomposable decomposition. If not, for definiteness let $\mathrm{c.u.dim}(N_1) = \mathrm{c.u.dim}(M)$. Then $M \cong N_1$, by Remark 3.2. Thus it contains an infinite direct sum of uniform modules, say $\oplus_{i=1}^{\infty} K_i$. Clearly, $\mathrm{c.u.dim}(\oplus_{i=1}^{\infty} K_i) \geq \omega$. Thus we have $M \cong \oplus_{i=1}^{\infty} K_i$. □

Remark 4.4 We do not know whether the above proposition holds for a module of arbitrary couniserial dimension. For infinite countable couniserial dimension one can show under some condition that the module can be represented as a direct sum of uniform modules.

Recall that a module M is called *Dedekind finite* if M is not isomorphic to any proper direct summand of itself. Clearly, every direct summand of a Dedekind finite module is a Dedekind finite module. Obviously, a Hopfian module is Dedekind finite. Since all finitely generated modules over a commutative ring are Hopfian (see [10]), they provide examples of Dedekind finite modules.

Theorem 4.5 *If M is a Dedekind finite module with couniserial dimension, then M has finite indecomposable decomposition.*

Proof The proof is by induction on $\mathrm{c.u.dim}(M) = \alpha$. The case $\alpha = 1$ is clear. Let $\alpha > 1$ and every Dedekind finite module with c.u.dim less than α be decomposed to finitely many indecomposable modules. If M is not indecomposable, then $M = M_1 \oplus M_2$. Since $M_i \ncong M$, using Remark 3.2, $\mathrm{c.u.dim}(M_i) < \mathrm{c.u.dim}(M)$ and so, by induction hypothesis, M_i have finite indecomposable decomposition. This completes the proof. □

A ring R is called a *von Neumann regular ring* if for each $x \in R$, there exists $y \in R$ such that $xyx = x$, equivalently, every principal right ideal is a direct summand. R is *unit regular ring* if for each $x \in R$, there exists a unit element $u \in R$ such that $x = xux$. As a consequence of the above theorem we have the following corollary.

Corollary 4.6 *Every Dedekind finite von Neumann regular ring (in particular, unit regular rings) with couniserial dimension is semisimple artinian.*

A ring R is called a PWD (*piecewise domain*) if it possesses a complete set $\{e_i | 0 \leq i \leq n\}$ of orthogonal idempotents such that $xy = 0$ implies $x = 0$ or $y = 0$ whenever $x \in e_i R e_k$ and $y \in e_k R e_j$. Note that the definition is left-right symmetric and all $e_i R e_i$ are domain, see [12].

An element x of R is called regular if its right and left annihilators are zero.

Proposition 4.7 *Let R be a semiprime right Goldie ring with couniserial dimension. If u.dim$(R_R) = n$, then R has a decomposition into n uniform modules. In particular, it is a piecewise domain.*

Proof We can assume that $n > 1$. Let $I_1 = U_1 \oplus \cdots \oplus U_n$ be an essential right ideal of R. Then, by [8, Proposition 6.13], I_1 contains a regular element x and thus $J_1 = xR$ is a right ideal of R which is R-isomorphic to R. So u.dim$(J_1) = n$ and it contains an essential right ideal I_2 of R such that it is a direct sum of n uniform right ideals. By continuing in this manner we obtain a descending chain $I_1 \geq J_1 \geq I_2 \geq \cdots$ of right ideals of R such that I_i are direct sum of n uniform right ideals and J_i are isomorphic to R. Since R has couniserial dimension, for some n, $I_n \cong R$. The last statement follows from [12, Pages 2–3]. This completes the proof. □

Remark 4.8 There exists an example of simple noetherian ring of uniform dimension 2 which has no non-trivial idempotents (c.f. [18, Example 7.16, page 441]). So by the above proposition this provides an example of prime right Goldie ring without couniserial dimension.

Lemma 4.9 *Let R be a right non-singular ring with maximal right quotient ring Q. Let M be a Q-module. If M is non-singular R-module, such that M_R has couniserial dimension, then M_Q has couniserial dimension.*

Proof Let $M \geq M_1 \geq M_2 \geq \cdots$ be a descending chain of Q-submodules of M. So it is a descending chain of R-submodules of M and thus, for some n, M_n is uniform R-module or $M_n \cong M_i$ as R-modules for all $i \geq n$. If M_n is uniform R-module, then it is also uniform Q-module. So let $M_n \cong M_i$ as R-modules and let φ_i be this isomorphism. If $q \in Q$ and $t \in M_n$ there exists an essential right ideal E of R such that $qE \leq R$. So $\varphi_i(tqE) = \varphi_i(tq)E$ and also $\varphi_i(tqE) = \varphi_i(t)qE$. Then $\varphi_i(tq)E = \varphi_i(t)qE$. Since Q is right non-singular, $\varphi_i(tq) = \varphi_i(t)q$. Thus φ_i is a Q-isomorphism. This completes the proof.

A ring R is semiprime (prime) right Goldie ring if and only if its classical right quotient ring is semisimple (simple) artinian ring [9, Theorems 3.35 and 3.36]. Semiprime right Goldie rings are non-singular. A semiprime right non-singular ring R is semiprime right Goldie ring if and only if u.dim(R_R) is finite, [9, Theorem 3.17]. Recall that a *right full linear ring* is the ring of all linear transformations (written on the left) of a right vector space over a division ring. If the dimension of the vector space is finite, a right full linear ring is exactly a simple artinian ring.

Theorem 4.10 *Let R be a semiprime right non-singular ring with maximal right quotient ring, Q. If Q as an R-module has couniserial dimension, then R is a semiprime right Goldie ring which is a finite product of prime Goldie rings, each of which is a piecewise domain.*

Proof It is enough to show that R has finite uniform dimension. Since Q_R has couniserial dimension, R_R has couniserial dimension and so every right ideal of R has couniserial dimension. Thus Lemma 3.6 implies that every right ideal contains a uniform submodule. Now by [9, Theorem 3.29] the maximal right quotient ring of R is a product of right full linear rings, say $Q = \prod_{i\in I} Q_i$, where Q_i are right full linear rings. Note that since R_R is right non-singular, Q_R is also non-singular and so, using Lemma 4.9, Q_Q has couniserial dimension. At first we claim each Q_i is endomorphism ring of a finite dimensional vector space. Assume the contrary. Then Q_j is the endomorphism ring of an infinite dimensional vector space, for some j. Thus $Q_j \cong Q_j \times Q_j$ and so if $\iota : Q_j \longrightarrow Q$ be the canonical embedding, then $\iota(Q_j)$ is a right ideal of Q and there exists a Q-isomorphism $Q \cong \iota(Q_j) \times Q$. Then there exist right ideals T_1 and T of Q such that $Q = T_1 \oplus T$, T_1 and Q are isomorphic as Q-modules and $T \cong \iota(Q_j)$ as Q-module. Because Q_j is the endomorphism ring of an infinite dimensional vector space, it has a right ideal which is not principal, for example its socle. So $\iota(Q_j)$ and thus T contains a non-cyclic right ideal of Q and thus since $T \cong Q/T_1$, there exists a non-cyclic right ideal of Q, say K_1 such that $Q \geq K_1 \geq T_1$. Now T_1 is isomorphic to Q. So we can have a descending chain $Q > K_1 > T_1 > K_2 > T_2 > \cdots$ of right ideals of Q such that T_i are cyclic but K_i are not cyclic. This is a contradiction. So all Q_i are endomorphism ring of finite dimensional vector spaces. Now to show R is semiprime right Goldie ring it is enough to show that the index set I is finite. If I is infinite, there exist infinite subsets I_1 and I_2 of I such that $I = I_1 \cup I_2$. and $I_1 \cap I_2$ is empty. Let $T_1 = \prod_{i\in I} N_i$ such that $N_i = Q_i$ for all $i \in I_1$ and $N_i = 0$ for all $i \in I_2$. Similarly let $T = \prod_{i\in I} M_i$ such that $M_i = Q_i$ for all $i \in I_2$ and $M_i = 0$ for all $i \in I_1$. Then T_1 and T are right ideals of Q and $Q = T_1 \oplus T$. T contains a right ideal of Q which is not cyclic, for example $\oplus_{i\in I} M_i$. Since $T \cong Q/T_1$, there exists a non-cyclic right ideal K_1 of Q such that $Q \geq K_1 \geq T_1$. Note that T_1 is a cyclic Q-module and because I_1 is infinite, the structure of T_1 is similar to that of Q. We can continue in this manner and find a descending chain of right ideals of Q such that K_i are non cyclic Q-modules and T_i are cyclic Q modules, which is a contradiction. Therefore I is finite and R must have finite uniform dimension. This shows R is semiprime right Goldie ring and so Proposition 4.7 and [12, Corollary 3] imply that it is a direct sum of prime right Goldie rings. □

We may point out that unlike a semiprime ring with right Krull dimension, a semiprime ring with couniserial dimension need not be a right Goldie ring. See Dubrovin [5] that contains an example of a primitive uniserial ring with non-zero nilpotent elements.

Next we show that the converse of the above theorem is not true, in general. In fact we show that there exists a prime right Goldie ring R such that $\text{c.u.dim}(R_R) = 2$ and Q_R does not have couniserial dimension. We need the following lemma to give the example.

Lemma 4.11 *For an ordinal number α, being of couniserial dimension α is a Morita invariant property for modules.*

Proof This is clear by the definition of couniserial dimension and [1, Proposition 21.7]. □

Example 4.12 Here we give an example of a prime right Goldie ring R with maximal right quotient ring Q such that Q_R does not have couniserial dimension. Take $R = M_2(\mathbb{Z})$, the 2×2 matrix ring over $\mathbb{Z}$. Then R is a prime right Goldie ring with maximal right quotient ring $Q = M_2(\mathbb{Q})$. Note that under the standard Morita equivalent between the ring $\mathbb{Z}$ and $R = M_2(\mathbb{Z})$, see [17, Theorem 17.20], R corresponds to $\mathbb{Z}\oplus\mathbb{Z}$ and so using the above lemma R has couniserial dimension 2. If $\{p_i | i \geq 1\}$ is the set of all prime numbers, then $\mathbb{Q}/\mathbb{Z} = \sum_{i=1}^{\infty} K_i/\mathbb{Z}$, where $K_i = \{m/p_i^n | n \geq 0$ and $m \in \mathbb{Z}\}$. Then take $Q_n = \sum_{i=n}^{\infty} K_i$. Then $M_2(Q_1) \geq M_2(Q_2) \geq \cdots$ is a descending chain of R-submodules of Q which are not uniform R-modules. Assume that for some n, $M_2(Q_n) \cong M_2(Q_{n+1})$ with an R-isomorphism ϕ. Let $\phi\begin{pmatrix} 1 & 0 \\ 0 & 1 \end{pmatrix} = \begin{pmatrix} m_1/t_1 & m_2/t_2 \\ m_3/t_3 & m_4/t_4 \end{pmatrix}$, where $m_i/t_i \in Q_{n+1}$. Suppose that $j \geq 1$ and $\phi\begin{pmatrix} 1/p_n^j & 0 \\ 0 & 1/p_n^j \end{pmatrix} = \begin{pmatrix} m_{1,j}/t_{1,j} & m_{2,j}/t_{2,j} \\ m_{3,j}/t_{3,j} & m_{4,j}/t_{4,j} \end{pmatrix}$, where p_n does not odd non of $t_{i,j}$ for all $1 \leq i \leq 4$. Then since ϕ is additive, we can easily see that $\begin{pmatrix} m_{1,j}p_n^j/t_{1,j} & m_{2,j}p_n^j/t_{2,j} \\ m_{3,j}p_n^j/t_{3,j} & m_{4,j}p_n^j/t_{4,j} \end{pmatrix} = \begin{pmatrix} m_1/t_1 & m_2/t_2 \\ m_3/t_3 & m_4/t_4 \end{pmatrix}$ and this implies that $p_n^j | m_i$ for all $j \geq 1$ and $0 \leq i \leq 4$ and so $m_i = 0$, a contradiction. So Q_R does not have couniserial dimension.

5 Some applications

A right R-module M which has a composition series is called a module of *finite length*. A right R-module M is of finite length if and only if M is both artinian and noetherian. The length of a composition series of M_R is said to be the length of M_R and is denoted by length(M). Clearly, by Corollary 3.4, a module of finite length has couniserial dimension. The next result shows a relation between couniserial dimension of a finite length module M and length(M).

Proposition 5.1 *Let M be a right R-module of finite length. Then the following statements hold:*

(1) *If N is a submodule of M, then* c.u.dim$(M/N) \leq$ c.u.dim(M).
(2) c.u.dim$(M) \leq$ length(M).

Proof (1) The proof is by induction on n, where length$(M) = n$. The case $n = 1$ is clear. Now, let $n > 1$ and assume that the assertion is true for all modules with length less than n. If N is a non-zero submodule of M, then the length$(M/N) < n$. Thus for every proper submodule K/N of M/N, by induction, c.u.dim$(K/N) \leq$ c.u.dim$(K)<$c.u.dim(M). Now, Remark 3.2 implies that c.u.dim$(M/N) \leq$ c.u.dim(M).

(2) The proof is by induction on length(M) $= n$. The case $n = 1$ is clear. Now if $n > 1$ and K is a proper submodule of M, then, by assumption, c.u.dim(K) $\leq$ length(K) $<$ length(M). Thus by Remark 3.2, c.u.dim(M) $\leq$ length(M). □

Recall that an R-module M is called *co-semisimple* if every simple R-module is M-injective, or equivalently, $\mathrm{Rad}(M/N) = 0$ for every submodule $N \leq M$ (See [23, Theorem 23.1]). The next proposition gives a condition as to when a module of finite length is semisimple. It may be of interest to state that for the finite length $\mathbb{Z}$-module $\mathbb{Z}_4$, $\oplus_{i=1}^{\infty}\mathbb{Z}_4$ does not possess couniserial dimension.

Proposition 5.2 *Let M be a non-zero right R-module of finite length. Then M is a semisimple R-module if and only if for every submodule N of M the right R-module $\oplus_{i=1}^{\infty}M/N$ has couniserial dimension.*

Proof ($\Rightarrow$) c.f. Remark 3.18.

($\Leftarrow$) For every submodule N of M the right R-module $\oplus_{i=1}^{\infty}M/N$ has couniserial dimension. Clearly, this also holds for any factor module of M. We will proof the result by induction on the length(M) $= n$. The case $n = 1$ is clear. Now assume that $n > 1$ and the result is true for all modules of length less than n. Let K be a non-zero submodule of M. Since length(M/K) $< n$, by the inductive hypothesis, M/K is semisimple. Therefore, for every non-zero submodule K of M, $\mathrm{Rad}(M/K) = 0$. If $\mathrm{Rad}(M) = 0$, then M is co-semisimple. Let S be a simple submodule of M. Consider the exact sequence $0 \longrightarrow S \longrightarrow M \longrightarrow M/S \longrightarrow 0$ which splits, because M is co-semisimple. Therefore, M is semisimple. Next suppose that $\mathrm{Rad}(M) \neq 0$. Let S be a simple submodule of M. Because by the above $\mathrm{Rad}(M/S) = 0$, we obtain $\mathrm{Rad}(M) \leq S$. This implies $\mathrm{Rad}(M) = S$ and so M has only one simple submodule. Thus $\mathrm{Rad}(M) = \mathrm{soc}(M) = S$ is a simple module. Suppose that M is not semisimple. Let N be a maximal submodule of M. Then for every submodule $K \leq N < M$, $\oplus_{i=1}^{\infty}N/K$ is a submodule of $\oplus_{i=1}^{\infty}M/K$ and thus $\oplus_{i=1}^{\infty}N/K$ has couniserial dimension. Since length(N) $< n$, we conclude that N is semisimple. Thus $N = \mathrm{soc}(M) = \mathrm{Rad}(M)$ is a simple module and so M is of length 2.

Now consider the descending chain

$$N \oplus (\oplus_{i=2}^{\infty}M) > N^{(2)} \oplus (\oplus_{i=3}^{\infty}M) > \cdots$$

of submodules of $\oplus_{i=1}^{\infty}M$. Using Proposition 3.3, there exists $k \geq 1$ such that $N^{(k)} \oplus (\oplus_{i=k+1}^{\infty}M) \cong N^{(k+1)} \oplus (\oplus_{i=k+2}^{\infty}M)$. Since $N^{(k+1)}$ is finitely generated, there exist $m \geq 0$ and an R-module T, such that $N^{(k)} \oplus M^m \cong N^{(k+1)} \oplus T$. N is simple and so it has cancellation property and thus $M^m \cong N \oplus T$. This implies $\mathrm{Rad}(T)$ is semisimple of length m and length($\mathrm{soc}(T)$) $= m - 1$, a contradiction. □

Recall that a ring R is called *right bounded* if every essential right ideal contains a two-sided ideal which is essential as a right ideal. A ring R is called right *fully bounded* if every prime factor ring is right bounded. A right noetherian right fully bounded ring is commonly abbreviated as a right FBN ring. Clearly all commutative noetherian rings are example of right FBN rings. Finite matrix rings over commutative noetherian rings are a large class of right FBN rings which are not commutative. In

[14, Theorem 2.11], Hirano and et.al. showed that a right FBN ring R is semisimple if and only if every right module of finite length is semisimple. As a consequence of the above proposition we have:

Corollary 5.3 *A right FBN ring R is semisimple if and only if for every finite length module M, the module $\oplus_{i=1}^{\infty} M$ has couniserial dimension.*

Proposition 5.4 *Let P be an anti-coHopfian projective right R-module. If $\oplus_{i=1}^{\infty} E(P)$ has couniserial dimension, then P is injective.*

Proof By part 1 of Lemma 3.20 it is enough to show P has cancellation property. Let $M = P \oplus B \cong P \oplus B'$. So there exist submodules P' and C of M such that $M = P \oplus B = P' \oplus C$ and $P' \cong P$ and $C \cong B'$. If p_1 is the projection map from $M = P \oplus B$ onto P with kernel B. Then considering the restriction of p_1 to C we have an exact sequence $0 \longrightarrow C \cap B \longrightarrow C \longrightarrow I \longrightarrow 0$, such that I is a submodule of P. Note that every submodule of P is projective, because it is anti-coHopfian. So I is projective and thus $C \cong (C \cap B) \oplus I$. Similarly by considering the map p_2 from $M = P' \oplus C$ to P' we have $B \cong C \cap B \oplus J$ for some submodule J of P'. Since $J \cong I \cong P$, we have $B \cong C$ and so $B \cong B'$. Then P has cancellation property. □

As a consequence of the above proposition we have the following corollary:

Corollary 5.5 *Let R be a principal right ideal domain with maximal right quotient ring Q (which is a division ring). If the right R-module $\oplus_{i=1}^{\infty} Q$ has couniserial dimension, then $R = Q$.*

We need the following lemmas to prove the next theorem. Using Proposition 3.3 we can see that:

Lemma 5.6 *Let I be a two sided ideal of R and M be an R/I-module. If M as R-module has couniserial dimension, then M as R/I-module has couniserial dimension.*

Lemma 5.7 *If all finitely generated right modules have couniserial dimension, then every right module contains a noetherian uniform module.*

Proof By Lemma 3.1 it is enough to show that every cyclic module contains an anti-coHopfian module. Let M be a non-zero cyclic right module which does not contain anti-coHopfian module and let S be a simple module. M is not anti-coHopfian, then M has a non-zero submodule $M_1 \ncong M$ and M_1 has a non-zero submodule M_2 such that $M_2 \ncong M_1$. By continuing in this manner we have a descending chain $S \oplus M \geq S \oplus M_1 \geq S \oplus M_2 \geq \cdots$ of submodules of $S \oplus M$. Since $S \oplus M$ is finitely generated, by Proposition 3.3, $S \oplus M_n \cong S \oplus M_{n+1}$ for some n. This implies that $M_n \cong M_{n+1}$ for some n, because S is cancellable and this is a contradiction. □

Theorem 5.8 *For a ring R the following are equivalent.*

(1) *R is a semisimple artinian ring.*
(2) *All right R-modules have couniserial dimension.*
(3) *All left R-modules have couniserial dimension.*

(4) *All right R-modules have uniserial dimension.*
(5) *All left R-modules have uniserial dimension.*

Proof For equivalence of (1), (4) and (5) refer [20, Theorem 2.6].

(1) $\Rightarrow$ (2). This is clear by Corollary 3.19.

(2) $\Rightarrow$ (1). At first we show R satisfies ascending chain condition on two sided deals. Let $I_1 \leq I_2 \leq \cdots$ be a chain of ideals of R. Since the right module $\oplus_{i=1}^{\infty} R/I_i$ has couniserial dimension, there exists n such that, for each $j \geq n$, $\oplus_{i=n}^{\infty} R/I_i \cong \oplus_{i=j}^{\infty} R/I_i$. Thus they have the same annihilators and so for each $j \geq n$, $I_n = I_j$. Suppose R is a non-semisimple ring. By Lemma 5.6 every module over a factor ring of R also has couniserial dimension. Thus by invoking the ascending chain condition on two sided ideals we may assume R is not semisimple artinian but every factor ring of R is semisimple artinian. Using Lemma 3.20, R is a right V-ring. First let us assume that R is primitive. So, by Theorem 4.10, R is a prime right Goldie ring. By [4, Theorem 5.16], a prime right V-ring right Goldie is simple. By Lemma 5.7, R has a right noetherian uniform submodule and so using [8, Corollary 7.25], R is right noetherian. Now we show that R is Morita equivalent to a domain. By [7, lemma 5.12], the endomorphism ring of every uniform right ideal of a prime right Goldie ring is a right Ore domain. So by [11, Theorem 1.2], it is enough to show that R has a uniform projective generator U. Let us assume that R is not uniform and u.dim$(R) = n$ and let U be a uniform right ideal of R. By [8, Corollary 7.25], U^n can be embedded in R and also R can be embedded in U^n. Then c.u.dim(R) = c.u.dim(U^n) and hence $R \cong U^n$, because R is not uniform. Thus U is a projective generator uniform right ideal of R. So R is Morita equivalent to a domain. Now Lemma 4.11, Lemma 3.11 and Corollary 5.5 show that R is simple artinian, a contradiction. So R is not primitive, but every primitive factor ring is artinian (indeed all proper factor rings are artinian). Then since R is a right V-ring, by [3], R is regular and Σ-V-ring. Also every right ideal contains a non-zero uniform right ideal, hence minimal. So R has non-zero essential soc(R). But R is Σ-V-ring and by Corollary 3.19 , we have only finitely many non-isomorphic simple modules. Thus soc(R) is injective. This implies R is semisimple, a contradiction. This completes the proof. □

6 Summary

This paper defines couniserial dimension of a module that measures how far a module is from being uniform. The results proved in the paper demonstrate its importance for studding the structure of modules and rings and is a beginning of a larger project to study its impact. We close with some open questions:

(1) Does a module with arbitrary couniserial dimension possesses indecomposable dimension?
(2) Is there a theory for modules with both finite uniserial and couniserial dimensions that parallels to Krull–Schmidt–Remak–Azumaya theorem?

Acknowledgments The authors would like to thank the anonymous referee for the careful reading of this article and a lot of useful suggestions. This paper was written when the third author was visiting Ohio University, United States during May–August 2014. She wishes to express her deepest gratitude to

Professor S. K. Jain for his kind guidance in her research project and Mrs. Parvesh Jain for the warm hospitality extended to her during her stay. She would also like to express her thanks to Professor E. Zelmanov for his kind invitation to visit the of University of Californian at San Diego and to give a talk.

References

1. Anderson, F.W., Fuller, K.R.: Rings and Categories of Modules, 2nd ed. Graduate Texts in Mathematics, vol. 13. Springer, Berlin (1992)
2. Baccella, G.: On flat factor rings and fully right idempotent rings. Ann. Univ. Ferrara Sez. **26**, 125–141 (1980)
3. Baccella, G.: Von Neumann regularity of V-rings with artinian primitive factor rings. Proc. Am. Math. Soc. **103**(3), 747–749 (1988)
4. Cozzens, J., Faith, C.: Simple Noetherian Rings. Cambridge University Press, Cambridge (1975)
5. Dubrovin, N.I.: An example of a chain primitive ring with nilpotent elements. Mat. Sb. (N.S.) 120(162), no. 3, pp. 441–447 (1983) (Russian)
6. Ghorbani, A., Nazemian, Z.: On commutative rings with uniserial dimension. J. Algebra Appl. **14**(1) (2015).
7. Goldie, A.W.: Rings with Maximum Condition, Lecture Notes. Yale University, New Haven (1961)
8. Goodearl, K.R., Warfield, R.B., Jr.: An Introduction to Non commutative noetherian Rings, London Math. Soc., vol. 61. Cambridge University Press, Cambridge (2004)
9. Goodearl, K.R.: Ring Theory. Nonsingular Rings and Modules. Dekker, New York (1976)
10. Goodearl, K.R.: Surjective endomorphisms of finitely generated modules. Commun. Algebra. **15**, 589–609 (1987)
11. Hart, R., Robson, J.C.: Simple rings and rings Morita equivalent to ore domains. Proc. Lond. Math. Soc. **21**(3), 232–242 (1970)
12. Gordon, R., Small, L.W.: Piecewise domains. J. Algebra **23**, 553–564 (1972)
13. Hirano, Y., Mogami, I.: Modules whose proper submodules are non-hopf kernels. Commun. Algebra. **15**(8), 1549–1567 (1987)
14. Hirano, Y., Poon, E., Tsutsui, H.: A generalization of complete reducibility. Commun. Algebra. **40**(6), 1901–1910 (2012)
15. Jain, S.K., Ashish Srivastava, K., Askar Tuganbaev, A.: Cyclic Modules and the Structure of Rings, Oxford Mathematical Monographs. Oxford University Press, Oxford (2012)
16. Lam, T.Y.: A crash course on stable range, cancellation, substitution, and exchange. J. Algebra Appl. **3**, 301–343 (2004)
17. Lam, T.Y.: Lectures on Modules and Rings, Graduate Texts in Mathematics, vol. 189. Springer, New York (1999)
18. McConnell, J.C., Robson, J.C.: Noncommutative Noetherian Rings. Wiley, New York (1987)
19. Michler, G.O., Villamayor, O.E.: On rings whose simple modules are injective. J. Algebra **25**, 185–201 (1973)
20. Nazemian, Z., Ghorbani, A., Behboodi, M.: Uniserial dimension of modules. J. Algebra **399**, 894–903 (2014)
21. Stoll, R.R.: Set Theory and Logic. Dover Publication Inc, New York (1961)
22. Varadarajan, K.: Anti hopfian and anti coHopfian modules. AMS Contemp. Math. Ser. **456**, 205–218 (2008)
23. Wisbauer, R.: Foundations of Module and Ring Theory. Gordon and Breach, Reading (1991)

8

Sharp interpolation inequalities for discrete operators and applications

Alexei Ilyin · Ari Laptev · Sergey Zelik

Abstract We consider interpolation inequalities for imbeddings of the l^2-sequence spaces over d-dimensional lattices into the l_0^∞ spaces written as interpolation inequality between the l^2-norm of a sequence and its difference. A general method is developed for finding sharp constants, extremal elements and correction terms in this type of inequalities. Applications to Carlson's inequalities and spectral theory of discrete operators are given.

Keywords Discrete operators · Sobolev inequality · Interpolation inequalities · Green's function · Sharp constants · Lieb–Thirring inequalities · Carlson inequality

Communicated by S.K. Jain.

A. Ilyin (✉) · S. Zelik
Keldysh Institute of Applied Mathematics, Moscow, Russia
e-mail: ilyin@keldysh.ru

A. Ilyin
Institute for Information Transmission Problems, Moscow, Russia

A. Laptev
Imperial College London, London, UK
e-mail: a.laptev@imperial.ac.uk

A. Laptev
Institute Mittag-Leffler, Djursholm, Sweden

S. Zelik
Department of Mathematics, University of Surrey, Surrey, UK
e-mail: s.zelik@surrey.ac.uk

1 Introduction

In this paper we study imbeddings of the sequence space $l^2(\mathbb{Z}^d)$ into $l_0^\infty(\mathbb{Z}^d)$ written in terms of a interpolation inequality involving the l^2-norms both of the sequence $u \in l^2(\mathbb{Z}^d)$, and the sequence of differences ∇u, where for $u \in l^2(\mathbb{Z})$ and $n \in \mathbb{Z}$

$$\mathrm{D}u(n) = u(n+1) - u(n),$$

and for $u \in l^2(\mathbb{Z}^d)$ and $n \in \mathbb{Z}^d$

$$\nabla u(n) = \{\mathrm{D}_1 u(n), \dots, \mathrm{D}_d u(n)\}, \quad \|\mathrm{D}u\|^2 = \|\nabla u\|^2 = \sum_{i=1}^{d} \|\mathrm{D}_i u\|^2.$$

Before we describe the content of the paper in greater detail we give a simple but important example [16], namely, let us prove the one-dimensional inequality

$$\sup_n u(n)^2 \le \|u\| \, \|\mathrm{D}u\|. \tag{1.1}$$

The proof repeats that in the continuous case. For an arbitrary $\gamma \in \mathbb{Z}$ we have

$$\begin{aligned} 2u^2(\gamma) &= \left(\sum_{n=-\infty}^{\gamma-1} - \sum_{n=\gamma}^{\infty} \right) \mathrm{D}u^2(n) \\ &= \left(\sum_{n=-\infty}^{\gamma-1} - \sum_{n=\gamma}^{\infty} \right) \big(u(n+1)\mathrm{D}u(n) + u(n)\mathrm{D}u(n) \big) \\ &\le \sum_{n=-\infty}^{\infty} \left(|u(n+1)\mathrm{D}u(n)| + |u(n)\mathrm{D}u(n)| \right) \le 2\|u\| \, \|\mathrm{D}u\|. \end{aligned}$$

Below we consider separately interpolation inequalities of the form

$$\sup_{\mathbf{n} \in \mathbb{Z}^d} u(\mathbf{n})^2 \le \mathrm{K}_d(\theta) \|u\|^{2\theta} \|\nabla u\|^{2(1-\theta)}, \quad 0 \le \theta \le 1 \tag{1.2}$$

in dimension $d = 1, 2$ and $d \ge 3$. By notational definition $\mathrm{K}_d(\theta)$ is the sharp constant in this inequality. This inequality clearly holds for $\theta = 1$ (with $\mathrm{K}_d(1) = 1$), and if it holds for a $\theta = \theta_* \in [0, 1)$, then it holds for $\theta \in [\theta_*, 1]$, when the 'weight' of the stronger norm $\|u\|$ is getting larger [see (1.11)].

For $d = 1$ we show that (1.2) holds for $1/2 \le \theta \le 1$ and find explicitly the corresponding sharp constant:

$$\mathrm{K}_1(\theta) = \frac{1}{2} \left(\frac{2}{\theta} \right)^\theta (2\theta - 1)^{\theta - 1/2}. \tag{1.3}$$

In the limiting case $\theta = 1/2$ we have $\mathrm{K}_1(1/2) = 1$, and we supplement inequality (1.1) (which is, in fact, sharp) with a refined inequality

$$u(0)^2 \le \frac{1}{2}\sqrt{4 - \frac{\|\mathrm{D}u\|^2}{\|u\|^2}}\,\|u\|\|\mathrm{D}u\|, \tag{1.4}$$

which for any $d \in (0,4)$ has a unique extremal sequence u^* with $\|\mathrm{D}u^*\|^2/\|u^*\|^2 = d$.

In the 2D case (1.2) holds for $0 < \theta \le 1$ and the sharp constant is given by

$$\mathrm{K}_2(\theta) = \frac{2}{\pi}\frac{1}{\theta^\theta(1-\theta)^{1-\theta}} \cdot \max_{\lambda>0} \frac{\lambda^\theta K\left(\frac{4}{4+\lambda}\right)}{4+\lambda}, \tag{1.5}$$

where K is the complete elliptic integral of the first kind, see (3.8). The constant $\mathrm{K}_2(\theta)$ logarithmically tends to ∞ as $\theta \to 0^+$, and for $\theta = 0$ we have the following limiting logarithmic inequality of Brezis–Gallouet type:

$$\begin{aligned} u(0,0)^2 \le \frac{1}{4\pi}\frac{\|\nabla u\|^2}{\|u\|^2}\left(1 - \frac{\|\nabla u\|^2}{8\|u\|^2}\right)&\left(\ln \frac{16}{\frac{\|\nabla u\|^2}{\|u\|^2}\left(8 - \frac{\|\nabla u\|^2}{\|u\|^2}\right)}\right. \\ &\left. + \ln\left(1 + \ln \frac{16}{\frac{\|\nabla u\|^2}{\|u\|^2}\left(8 - \frac{\|\nabla u\|^2}{\|u\|^2}\right)}\right) + 2\pi\right), \end{aligned} \tag{1.6}$$

where the constants in front of logarithms and 2π are sharp. The inequality saturates for $u = \delta$, otherwise the inequality is strict.

Finally, in dimension three and higher the inequality holds for the limiting exponent $\theta = 0$:

$$u(0)^2 \le \mathrm{K}_d\|\nabla u\|^2, \tag{1.7}$$

where the sharp constant is given by

$$\mathrm{K}_d = \frac{1}{4(2\pi)^d}\int_0^{2\pi}\cdots\int_0^{2\pi}\frac{dx_1\cdots dx_d}{\sin^2\frac{x_1}{2} + \cdots + \sin^2\frac{x_d}{2}}. \tag{1.8}$$

In the three dimensional case the constant K_3 can be evaluated in closed form since it is expressed in terms of the so-called third Watson's triple integral:

$$\mathrm{K}_3 = \frac{1}{2}W_S = 0.2527\ldots, \tag{1.9}$$

where (see [3] and the references therein)

$$W_S := \frac{1}{\pi^3}\int_0^\pi\int_0^\pi\int_0^\pi \frac{dxdydz}{3-\cos x-\cos y-\cos z} = \frac{\sqrt{6}}{12(2\pi)^3}\Gamma(\tfrac{1}{24})\Gamma(\tfrac{5}{24})\Gamma(\tfrac{7}{24})\Gamma(\tfrac{11}{24}). \tag{1.10}$$

It is natural to compare interpolation inequalities for differences and inequalities for derivatives in the continuous case. While in the continuous case the L_∞-norm is the strongest (at least locally), in the discrete case the l^1-norm is the strongest. Obviously, $\|u\|_{l^\infty} \le \|u\|_{l^p}$ for $p \ge 1$, and therefore $\|u\|_{l^p} \le \|u\|_{l^q}$ for $q \le p$:

$$\|u\|_{l^p}^p \le \|u\|_{l^\infty}^{p-q}\|u\|_{l^q}^q \le \|u\|_{l^p}^{p-q}\|u\|_{l^q}^q.$$

Also, unlike the continuous case, the difference operator is *bounded*:

$$\|\mathrm{D}u\|^2_{l^2(\mathbb{Z}^d)} \le 4d\|u\|^2_{l^2(\mathbb{Z}^d)}. \tag{1.11}$$

Roughly speaking, the situation (at least in the one-dimensional case) is as follows. The discrete inequality (1.2) for $d = 1$ holds for $\theta \in [1/2, 1]$, while the corresponding continuous inequality

$$\|f\|^2_\infty \le \mathrm{C}_1(\theta)\|f\|^{2\theta}\|f'\|^{2(1-\theta)}, \quad f \in H^1(Q)$$

holds only for $\theta = 1/2$ in case when $Q = \mathbb{R}$, and for $\theta \in [0, 1/2]$ for periodic function with zero mean, $Q = \mathbb{T}^1$. Hence, it makes sense to compare the constants at a unique common point $\theta_* = 1/2$ where both constants are equal to 1. For n-order derivatives and differences, $n > 1$, the constants in the discrete inequalities are strictly greater than those in the continuous case, the corresponding $\theta_* = 1 - 1/(2n)$.

For example, the second-order inequality on the line $\mathbb{R}$ and the corresponding discrete inequality are as follows

$$\|f\|^2_{L_\infty(\mathbb{R})} \le \frac{\sqrt{2}}{\sqrt[4]{27}}\|f\|^{3/2}\|f''\|^{1/2}, \quad f \in H^2(\mathbb{R}),$$
$$\|u\|^2_{l^\infty(\mathbb{Z})} \le \frac{\sqrt{2}}{2}\|u\|^{3/2}\|\Delta u\|^{1/2}, \quad u \in l^2(\mathbb{Z}).$$

Both constants are sharp, the second one is strictly greater than the first. Up to a constant factor (and shift of the origin) the family of extremal functions in the first inequality is produced by scaling $x \to \lambda x$, $\lambda > 0$ of the extremal $f_*(x)$, where

$$\int_{-\infty}^{\infty} \frac{e^{-ixy}\,dx}{x^4+1} = \frac{\pi\sqrt{2}}{2} f_*\Big(\frac{y}{\sqrt{2}}\Big), \quad f_*(x) = e^{-|x|}(\cos x + \sin|x|),$$

In the discrete inequality the unique extremal sequence is $\{u_*(n)\}_{n=-\infty}^{\infty}$,

$$u_*(n) = \int_0^{\pi} \frac{\cos nx\, dx}{\lambda_* + 16\sin^4 \frac{x}{2}}, \quad \text{where } \lambda_* = \frac{16}{3};$$

see (5.7) for the explicit formula for $u_*(n)$.

In two dimensions in the continuous case the imbedding $H^1 \subset L_\infty$ holds only with a logarithmic correction term involving higher Sobolev norms (and $\theta = 0$), which is the well-known Brezis–Gallouet inequality. On the contrary, in the $2D$ discrete case inequality (1.2) holds for $\theta \in (0, 1]$ and also requires a logarithmic correction for $\theta = 0$, see (1.6).

In higher dimensional case $d \geq 3$ the imbedding $H^1 \subset L_\infty$ fails at all, while inequality (1.2) holds for all $\theta \in [0, 1]$.

Next, we consider applications of discrete interpolation inequalities. Using the discrete Fourier transform and Parseval's identities we show that each discrete interpolation inequality is equivalent to an integral Carlson-type inequality. For example, in the 1D case, setting for a function $g \in L_2(0, 2\pi)$

$$I_1 := \int_0^{2\pi} g(x)dx, \quad I_2^2 := \int_0^{2\pi} g(x)^2 dx, \quad \hat{I}_2^2 := \int_0^{2\pi} 4\sin^2 \tfrac{x}{2}\, g(x)^2 dx,$$

we obtain that inequality (1.1) is equivalent to the sharp inequality

$$I_1^2 \leq 2\pi I_2 \hat{I}_2,$$

with no extremal functions, while the refined inequality (1.4) is equivalent to the inequality

$$I_1^2 \leq \pi \sqrt{4 - \frac{\hat{I}_2^2}{I_2^2}}\, I_2 \hat{I}_2,$$

saturating for each $\lambda \in (-\infty, -4) \cup (0, \infty)$ at

$$g_\lambda(x) = \frac{1}{\lambda + 4\sin^2 \frac{x}{2}}.$$

Developing further this approach we prove a Sobolev l^q-type discrete inequality for a non-limiting exponent

$$\|u\|_{l^q(\mathbb{Z}^d)} \leq \mathrm{C}(q, d)\|\nabla u\| \quad \text{for } q > 2d/(d-2). \tag{1.12}$$

Our explicit estimate for the constant $\mathrm{C}(q, d)$ is non-sharp, moreover, it blows up as $q \to 2d/(d-2)$ however, it is sharp in the limit $q \to \infty$.

Finally, we apply the results on discrete inequalities to the estimates of negative eigenvalues of discrete Schrödinger operators

$$-\Delta - V \tag{1.13}$$

acting in $l^2(\mathbb{Z}^d)$. Here $-\Delta := \mathrm{D}^*\mathrm{D}$ and $V(n) \geq 0$. Each discrete interpolation inequality for the imbedding into $l^\infty(\mathbb{Z}^d)$ produces by the method of [7] a collective inequality for families of orthonormal sequences, which, in turn, is equivalent to a Lieb–Thirring estimate for the negative trace. For example, we deduce from (1.7) the estimate

$$\sum_{\lambda_j<0} |\lambda_j| \leq \frac{\mathrm{K}_d}{4} \sum_{\alpha \in \mathbb{Z}^d} V^2(\alpha),$$

which holds for $d \geq 3$.

We finally point out that in the continuous case the classical Lieb–Thirring inequality for the negative trace of operator (1.13) in $L_2(\mathbb{R}^d)$ is as follows (see [6,13,14])

$$\sum_{\lambda_j<0} |\lambda_j| \leq \mathrm{L}_{1,d} \int_{\mathbb{R}^d} V^{1+d/2}(x)dx.$$

2 1D case

Since $u_n \to 0$ as $|n| \to \infty$, without loss of generality we can assume that $\sup_n u(n)^2 = u(0)^2$.

We consider a more general problem of finding sharp constants, existence of extremals and possibly correction terms in the inequalities of the type

$$u(0)^2 \leq \mathrm{K}_1(\theta)\|u\|^{2\theta}\|\mathrm{D}u\|^{2(1-\theta)}, \quad 0 \leq \theta \leq 1, \tag{2.1}$$

including, to begin with, the problem of finding those θ for which (2.1) holds at all. Here

$$\|u\|^2 = \sum_{k=-\infty}^{\infty} u(k)^2, \quad \|\mathrm{D}u\|^2 = \sum_{k=-\infty}^{\infty} \mathrm{D}u(k)^2.$$

Since $|a-b| \geq ||a|-|b||$, we have

$$\|\mathrm{D}|u|\| \leq \|\mathrm{D}u\|, \quad \text{where } |u| := \{|u(n)|\}_{n=-\infty}^{\infty}, \tag{2.2}$$

and we could have further reduced our treatment to the case when $u(n) \geq 0$. However, we shall be dealing below with a more general problem (2.4) which has both sing-definite and non-sign-definite extremals. We have the following 'reverse' Poincaré inequality:

$$\|\mathrm{D}u\|^2 \leq \begin{cases} 2\|u\|^2, & u \text{ is sign definite;} \\ 4\|u\|^2, & \text{otherwise.} \end{cases} \tag{2.3}$$

The adjoint to D is the operator:

$$\mathrm{D}^*u(n) = -(u(n) - u(n-1)),$$

and

$$\mathrm{D}^*\mathrm{D}u(n) = \mathrm{D}\mathrm{D}^*u(n) = -\big(u(n+1) - 2u(n) + u(n-1)\big).$$

To find the sharp constant $\mathrm{K}_1(\theta)$ in (2.1) we consider a more general problem: find $\mathbb{V}(d)$, where $\mathbb{V}(d)$ is the solution of the following maximization problem:

$$\mathbb{V}(d) := \sup\big\{u(0)^2 : \ u \in l^2(\mathbb{Z}), \ \|u\|^2 = 1, \ \|\mathrm{D}u\|^2 = d\big\}, \tag{2.4}$$

where $0 < d < 4$.

Its solution is found in terms of the Green's function of the corresponding second-order self-adjoint positive operator, see [1,18]. The spectrum of the operator $-\Delta = \mathrm{D}^*\mathrm{D}$ is the closed interval [0, 4], and we set

$$\mathbb{A}(\lambda) = \begin{cases} \mathrm{D}^*\mathrm{D} + \lambda, & \text{for } \lambda > 0; \\ -\mathrm{D}^*\mathrm{D} - \lambda, & \text{for } \lambda < -4. \end{cases} \tag{2.5}$$

Then $\mathbb{A}(\lambda)$ is positive definite

$$(\mathbb{A}(\lambda)u, u) = \begin{cases} \|\mathrm{D}u\|^2 + \lambda\|u\|^2 > \lambda\|u\|^2, & \text{for } \lambda > 0; \\ -\|\mathrm{D}u\|^2 - \lambda\|u\|^2 > (-\lambda - 4)\|u\|^2, & \text{for } \lambda < -4. \end{cases}$$

Let δ be the delta-sequence: $\delta(0) = 1$, $\delta(n) = 0$ for $n \neq 0$, and let $G_\lambda = \{G_\lambda(n)\}_{n=-\infty}^{\infty} \in l^2(\mathbb{Z})$ be the Green's function of operator (2.5), that is, the solution of the equation:

$$\mathbb{A}(\lambda)G_\lambda = \delta. \tag{2.6}$$

Then we have by the Cauchy–Schwartz inequality

$$\begin{aligned} u(0)^2 &= (\delta, u)^2 = (\mathbb{A}(\lambda)G_\lambda, u)^2 \\ &= (\mathbb{A}(\lambda)^{1/2}G_\lambda, \mathbb{A}(\lambda)^{1/2}u)^2 \leq (\mathbb{A}(\lambda)G_\lambda, G_\lambda)(\mathbb{A}(\lambda)u, u) \\ &= G_\lambda(0)(\mathbb{A}(\lambda)u, u). \end{aligned} \tag{2.7}$$

Furthermore, this inequality is sharp and turns into equality if and only if $u = \mathrm{const}\cdot G_\lambda$.

We find in Lemma 2.2 explicit formulas for $\mathbb{V}(d)$ and $G_\lambda(n)$. Nevertheless, we now independently prove the following two symmetry properties of $\mathbb{V}(d)$ and $G_\lambda(n)$, especially since their counterparts will be useful in the two-dimensional case below.

Proposition 2.1 *For* $d \in (0, 4)$

$$\mathbb{V}(d) = \mathbb{V}(4-d). \tag{2.8}$$

For $\lambda > 0$ *and* $n \in \mathbb{Z}$

$$0 < G_\lambda(n) = (-1)^{|n|} G_{-4-\lambda}(n). \tag{2.9}$$

Proof For $u \in l^2(\mathbb{Z})$ we define the orthogonal operator T

$$Tu = u^\star := \{(-1)^{|n|} u(n)\}_{n=-\infty}^{\infty}.$$

Then clearly $\|u\|^2 = \|u^\star\|^2$ and, in addition,

$$\|\mathrm{D}u^\star\|^2 = 4\|u\|^2 - \|\mathrm{D}u\|^2. \tag{2.10}$$

Therefore if for a fixed d and $u = u_d$ we have

$$\mathbb{V}(d) = u(0)^2, \quad \|u\|^2 = 1 \quad \text{and} \quad \|\mathrm{D}u\|^2 = d,$$

then for $u^* = Tu$ it holds

$$u^\star(0)^2 = u(0)^2 = \mathbb{V}(d), \quad \|u^\star\|^2 = 1 \quad \text{and} \quad \|\mathrm{D}u^\star\|^2 = 4 - d,$$

which gives that $\mathbb{V}(4-d) \geq \mathbb{V}(d)$. However, the strict inequality here is impossible, since otherwise by repeating this procedure we would have found that $\mathbb{V}(d) > \mathbb{V}(d)$. This proves (2.8).

Turning to (2.9) we note that $T^{-1} = T^* = T$ and we see from (2.10) that

$$(\mathrm{D}^*\mathrm{D}u, u) = (T(-\mathrm{D}^*\mathrm{D} + 4)Tu, u),$$

and, consequently,

$$\mathrm{D}^*\mathrm{D} = T(-\mathrm{D}^*\mathrm{D} + 4)T.$$

Therefore, if for $\lambda > 0$, G_λ solves

$$\mathbb{A}(\lambda)G_\lambda = (\mathrm{D}^*\mathrm{D} + \lambda)G_\lambda = \delta,$$

then

$$T(-\mathrm{D}^*\mathrm{D} + 4 + \lambda)TG_\lambda = \delta.$$

Since $T^{-1}\delta = \delta$, using definition (2.5) we obtain

$$(-\mathrm{D}^*\mathrm{D} + 4 + \lambda)TG_\lambda = \mathbb{A}(-4 - \lambda)TG_\lambda = \delta,$$

which gives

$$TG_\lambda = G_{-4-\lambda},$$

and proves the equality in (2.9).

It remains to show that for $\lambda > 0$ $G_\lambda(n) > 0$ for all n. Since $\mathbb{A}(\lambda)$ is positive definite, it follows that $G_\lambda(0) = (\mathbb{A}(\lambda)G_\lambda, G_\lambda) > 0$. We use the maximum principle and suppose that for some $n \neq 1$, $G_\lambda(n) < 0$. Since $G_\lambda(n) \to 0$ as $n \to \infty$ and $G_\lambda(0) > 0$, it follows that G_λ attains a global strictly negative minimum at some point $n > 1$ (the case $n < -1$ is similar). Then the sum of the first three terms in (2.20) is non-positive and the fourth term is strictly negative, which contradicts $\delta(n) = 0$. This proves that $G_\lambda(n) \geq 0$ for all n. Finally, to prove strict positivity, we suppose that $G_\lambda(n) = 0$ for some $n > 1$. Then we see from (2.20) that $G_\lambda(n-1)+G_\lambda(n+1) = 0$, and what has already been proved gives $G_\lambda(n-1) = G_\lambda(n+1) = 0$. Repeating this we reach $n = 1$ giving that $G_\lambda(0) = 0$, which is a contradiction. □

To denote the three norms of G_λ we set

$$f(\lambda) := G_\lambda(0), \quad g(\lambda) := \|G_\lambda\|^2, \quad h(\lambda) := \|\mathrm{D}G_\lambda\|^2. \tag{2.11}$$

Lemma 2.1 *The functions f, g and h satisfy*

$$g(\lambda) = -\operatorname{sign}(\lambda)f'(\lambda), \quad h(\lambda) = \operatorname{sign}(\lambda)(f(\lambda) + \lambda f'(\lambda)). \tag{2.12}$$

Proof Let $\lambda > 0$. Then $\mathbb{A}(\lambda) = \mathrm{D}^*\mathrm{D} + \lambda$. Taking the scalar product of (2.6) with G_λ we have

$$f(\lambda) = G_\lambda(0) = \|\mathrm{D}G_\lambda\|^2 + \lambda\|G_\lambda\|^2 = h(\lambda) + \lambda g(\lambda). \tag{2.13}$$

Differentiating this formula with respect to λ we obtain

$$\begin{aligned} f'(\lambda) &= 2(\mathbb{A}(\lambda)G'_\lambda, G_\lambda) + g(\lambda) \\ &= -2(G_\lambda, G_\lambda) + g(\lambda) = -g(\lambda), \end{aligned} \tag{2.14}$$

where we used that $G_\lambda + \mathbb{A}(\lambda)G'_\lambda = 0$, which, in turn, follows from (2.6). The case $\lambda < -4$ is treated similarly taking into account that now $\mathbb{A}(\lambda) = -\mathrm{D}^*\mathrm{D} - \lambda$. □

Corollary 2.1 *The function $d(\lambda)$ defined as follows*

$$d(\lambda) := \frac{\|\mathrm{D}G_\lambda\|^2}{\|G_\lambda\|^2} = \frac{h(\lambda)}{g(\lambda)} \tag{2.15}$$

satisfies the functional equation

$$d(-4-\lambda) = 4 - d(\lambda). \tag{2.16}$$

Proof It follows from (2.9) and (2.11) that

$$f(-4-\lambda) = f(\lambda).$$

Hence, $f'(\lambda) = -f'(-4-\lambda)$ and we obtain from (2.12)

$$\begin{aligned} d(-4-\lambda) &= \frac{h(-4-\lambda)}{g(-4-\lambda)} = \frac{f(-4-\lambda) + (-4-\lambda)f'(-4-\lambda)}{-f'(-4-\lambda)} \\ &= \frac{f(\lambda) + (4+\lambda)f'(\lambda)}{f'(\lambda)} = \frac{f(\lambda) + \lambda f'(\lambda)}{f'(\lambda)} + 4 = 4 - d(\lambda). \end{aligned}$$

□

Next, we find explicit formulas for f, g and h.

Lemma 2.2 *The Green's function G_λ belongs to $l^2(\mathbb{Z})$, and both for $\lambda \in (-\infty, -4)$ and $\lambda \in (0, \infty)$*

$$f(\lambda) = \frac{1}{\sqrt{\lambda(\lambda+4)}}, \quad g(\lambda) = \frac{\lambda+2}{(\lambda+4)\sqrt{\lambda^3(\lambda+4)}}, \quad h(\lambda) = \frac{2}{\sqrt{\lambda(\lambda+4)^3}}. \tag{2.17}$$

Furthermore, the elements $G_\lambda(n)$ can be found explicitly: for $\lambda > 0$

$$G_\lambda(n) = \frac{1}{\pi}\int_0^\pi \frac{\cos nx\, dx}{\lambda + 4\sin^2\frac{x}{2}} = \frac{1}{\sqrt{\lambda(\lambda+4)}}\left(\frac{\lambda+2-\sqrt{\lambda(\lambda+4)}}{2}\right)^{|n|}, \tag{2.18}$$

for $\lambda < -4$

$$G_\lambda(n) = -\frac{1}{\pi}\int_0^\pi \frac{\cos nx\, dx}{\lambda + 4\sin^2\frac{x}{2}} = \frac{1}{\sqrt{\lambda(\lambda+4)}}\left(\frac{\lambda+2+\sqrt{\lambda(\lambda+4)}}{2}\right)^{|n|}. \tag{2.19}$$

Proof In view of (2.12), for the proof of (2.17) it suffices to find only $f(\lambda) = G_\lambda(0)$. We consider two cases: $\lambda > 0$ and $\lambda < -4$. For $\lambda > 0$ the sequence G_λ solves (2.6), which takes the form $(\mathrm{D}^*\mathrm{D} + \lambda)G_\lambda = \delta$, or component-wise

$$-G_\lambda(n+1) + 2G_\lambda(n) - G_\lambda(n-1) + \lambda G_\lambda(n) = \delta(n). \tag{2.20}$$

We multiply each equation by e^{inx} and sum the results from $n = -\infty$ to ∞. Setting

$$\widehat{g}_\lambda(x) := \sum_{n=-\infty}^{\infty} G_\lambda(n)e^{inx},$$

we obtain

$$-\sum_{n=-\infty}^{\infty} e^{inx}G_\lambda(n+1) + (2+\lambda)\sum_{n=-\infty}^{\infty} e^{inx}G_\lambda(n) - \sum_{n=-\infty}^{\infty} e^{inx}G_\lambda(n-1) = 1$$

or

$$\begin{aligned}1 &= \widehat{g}_\lambda(x)(\lambda - e^{-ix} + 2 - e^{ix})) \\ &= \widehat{g}_\lambda(x)\big(\lambda - (e^{ix/2} - e^{-ix/2})^2\big) = \widehat{g}_\lambda(x)\left(\lambda + 4\sin^2\frac{x}{2}\right),\end{aligned}$$

which gives $\widehat{g}_\lambda(x) = 1/(\lambda + 4\sin^2\frac{x}{2})$.

In the case when $\lambda < -4$ Eq. (2.6) becomes $(\mathrm{D}^*\mathrm{D} + \lambda)G_\lambda = -\delta$ and we merely have to change the sign of $\widehat{g}_\lambda(x)$ and we obtain:

$$\widehat{g}_\lambda(x) = \begin{cases} \frac{1}{\lambda+4\sin^2\frac{x}{2}}, & \lambda > 0; \\ -\frac{1}{\lambda+4\sin^2\frac{x}{2}}, & \lambda < -4, \end{cases} \tag{2.21}$$

and

$$G_\lambda(n) = \frac{1}{2\pi}\int_{-\pi}^{\pi} \widehat{g}_\lambda(x)\cos nx\,dx. \tag{2.22}$$

Since $\widehat{g}_\lambda \in L_2(0, 2\pi)$, it follows that $G_\lambda \in l^2(\mathbb{Z})$.

Using the integral

$$\int_{-\pi}^{\pi} \frac{dx}{b + \sin^2\frac{x}{2}} = \begin{cases} \frac{2\pi}{\sqrt{b(b+1)}}, & b > 0; \\ -\frac{2\pi}{\sqrt{b(b+1)}}, & b < -1. \end{cases} \tag{2.23}$$

we finally obtain both for $\lambda > 0$, and $\lambda < -4$

$$f(\lambda) = G_\lambda(0) = \frac{1}{2\pi}\int_{-\pi}^{\pi} \widehat{g}_\lambda(x)dx = \frac{1}{\sqrt{\lambda(\lambda+4)}}. \tag{2.24}$$

Finally, to obtain the explicit formula (2.18) (which will not be used below) we observe that the Eq. (2.20) for positive (and negative) n is a homogeneous linear recurrence relation with constant coefficients. The characteristic equation is

$$q^2 - (2+\lambda)q + 1 = 0$$

with roots

$$q_1(\lambda) = \frac{\lambda + 2 - \sqrt{\lambda(\lambda+4)}}{2}, \quad 0 < q_1(\lambda) < 1 \text{ for } \lambda > 0,$$
$$q_2(\lambda) = \frac{\lambda + 2 + \sqrt{\lambda(\lambda+4)}}{2}, \quad -1 < q_2(\lambda) < 0 \text{ for } \lambda < -4.$$

For $\lambda > 0$ the general l^2-solution of (2.20) is $G_\lambda(n) = c_1(\lambda)q_1(\lambda)^n$ for $n > 0$ and $G_\lambda(n) = c_2(\lambda)q_1(\lambda)^{|n|}$ for $n < 0$. Since we already know that $G_\lambda(n) = G_\lambda(-n)$, it follows that $c_1(\lambda) = c_2(\lambda) =: a(\lambda)$. Substituting $G_\lambda(n) = a(\lambda)q_1(\lambda)^{|n|}$ into (2.20) with $n = 0$ we obtain $-2a(\lambda)q_1(\lambda) + (2+\lambda)a(\lambda) = 1$, which gives

$$a(\lambda) = \frac{1}{2 + \lambda - 2q_1(\lambda)} = \frac{1}{\sqrt{\lambda(\lambda+4)}}$$

and proves (2.18). The proof of (2.19) in the case $\lambda < -4$ is totally similar, we only have to use the second root $q_2(\lambda)$ with $|q_2(\lambda)| < 1$.

We finally point out that the equality (2.9) can now be also verified by a direct calculation: $q_2(-4-\lambda) = -q_1(\lambda)$. □

We can now give the solution to the problem (2.4).

Theorem 2.1 *For any* $0 < d < 4$ *the solution of the maximization problem* (2.4) *is given by*

$$\mathbb{V}(d) = \frac{1}{2}\sqrt{d(4-d)}. \tag{2.25}$$

The supremum in (2.4) *is the maximum that is attained at a unique sequence* $u^*_{\lambda(d)} = \|G_{\lambda(d)}\|^{-1}G_{\lambda(d)}$, *where*

$$\lambda(d) = \frac{2d}{2-d} \tag{2.26}$$

for $d \neq 2$; *for* $d = 2$, $u^* = \delta$.

Proof It follows from (2.7) that for any $u \in l^2(\mathbb{Z})$

$$u(0)^2 \leq G_\lambda(0)(\mathbb{A}(\lambda)u, u),$$

and, furthermore, for

$$u^*_\lambda := \frac{1}{\|G_\lambda\|} \cdot G_\lambda$$

with $\|u^*_\lambda\|^2 = 1$ the above inequality turns into equality.

Next, using (2.17) we find the formula for the function $d(\lambda)$ defined in (2.15)

$$d(\lambda) = \frac{\|\mathrm{D}G_\lambda\|^2}{\|G_\lambda\|^2} = \frac{h(\lambda)}{g(\lambda)} = \frac{2\lambda}{2+\lambda}, \quad d : \begin{cases} (0, \infty) \to (0, 2), \\ (-\infty, -4) \to (2, 4). \end{cases}$$

The inverse function $\lambda(d)$ is given by (2.26) and with this $\lambda(d)$ we have

$$\frac{\|\mathrm{D}u^*_{\lambda(d)}\|^2}{\|u^*_{\lambda(d)}\|^2} = \frac{h(\lambda(d))}{g(\lambda(d))} = \frac{2\lambda(d)}{2+\lambda(d)} = d.$$

Therefore $u^*_{\lambda(d)}$ is the extremal sequence in (2.4) and its solution is

$$\mathbb{V}(d) = u^*_{\lambda(d)}(0)^2 = \frac{f(\lambda(d))^2}{g(\lambda(d))} = \frac{1}{2}\sqrt{d(4-d)}.$$

□

Remark 2.1 It is worth pointing out that in accordance with Proposition 2.1 and Corollary 2.1 we directly see here that $\mathbb{V}(d) = \mathbb{V}(4-d)$ and $d(-4-\lambda) = 4 - d(\lambda)$. For the inverse function $\lambda(d)$ we have the functional equation $\lambda(4-d) = -4 - \lambda(d)$.

Corollary 2.2 *For any $u \in l^2(\mathbb{Z})$ inequality* (1.1) *holds, the constant* 1 *is sharp and no extremals exist. The following refined inequality holds:*

$$u(0)^2 \le \frac{1}{2}\sqrt{4 - \frac{\|\mathrm{D}u\|^2}{\|u\|^2}}\,\|u\|\,\|\mathrm{D}u\|. \tag{2.27}$$

For any $0 < d < 4$ the inequality saturates for $u^*_{\lambda(d)} = G_{\lambda(d)}$, where $\lambda(d) = \frac{2d}{2-d}$ [*see* (2.26)] *with $\|\mathrm{D}u^*_{\lambda(d)}\|^2/\|u^*_{\lambda(d)}\|^2 = d$. For $d = 2$, $u^* = \delta$.*

Proof Inequality (2.27) follows from (2.25) by homogeneity.

Since $\mathbb{V}(d) < \sqrt{d}$, we obtain inequality (1.1), and since $\mathbb{V}(d)/\sqrt{d} \to 1$ as $d \to 0$ the constant 1 is sharp. In view of the refined inequality (2.27) there can be no extremals in the original inequality (1.1). □

We now consider (2.1) for $\theta \neq 1/2$.

Theorem 2.2 *Inequality* (2.1) *holds only for $1/2 \le \theta \le 1$. The sharp constant $\mathrm{K}_1(\theta)$ is*

$$\mathrm{K}_1(\theta) = \frac{1}{2}\left(\frac{2}{\theta}\right)^{\theta}(2\theta - 1)^{\theta - 1/2}. \tag{2.28}$$

For each $1/2 < \theta \le 1$ there exists a unique extremal sequence.

Proof The proof is similar to the proof of Theorem 2.5 in [18] where the classical Sobolev spaces were considered. For convenience we include some details.

We first observe that inequality (2.1) cannot hold for $\theta < 1/2$, since otherwise we would have found that $\mathbb{V}(d) \le cd^{\eta}$, $\eta = 1 - \theta > 1/2$, a contradiction with (2.25): $\mathbb{V}(d) \sim d^{1/2}$ as $d \to 0$.

The case $\theta = 1/2$ was treated above and we assume in what follows that $\theta > 1/2$. We set

$$\lambda := \frac{\theta}{1-\theta}\frac{\|\mathrm{D}u\|^2}{\|u\|^2}. \tag{2.29}$$

Then, using (2.7), we have

$$\begin{aligned}
u(0)^2 &\le G_\lambda(0)\|u\|^2\left(\frac{\|\mathrm{D}u\|^2}{\|u\|^2} + \lambda\right) = \frac{1}{\theta}G_\lambda(0)\lambda^\theta\lambda^{1-\theta}\|u\|^2\\
&= \frac{1}{\theta}\lambda^\theta G_\lambda(0)\left(\frac{\theta}{1-\theta}\frac{\|\mathrm{D}u\|^2}{\|u\|^2}\right)^{1-\theta}\|u\|^2\\
&= \frac{1}{\theta^\theta(1-\theta)^{1-\theta}}\cdot\lambda^\theta G_\lambda(0)\|u\|^{2\theta}\|\mathrm{D}u\|^{2(1-\theta)}\\
&\le \frac{1}{\theta^\theta(1-\theta)^{1-\theta}}\cdot\sup_{\lambda>0}\left\{\lambda^\theta G_\lambda(0)\right\}\|u\|^{2\theta}\|\mathrm{D}u\|^{2(1-\theta)}\\
&= \mathrm{K}_1(\theta)\|u\|^{2\theta}\|\mathrm{D}u\|^{2(1-\theta)}.
\end{aligned} \tag{2.30}$$

We have taken into account in the last equality that

$$G_\lambda(0) = f(\lambda) = \frac{1}{\sqrt{\lambda(\lambda+4)}}.$$

Hence, the supremum in the above formula is a (unique) maximum on $\lambda \in \mathbb{R}^+$ of the function

$$\lambda^\theta f(\lambda) = \frac{\lambda^{\theta-1/2}}{\sqrt{\lambda+4}}$$

attained at $\lambda_* = (4\theta-2)/(1-\theta)$, which gives (2.28). To see that the constant $\mathrm{K}_1(\theta)$ is sharp we use that $\frac{d}{d\lambda}(\lambda^\theta f(\lambda))|_{\lambda=\lambda_*} = 0$, and $\lambda_* f'(\lambda_*) + \theta f(\lambda_*) = 0$. In view of (2.12) this gives

$$d_* := d(\lambda_*) = \frac{\|\mathrm{D}G_{\lambda_*}\|^2}{\|G_{\lambda_*}\|^2} = \frac{h(\lambda_*)}{g(\lambda_*)} = -\frac{f(\lambda_*) + \lambda_* f'(\lambda_*)}{f'(\lambda_*)} = \frac{1-\theta}{\theta}\lambda_*.$$

Hence (2.29) is satisfied for $u_* = G_{\lambda_*}$ the two inequalities in (2.30) become equalities, and u_* is the unique extremal. □

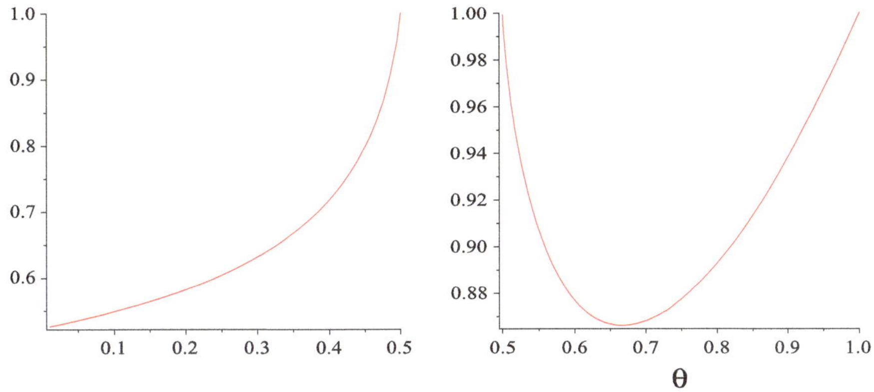

Fig. 1 Graphs of sharp constants in one-dimensional first-order inequalities on complementary intervals: periodic functions (*left*) (5.18), discrete case (*right*) (2.28)

The graph of the function $\mathrm{K}_1(\theta)$ is shown in Fig. 1 on the right. Here $\mathrm{K}_1(1/2) = 1$ corresponds to (1.1), and $\mathrm{K}_1(1) = 1$ corresponds to the trivial inequality $u(0)^2 \le \|u\|^2$ with extremal $u = \delta$.

Remark 2.2 In this theorem we do not use the formula (2.25) for $\mathbb{V}(d)$. However, if we do, then finding $\mathrm{K}_1(\theta)$ for $\theta \in [1/2, 1]$ becomes very easy. In fact, by the definition of $\mathbb{V}(d)$ and homogeneity, $\mathrm{K}_1(\theta)$ is the smallest constant for which $\mathbb{V}(d) \le \mathrm{K}_1(\theta) d^{1-\theta}$ for all $d \in [0, 4]$. Therefore

$$\mathrm{K}_1(\theta) = \max_{d\in[0,4]} \mathbb{V}(d)/d^{1-\theta} = \max_{d\in[0,4]} \frac{1}{2} d^{\theta-1/2}(4-d)^{1/2} = \text{r.h.s.}(2.28).$$

The corresponding $d_* = (4\theta - 2)/\theta \le 2$ and $\lambda(d_*) > 0$, see (2.26). This also explains why the region of negative λ does not play a role in Theorem 2.2.

3 2D case

In this section we consider the two-dimensional inequalities

$$u(0,0)^2 \le \mathrm{K}_2(\theta)\|u\|^{2\theta}\|\nabla u\|^{2(1-\theta)}, \quad 0 \le \theta \le 1, \tag{3.1}$$

and address the same problems as in the previous section.

We set

$$\Delta = -\mathrm{D}_1^*\mathrm{D}_1 - \mathrm{D}_2^*\mathrm{D}_2.$$

Then $(-\Delta u, u) = \|\nabla u\|^2 \le 8\|u\|^2$ for $u \in l_2(\mathbb{Z}^2)$. As in the 1D case we shall be dealing with the following extremal problem:

$$\mathbb{V}(d) := \sup\{u(0,0)^2 : u \in l^2(\mathbb{Z}^2),\ \|u\|^2 = 1,\ \|\mathrm{D}u\|^2 = d\}, \tag{3.2}$$

where $0 < d < 8$.

The resolvent set of $-\Delta + \lambda$ is $(-\infty, -8) \cup (0, \infty)$ and as before we consider the positive self-adjoint operator operator

$$\mathbb{A}(\lambda) = \begin{cases} -\Delta + \lambda, & \text{for } \lambda > 0; \\ \Delta - \lambda, & \text{for } \lambda < -8. \end{cases}$$

Our main goal is to find the Green's function of it:

$$\mathbb{A}(\lambda) G_\lambda = \delta, \tag{3.3}$$

more precisely, $G_\lambda(0, 0)$.

Proposition 3.1 *For $d \in (0, 8)$*

$$\mathbb{V}(d) = \mathbb{V}(8 - d). \tag{3.4}$$

For $\lambda > 0$ and $(n, m) \in \mathbb{Z}^2$

$$0 < G_\lambda(n, m) = (-1)^{|n+m|} G_{-8-\lambda}(n, m). \tag{3.5}$$

Finally, the function $d(\lambda) = \frac{\|\nabla G_\lambda\|^2}{\|G_\lambda\|^2}$ satisfies

$$d(-8 - \lambda) = 8 - d(\lambda). \tag{3.6}$$

Proof The proof is completely analogous to that of Proposition 2.1 and Corollary 2.1, where the functions $f(\lambda)$, $g(\lambda)$ and $h(\lambda)$ have the same meaning as in (2.11) and satisfy (2.12). The operator T is as follows

$$Tu(n, m) = (-1)^{|n+m|} u(n, m).$$

□

Lemma 3.1 *For $\lambda \in (-\infty, -8) \cup (0, \infty)$ the Green's function $G_\lambda \in l^2(\mathbb{Z}^2)$ and*

$$G_\lambda(0, 0) = \frac{2}{\pi} \frac{K\left(\frac{4}{4+\lambda}\right)}{|4 + \lambda|}, \tag{3.7}$$

where $K(k)$ is the complete elliptic integral of the first kind:

$$K(k) = \int_0^1 \frac{dt}{\sqrt{(1 - t^2)(1 - k^2 t^2)}}. \tag{3.8}$$

Proof Setting

$$\widehat{g}_\lambda(x, y) := \sum_{k,l=-\infty}^{\infty} G_\lambda(k, l)e^{ikx+ily},$$

and acting as in Lemma 2.2 we find that

$$\widehat{g}_\lambda(x, y) = \frac{\operatorname{sign}(\lambda)}{\lambda + 4(\sin^2 \frac{x}{2} + \sin^2 \frac{y}{2})}$$

and

$$G_\lambda(n, m) = \frac{\operatorname{sign}(\lambda)}{4\pi^2} \int_0^{2\pi} \int_0^{2\pi} \frac{\cos(nx + my)dxdy}{\lambda + 4(\sin^2 \frac{x}{2} + \sin^2 \frac{y}{2})}. \tag{3.9}$$

Therefore for $\lambda > 0$, using (2.23)

$$\begin{aligned} G_\lambda(0, 0) &= \frac{1}{4}\frac{1}{4\pi^2} \int_0^{2\pi} dx \int_0^{2\pi} \frac{dy}{(\frac{\lambda}{4} + \sin^2 \frac{x}{2}) + \sin^2 \frac{y}{2}} \\ &= \frac{1}{16\pi^2} \int_0^{2\pi} \frac{2\pi dx}{\sqrt{(\frac{\lambda}{4} + \sin^2 \frac{x}{2})(\frac{\lambda}{4} + 1 + \sin^2 \frac{x}{2})}} \\ &= \frac{1}{4\pi} \int_0^{\pi} \frac{dx}{\sqrt{(\frac{\lambda}{4} + \sin^2 \frac{x}{2})(\frac{\lambda}{4} + 1 + \sin^2 \frac{x}{2})}} \\ &= \frac{1}{4\pi} \int_0^{1} \frac{dt}{\sqrt{t(1-t)(\frac{\lambda}{4} + t)(\frac{\lambda}{4} + 1 + t)}} \\ &= \frac{1}{4\pi} \cdot \frac{2K\left(\frac{1}{\frac{\lambda}{4}+1}\right)}{\frac{\lambda}{4} + 1}, \end{aligned} \tag{3.10}$$

where the last integral was calculated by transforming general elliptic integrals to the standard form (see formula 3.147.7 in [8]).

Since $K(k)$ is even, we see from (3.5) that formula (3.7) works both for $\lambda > 0$ and $\lambda < -8$. □

Remark 3.1 The equality in (3.5) also follows from (3.9) by changing the variables $(x, y) \to (x' + \pi, y' + \pi)$ and using the fact that the integrand is even.

Theorem 3.1 *The inequality*

$$u(0, 0)^2 \leq \mathrm{K}_2(\theta)\|u\|^{2\theta}\|\nabla u\|^{2(1-\theta)}$$

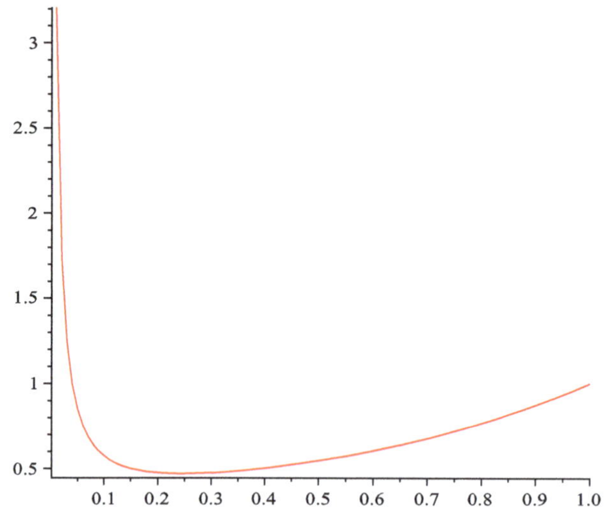

Fig. 2 Graph of $\mathrm{K}_2(\theta)$ on $\theta \in (0, 1]$

holds for $\theta \in (0, 1]$. *For* $\theta \in (0, 1)$ *the sharp constant* $\mathrm{K}_2(\theta)$ *is*

$$\mathrm{K}_2(\theta) = \frac{1}{\theta^\theta (1-\theta)^{1-\theta}} \cdot \max_{\lambda>0}(\lambda^\theta G_\lambda(0,0)), \tag{3.11}$$

and for each $0 < \theta < 1$ *there exists a unique extremal sequence*

$$u_{\lambda_*} = G_{\lambda_*}, \quad \textit{where } \lambda_* = \operatorname{argmax}(\lambda^\theta G_\lambda(0,0)). \tag{3.12}$$

Finally, $\mathrm{K}_2(1) = 1$ *with* $u_* = \delta$, *and*

$$\mathrm{K}_2(\theta) = \frac{1}{4\pi e \theta} + o\left(\frac{1}{\theta}\right) \quad \textit{as } \theta \to 0^+. \tag{3.13}$$

The graph of the function $\mathrm{K}_2(\theta)$ *is shown in Fig.* 2.

Proof Similarly to Theorem 2.2, we have

$$\mathrm{K}_2(\theta) = \frac{1}{\theta^\theta (1-\theta)^{1-\theta}} \cdot \sup_{\lambda>0}(\lambda^\theta G_\lambda(0,0)), \tag{3.14}$$

where, of course, $G_\lambda(0,0)$ is given by (3.7). We have the following asymptotic expansions

$$\begin{aligned} G_\lambda(0,0) &= \frac{1}{4\pi}(2\log(4\sqrt{2}) + \log(\tfrac{1}{\lambda})) + O(\lambda \log(\tfrac{1}{\lambda})), \quad \text{as } \lambda \to 0, \\ G_\lambda(0,0) &= \frac{1}{\lambda} - \frac{4}{\lambda^2} + O(\tfrac{1}{\lambda^3}), \quad \text{as } \lambda \to \infty. \end{aligned} \tag{3.15}$$

Hence, for $0<\theta<1$ we see that $\lambda^\theta G_\lambda(0,0)=0$ both at $\lambda=0$ and $\lambda=\infty$, and the supremum in (3.14) is the maximum, which proves (3.11) and (3.12).

We also see from the first formula that for small positive θ the leading term in the second factor in (3.14) is

$$\max_{0<\lambda<1}\frac{1}{4\pi}\lambda^\theta\log(\tfrac{1}{\lambda})=\frac{1}{4\pi e\theta},$$

while the first factor tends to 1. This proves (3.13). For example,

$$\mathrm{K}_2(0.01)=3.205\ldots,\quad\text{while}\ \frac{1}{0.99^{0.99}0.01^{0.01}}\frac{1}{4\pi e\cdot 0.01}=3.096\ldots.$$

□

In the limiting case $\theta=0$ inequality (3.1) holds with a logarithmic correction term of Brezis–Gallouet type [1,4].

The solution of the extremal problem (3.2) is given in terms of the functions $f(\lambda)$, $g(\lambda)$ and $h(\lambda)$:

$$\begin{aligned}
f(\lambda)&=G_\lambda(0,0)=\frac{2}{\pi}\frac{K(\frac{4}{4+\lambda})}{|4+\lambda|},\\
g(\lambda)&=\|G_\lambda\|^2=-\mathrm{sign}(\lambda)f'(\lambda)=\frac{2E(\frac{4}{4+\lambda})}{\pi\lambda(\lambda+8)},\\
h(\lambda)&=\|\nabla G_\lambda\|^2=\mathrm{sign}(\lambda)\big(f(\lambda)+\lambda f'(\lambda)\big)=\frac{2}{\pi}\frac{K(\frac{4}{4+\lambda})}{4+\lambda}-\frac{2E(\frac{4}{4+\lambda})}{\pi(\lambda+8)},
\end{aligned}\tag{3.16}$$

where $E(k)$ is the complete elliptic integral of the second kind:

$$E(k)=\int_0^1\frac{\sqrt{1-k^2t^2}}{\sqrt{1-t^2}}dt,$$

and where we used $\frac{\mathrm{d}K(k)}{\mathrm{d}k}=\frac{E(k)}{k(1-k^2)}-\frac{K(k)}{k}$.

Theorem 3.2 *The solution* $\mathbb{V}(d)$ *of problem* (3.2) *is*

$$\mathbb{V}(d)=u^*_{\lambda(d)}(0)^2=\frac{f^2(\lambda(d))}{g(\lambda(d))}=\frac{2K(\frac{4}{4+\lambda(d)})^2\lambda(d)(\lambda(d)+8)}{\pi(4+\lambda(d))^2E(\frac{4}{4+\lambda(d)})},\tag{3.17}$$

where $\lambda(d)$ *is the inverse function of the function* $d(\lambda)$*:*

$$d(\lambda)=\frac{h(\lambda)}{g(\lambda)}=\frac{\lambda(\lambda+8)}{(\lambda+4)}\frac{K(\frac{4}{\lambda+4})}{E(\frac{4}{\lambda+4})}-\lambda,\tag{3.18}$$

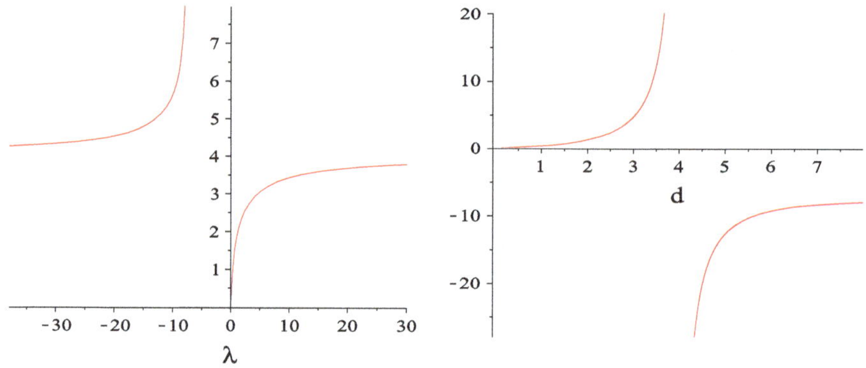

Fig. 3 Graphs of $d(\lambda)$ and $\lambda(d)$

and where $u^*_{\lambda(d)} = G_\lambda/\|G_\lambda\|$. *Here* $d(\lambda)$ *is defined on* $(-\infty, -8) \cup (0, \infty)$, *satisfies* (3.6) *and monotonically increases from* $d(-\infty) = 4$ *to* $d(-8) = 8$ *and then from* $d(0) = 0$ *to* $d(\infty) = 4$. *The inverse function* $\lambda(d)$ *is defined on* $d \in [0, 8]\backslash\{0\}$ *and satisfies*

$$\lambda(8 - d) = -8 - \lambda(d).$$

Their graphs are shown in Fig. 3. *Finally,* $\mathbb{V}(4) = 1$ *and* $u^* = \delta$.

Proof We act as in Theorem 2.1, the essential difference being that we now do not have a formula for the inverse function $\lambda(d)$, by means of which we construct the extremal element for each d. Although $d(\lambda)$ is given explicitly, the monotonicity of it required for the existence of the inverse function is a rather general fact and can be verified as in [18, Theorem 2.1], where the continuous case was considered. □

We now find an explicit majorant $\mathbb{V}_0(d)$ for the implicitly defined solution $\mathbb{V}(d)$. In view of the symmetry (3.4) it suffices to study the case $d \to 0$ only and then, by replacing $d \to d(8-d)/8$ we get the symmetric expansions valid for both singularities. We have the following expansions

$$\begin{aligned} d(\lambda) &= (5\ln 2 - \ln\lambda - 1)\lambda + O_{\lambda\to 0}((\lambda\ln\lambda)^2), \\ d(\lambda) &= 4 - \frac{8}{\lambda} + O_{\lambda\to\infty}(1/\lambda^2). \end{aligned} \tag{3.19}$$

Truncating the first expansion and solving $d = (5\ln 2 - \ln\lambda - 1)\lambda$, we have

$$\lambda = -\frac{d}{W_{-1}(-\frac{1}{32}ed)}$$

where $W_{-1}(z)$ is the -1th branch of the Lambert function. Using the known asymptotic expansions for the Lambert function, we get the following expression for $\lambda(d)$

$$\lambda(d) = \frac{d}{5\ln 2 - 1 - \ln(d) + \ln(5\ln(2) - 1 - \ln d) + O(\frac{\ln(-\ln d)}{\ln d})}. \tag{3.20}$$

Using

$$\frac{f^2(\lambda)}{g(\lambda)} = \frac{(\frac{5}{2}\ln 2 - \frac{1}{2}\ln\lambda)^2}{\pi}\lambda + O_{\lambda\to 0}(\lambda^2(\ln\lambda)^3),$$
$$\frac{f^2(\lambda)}{g(\lambda)} = 1 - \frac{4}{\lambda^2} + O_{\lambda\to\infty}\left(\frac{1}{\lambda^3}\right) \tag{3.21}$$

and substituting (3.20) into the first expansion we get

$$\mathbb{V}(d) = \frac{1}{4\pi}d(-\ln d + \ln(1 - \ln d) + \gamma + o_{d\to 0}(1))$$

where $\gamma = 5\ln(2) + 1 < 2\pi$. This justifies our choice of the approximation to $\mathbb{V}(d)$:

$$\mathbb{V}_0(d) = \frac{1}{4\pi}\frac{d(8-d)}{8}\left(\ln\frac{16}{d(8-d)} + \ln\left(1 + \ln\frac{16}{d(8-d)}\right) + 2\pi\right)$$

The constant 2π instead of γ (and the numerator 16) are chosen so that for $d = 4$ we have $\mathbb{V}(4) = \mathbb{V}_0(4) = 1$.

The asymptotic expansion of $\mathbb{V}_0(d)$ at $d = 0$ shows that $\mathbb{V}(d) < \mathbb{V}_0(d)$ for $0 < d \le d_0$, where d_0 is sufficiently small.

Using the expansions at $\lambda = \infty$ in (3.19) and (3.21) we find that

$$\mathbb{V}(d) = 1 - \frac{(4-d)^2}{16} + O_{d\to 4}((4-d)^3).$$

Since

$$\mathbb{V}_0(d) = 1 - \left(1 - \frac{1}{\pi}\right)\frac{(4-d)^2}{16} + O_{d\to 4}((4-d)^3),$$

it follows that $\mathbb{V}(d) \le V_0(d)$ for $d \in [4 - d_1, 4]$ for a small $d_1 > 0$. Corresponding to $[d_0, d_1]$ is the *finite* interval $[\lambda_0, \lambda_1]$ on which computer calculations show that the inequality $\mathbb{V}(d) \le V_0(d)$ still holds. This gives that

$$\mathbb{V}(d) \le V_0(d)$$

for all $d \in [0, 4]$ and hence, by symmetry, for $d \in [0, 8]$.

Thus, we have proved the following inequality.

Theorem 3.3 *For* $u \in l_2(\mathbb{Z}^2)$

$$u(0,0)^2 \le \frac{1}{4\pi}\frac{\|\nabla u\|^2}{\|u\|^2}\left(1-\frac{\|\nabla u\|^2}{8\|u\|^2}\right)\left(\ln\frac{16}{\frac{\|\nabla u\|^2}{\|u\|^2}\left(8-\frac{\|\nabla u\|^2}{\|u\|^2}\right)}\right.$$
$$\left. + \ln\left(1+\ln\frac{16}{\frac{\|\nabla u\|^2}{\|u\|^2}\left(8-\frac{\|\nabla u\|^2}{\|u\|^2}\right)}\right)+2\pi\right), \tag{3.22}$$

where the constants in front of logarithms and 2π *are sharp. The inequality saturates for* $u=\delta$, *otherwise the inequality is strict.*

4 3D case

In the three-dimensional case the following result holds which is somewhat similar to the classical Sobolev inequality for the limiting exponent.

Theorem 4.1 *Let* $u \in l^2(\mathbb{Z}^3)$. *Then for any* $\theta \in [0,1]$

$$u(0,0,0)^2 \le \mathrm{K}_3(\theta)\|u\|^{2\theta}\|\nabla u\|^{2(1-\theta)}, \tag{4.1}$$

where $\mathrm{K}_3(\theta) < \infty$ *for* $\theta \in [0,1]$, *and its sharp value for* $\theta \in (0,1)$ *is given by*

$$\mathrm{K}_3(\theta) = \frac{1}{2\pi^2}\frac{1}{\theta^\theta(1-\theta)^{1-\theta}}\max_{\lambda>0}\lambda^\theta\int_0^\pi \frac{K\left(\frac{1}{\frac{\lambda}{4}+1+\sin^2\frac{x}{2}}\right)}{\frac{\lambda}{4}+1+\sin^2\frac{x}{2}}dx, \tag{4.2}$$

and there exists a unique extremal element, which belongs to $l^2(\mathbb{Z}^3)$.

In the limiting case $\theta = 0$ *inequality* (4.1) *still holds:*

$$u(0,0,0)^2 \le \mathrm{K}_3(0)\|\nabla u\|^2, \tag{4.3}$$

where

$$\mathrm{K}_3(0) = \frac{1}{(2\pi)^3}\int_{\mathbb{T}^3}\frac{dx}{4(\sin^2\frac{x_1}{2}+\sin^2\frac{x_2}{2}+\sin^2\frac{x_3}{2})}$$
$$= \frac{1}{2\pi^2}\cdot\int_0^\pi \frac{K\left(\frac{1}{1+\sin^2\frac{x}{2}}\right)}{1+\sin^2\frac{x}{2}}dx = \frac{4.9887\ldots}{2\pi^2} = 0.2527\ldots. \tag{4.4}$$

The constant is sharp and there exists a unique extremal element, which does not lie in $l^2(\mathbb{Z}^3)$, *but rather in* $l_0^\infty(\mathbb{Z}^3)$, *but whose gradient does belong to* $l^2(\mathbb{Z}^3)$. *Furthermore, as we already mentioned in Sect.* 1, *we have the closed form formula for* $\mathrm{K}_3(0)$ *(see* [3])

$$\mathrm{K}_3(0) = \frac{\sqrt{6}}{24(2\pi)^3}\Gamma(\tfrac{1}{24})\Gamma(\tfrac{5}{24})\Gamma(\tfrac{7}{24})\Gamma(\tfrac{11}{24}).$$

Proof We have to find the fundamental solution $G_\lambda(k,l,m)$ of the equation

$$\mathbb{A}(\lambda)G_\lambda = (\mathrm{D}_1^*\mathrm{D}_1 + \mathrm{D}_2^*\mathrm{D}_2 + \mathrm{D}_3^*\mathrm{D}_3 + \lambda)G_\lambda = \delta. \tag{4.5}$$

Similarly to the 1D and 2D cases we find that the function

$$\widehat{g}_\lambda(x,y,z) := \sum_{k,l,m=-\infty}^{\infty} G_\lambda(k,l,m)e^{ikx+ily+imz},$$

satisfies

$$\widehat{g}_\lambda(x,y,z) = \frac{\frac{1}{4}}{\frac{\lambda}{4} + \sin^2\frac{x}{2} + \sin^2\frac{y}{2} + \sin^2\frac{z}{2}}.$$

As before we have the inequality

$$u(0,0,0)^2 \le G_\lambda(0,0,0)(\|\nabla u\|^2 + \lambda\|u\|^2), \tag{4.6}$$

which saturates for $u = \mathrm{const}\cdot G_\lambda$.

For $\lambda > 0$ as in the 1D and 2D cases we have $\widehat{g}_\lambda \in L_2(\mathbb{T}^3)$, and, hence, $G_\lambda \in l^2(\mathbb{Z}^3)$ for $\lambda > 0$. In particular, using (3.10) we find

$$\begin{aligned} G_\lambda(0,0,0) &= \frac{1}{8\pi^3}\int_{-\pi}^{\pi}\int_{-\pi}^{\pi}\int_{-\pi}^{\pi}\widehat{g}_\lambda(x,y,z)dxdydz \\ &= \frac{1}{2\pi^2}\int_0^\pi \frac{K\left(\frac{1}{\frac{\lambda}{4}+1+\sin^2\frac{x}{2}}\right)}{\frac{\lambda}{4}+1+\sin^2\frac{x}{2}}dx. \end{aligned} \tag{4.7}$$

However, unlike the previous two cases, now $\widehat{g}_\lambda$ is integrable for all $\lambda \ge 0$ including $\lambda = 0$: $\widehat{g}_\lambda \in L_1(\mathbb{T}^3)$ for $\lambda \ge 0$. Therefore the Green's function G_0 is well defined and belongs to $l_0^\infty(\mathbb{Z}^3)$. We point out, however, that since $\widehat{g}_0 \notin L_2(\mathbb{T}^3)$, it follows that $G_0 \notin l^2(\mathbb{Z}^3)$.

For $\lambda = 0$, the integrand has only a logarithmic singularity at $x = 0$ and we obtain

$$G_0(0,0,0) = \frac{1}{2\pi^2}\int_0^\pi \frac{K\left(\frac{1}{1+\sin^2\frac{x}{2}}\right)}{1+\sin^2\frac{x}{2}}dx.$$

We now see that $f(\lambda) := G_\lambda(0,0,0)$ is continuous on $\lambda \in [0,\infty)$ and is of the order $1/\lambda$ at infinity. This gives that for $\theta \in (0,1)$ the function $\lambda^\theta f(\lambda)$ vanishes both at the origin and at infinity. Hence, it attains its maximum at a (generically) unique point $\lambda_*(\theta)$, and the claim of the theorem concerning the case $\theta \in (0,1)$ follows in exactly the same way as in Theorem 2.2.

Setting $\lambda = 0$ in (4.6) we obtain (4.3) with (4.4). It remains to verify that $\nabla G_0 \in l^2(\mathbb{Z}^3)$. To see this we use notation (2.11) and Lemma 2.1. We obtain

$$\begin{aligned}
\|\nabla G_\lambda\|^2 &= h(\lambda) = f(\lambda) + \lambda f'(\lambda) \\
&= \frac{1}{8\pi^3}\int_{-\pi}^{\pi}\int_{-\pi}^{\pi}\int_{-\pi}^{\pi} (\widehat{g}_\lambda(x,y,z) + \lambda \widehat{g}_\lambda(x,y,z)'_\lambda)\,dxdydz \\
&= \frac{1}{8\pi^3}\int_{-\pi}^{\pi}\int_{-\pi}^{\pi}\int_{-\pi}^{\pi} \frac{\frac{1}{4}(\sin^2\frac{x}{2} + \sin^2\frac{y}{2} + \sin^2\frac{z}{2})}{(\frac{\lambda}{4} + \sin^2\frac{x}{2} + \sin^2\frac{y}{2} + \sin^2\frac{z}{2})^2}\,dxdydz.
\end{aligned}$$

Since the integral on right-hand side is bounded for $\lambda = 0$ we have $\|\nabla G_0\|^2 < \infty$. Finally, G_λ has strictly positive elements for $\lambda \geq 0$, since we have as before the maximum principle. In the case when $\lambda = 0$ we use, in addition, the fact that $G_0 \in l_0^\infty$. The proof is complete. □

The graph of $K_3(\theta)$ is shown in Fig. 4.

Remark 4.1 Higher dimensional cases are treated similarly, in particular, for $d \geq 3$ and $\theta = 0$

$$u(0)^2 \leq K_d(0)\|\nabla u\|^2, \quad K_d(0) = \frac{1}{(2\pi)^d}\int_{\mathbb{T}^d} \frac{dx}{4(\sin^2\frac{x_1}{2} + \cdots + \sin^2\frac{x_d}{2})}. \tag{4.8}$$

In Sect. 6 we give an independent elementary proof of this inequality.

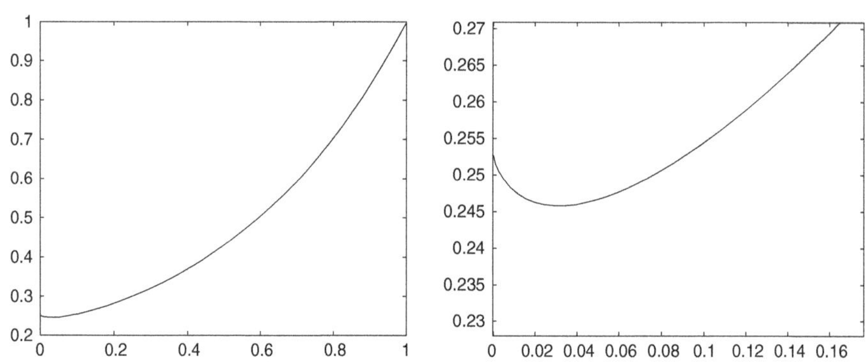

Fig. 4 Graph of $K_3(\theta)$ on the interval $\theta \in [0,1]$ (*left*) with a closer look at its behavior near $\theta = 0$ (*right*)

5 Higher order difference operators

The method developed above admits a straight forward generalization to higher order difference operators. We consider the second-order operator in the one dimensional case:

$$u(0)^2 \le \mathrm{K}_{1,2}(\theta)\|u\|^{2\theta}\|\Delta u\|^{2(1-\theta)}, \tag{5.1}$$

where

$$-\Delta u(n) := \mathrm{D}^*\mathrm{D}u(n) = -\big(u(n+1) - 2u(n) + u(n-1)\big).$$

Accordingly, the operator $\mathbb{A}(\lambda)$ is

$$\mathbb{A}(\lambda) = \begin{cases} \Delta^2 + \lambda, & \text{for } \lambda > 0; \\ -\Delta^2 - \lambda, & \text{for } \lambda < -16. \end{cases} \tag{5.2}$$

Here

$$\Delta^2 u(n) = u(n+2) - 4u(n+1) + 6u(n) - 4u(n-1) + u(n-2).$$

As before, we have to find the Green's function G_λ solving $\mathbb{A}(\lambda)G_\lambda = \delta$. Furthermore, for finding $\mathrm{K}_{1,2}(\theta)$ it suffices to solve this equation for $\lambda > 0$. Setting

$$\widehat{g}_\lambda(x) := \sum_{n=-\infty}^{\infty} G_\lambda(n)e^{inx},$$

and arguing as in Lemma 2.2 we get from (5.2)

$$\begin{aligned} 1 &= \widehat{g}_\lambda(x)(\lambda + e^{-i2x} - 4e^{-ix} + 6 - 4e^{ix} + e^{i2x}) \\ &= \widehat{g}_\lambda(x)\big(\lambda + (e^{ix/2} - e^{-ix/2})^4\big) = \widehat{g}_\lambda(x)\left(\lambda + 16\sin^4\frac{x}{2}\right), \end{aligned} \tag{5.3}$$

so that

$$G_\lambda(0) = \frac{1}{2\pi}\int_{-\pi}^{\pi}\frac{dx}{\lambda + 16\sin^4\frac{x}{2}} = \frac{\sqrt{2}}{2}\frac{1}{\lambda^{3/4}}\sqrt{\frac{\sqrt{\lambda+16}+\sqrt{\lambda}}{\lambda+16}}. \tag{5.4}$$

Now a word for word repetition of the argument in Theorem 2.2 gives that

$$\mathrm{K}_{1,2}(\theta) = \frac{1}{\theta^\theta(1-\theta)^{1-\theta}} \cdot \sup_{\lambda>0}\lambda^\theta G_\lambda(0).$$

Therefore we see from (5.4) that $\mathrm{K}_{1,2}(\theta) < \infty$ if and only if

$$\frac{3}{4} \le \theta \le 1.$$

For example, for $\theta = 3/4$ supremum is the maximum attained at $\lambda_*(3/4) = 16/3$, giving

$$\mathrm{K}_{1,2}(3/4) = \frac{4}{3^{3/4}} \cdot \lambda^{3/4} G_\lambda(0)|_{\lambda_*(3/4)=\frac{16}{3}} = \frac{\sqrt{2}}{2}.$$

We only mention that in the general case

$$\lambda_*(\theta) = \mathrm{argmax}_{\lambda>0}\, \lambda^{\theta-3/4} \sqrt{\frac{\sqrt{\lambda+16}+\sqrt{\lambda}}{\lambda+16}}$$
$$= \frac{64\theta - 32\theta^2 - 29 + \sqrt{32\theta - 23}}{2\theta^2 - 5\theta + 3}, \tag{5.5}$$

however, the corresponding substitution produces a long (but explicit) formula for $\mathrm{K}_{1,2}(\theta)$, and instead we present in Fig. 5 the graph of the sharp constant $\mathrm{K}_{1,2}(\theta)$, where $\mathrm{K}_{1,2}(3/4) = \sqrt{2}/2$ and $\mathrm{K}_{1,2}(1) = 1$.

Finally, it is possible to find $G_\lambda(n)$ explicitly. In fact, the free recurrence relation $\Delta^2 G_\lambda + \lambda G_\lambda = 0$ has the characteristic equation

$$q^2 - 4q + (6+\lambda) - 4q^{-1} + q^{-2} = 0,$$

or $(q^{1/2} - q^{-1/2})^4 = -\lambda$, which decomposes into two quadratic equations

$$q + \frac{1}{q} - 2 = i\sqrt{\lambda} \quad \text{and} \quad q + \frac{1}{q} - 2 = -i\sqrt{\lambda},$$

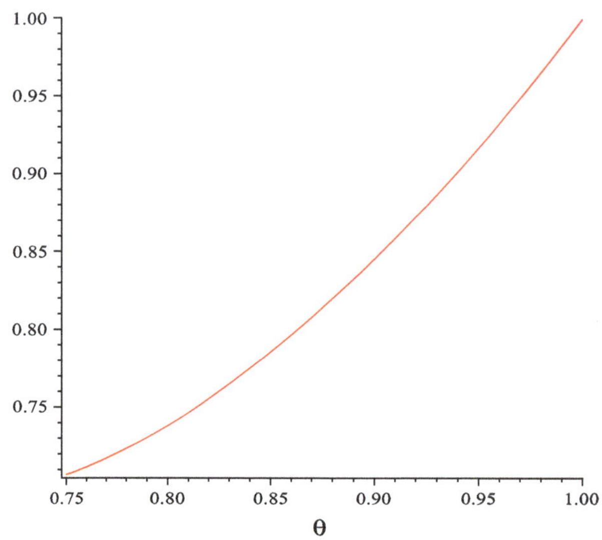

Fig. 5 Graph of $\mathrm{K}_{1,2}(\theta)$ on $\theta \in [3/4, 1]$

with four roots q_1, q_2, q_3, q_4, where $q_2 = 1/q_1$, $q_3 = \bar{q}_2$, $q_4 = \bar{q}_1$, where

$$q_1 = q(\lambda) = 1 - \frac{\lambda^{1/4}\sqrt{\sqrt{\lambda+16}-\sqrt{\lambda}}}{2\sqrt{2}} + i\left(\frac{\sqrt{\lambda}}{2} - \frac{\lambda^{1/4}\sqrt{\sqrt{\lambda+16}+\sqrt{\lambda}}}{2\sqrt{2}}\right). \tag{5.6}$$

Since $|q(\lambda)| < 1$ for $\lambda > 0$, it follows that any symmetric l_2-solution of (5.2) is of the form $a(\lambda)q(\lambda)^{|n|} + b(\lambda)\bar{q}(\lambda)^{|n|}$, and since, in addition $G_\lambda(n)$ is real, we have

$$G_\lambda(n) = a(\lambda)q(\lambda)^{|n|} + \bar{a}(\lambda)\bar{q}(\lambda)^{|n|}.$$

Setting $n = 0$ and $n = 1$ we obtain a linear system for $a(\lambda)$

$$\begin{aligned} G_\lambda(0) &= 2\mathrm{Re}\,a(\lambda) \\ G_\lambda(1) &= a(\lambda)q(\lambda) + \bar{a}(\lambda)\bar{q}(\lambda), \end{aligned}$$

where $G_\lambda(0)$ is given in (5.4) and

$$G_\lambda(1) = \frac{1}{\pi}\int_0^\pi \frac{\cos x\, dx}{\lambda + 16\sin^4\frac{x}{2}} = \frac{\sqrt{2}}{2}\frac{1}{\lambda^{3/4}}\frac{\sqrt{\lambda+16}-\sqrt{\lambda}}{\sqrt{\sqrt{\lambda+16}+\sqrt{\lambda}}\sqrt{\lambda+16}}.$$

Solving this system we find $a(\lambda)$:

$$a(\lambda) = \frac{\sqrt{2}}{4}\frac{1}{\lambda^{3/4}\sqrt{\lambda+16}\sqrt{\sqrt{\lambda+16}+\sqrt{\lambda}}}\left(\sqrt{\lambda+16}+\sqrt{\lambda}+4i\right),$$

and, consequently, the formula for $G_\lambda(n)$ with $\lambda > 0$:

$$G_\lambda(n) = \frac{1}{\pi}\int_0^\pi \frac{\cos nx\, dx}{\lambda + 16\sin^4\frac{x}{2}} = \frac{\mathrm{Re}\left[\left(\sqrt{\lambda+16}+\sqrt{\lambda}+4i\right)\cdot q(\lambda)^{|n|}\right]}{\sqrt{2}\,\lambda^{3/4}\sqrt{\lambda+16}\sqrt{\sqrt{\lambda+16}+\sqrt{\lambda}}}, \tag{5.7}$$

where $q(\lambda)$ is given in (5.6).

Thus, we obtain the following result.

Theorem 5.1 *Inequality* (5.1) *holds for* $\theta \in [3/4, 1]$. *In particular, in the limiting case* $\theta = 3/4$

$$u(0)^2 \le \frac{\sqrt{2}}{2}\|u\|^{3/2}\|\Delta u\|^{1/2}. \tag{5.8}$$

In the general case,

$$\mathrm{K}_{1,2}(\theta) = \frac{1}{\theta^\theta(1-\theta)^{1-\theta}}\cdot\lambda_*(\theta)^\theta G_{\lambda_*(\theta)}(0),$$

where $\lambda_*(\theta)$ is given in (5.5) *and $G_\lambda(0)$ in* (5.4) (*see also* (5.7)). *For $\theta \in [3/4, 1)$ the unique extremal is $u_{\lambda_*(\theta)} = G_{\lambda_*(\theta)}$. For $\theta = 1$, $\lambda_*(1) = \infty$ and $u_* = \delta$.*

Remark 5.1 It is not difficult to find the function $\mathbb{V}(d)$, that is, the solution of the maximization problem

$$\mathbb{V}(d) := \sup\{u(0)^2 : u \in l^2(\mathbb{Z}),\ \|u\|^2 = 1,\ \|\Delta u\|^2 = d\}, \tag{5.9}$$

where $d \in [0, 16]$. For this purpose we also need the expression for the Green's function $G_\lambda(0)$ in the region $\lambda \le -16$, which is as follows

$$G_\lambda(0) = -\frac{1}{2\pi}\int_{-\pi}^{\pi}\frac{dx}{\lambda + 16\sin^4\frac{x}{2}} = \frac{1}{2(-\lambda)^{3/4}}\frac{\sqrt{\sqrt{-\lambda}+4}+\sqrt{\sqrt{-\lambda}-4}}{\sqrt{-\lambda-16}}. \tag{5.10}$$

Using (5.4), (5.10) we can write down a parametric representation of $\mathbb{V}(d)$ as in Theorem 3.2, but instead we merely show its graph in Fig. 7.

This time we do not have the maximum principle, and the Green's function $G_\lambda(n)$ is not positive for all n, but is rather oscillating with exponentially decaying amplitude, see Fig. 6. Nor do we have the symmetry $\mathbb{V}(d) = \mathbb{V}(16 - d)$ in Fig. 7 that we have seen in the first-order inequalities in the one- and two-dimensional cases, see (2.8) and (3.4). The maximum is attained at $d = 6$ corresponding to $u = \delta$. The component $\lambda \in (0, \infty)$ of the resolvent set corresponds to $d \in (0, 6)$ and $\lambda \in (-\infty, -16)$ corresponds to $d \in (6, 16)$.

It is worth to compare the results so obtained in the discrete case with the corresponding interpolation inequalities for Sobolev spaces in the continuous case. It is well known that the interpolation inequality on the whole line $\mathbb{R}$

$$\|f\|^2_{L_\infty} \le \mathrm{C}_{1,n}(\theta)\|f\|^{2\theta}\|f^{(n)}\|^{2(1-\theta)}, \tag{5.11}$$

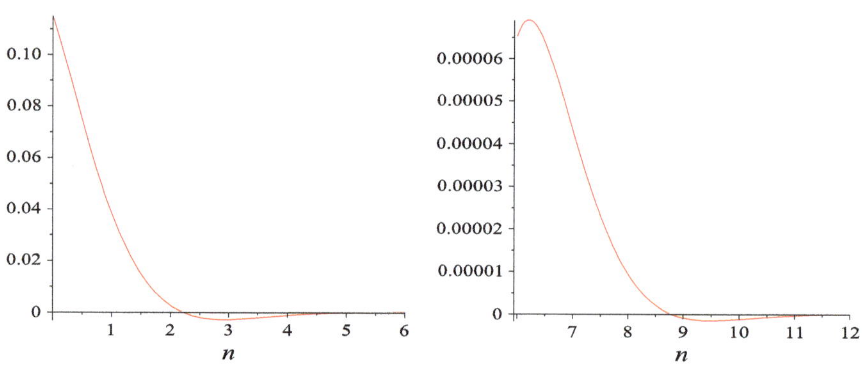

Fig. 6 Graph of the maximizer $G_{\frac{16}{3}}(n)$ for $n \in [0, 6]$ (*left*) and $n \in [6, 12]$ (*right*)

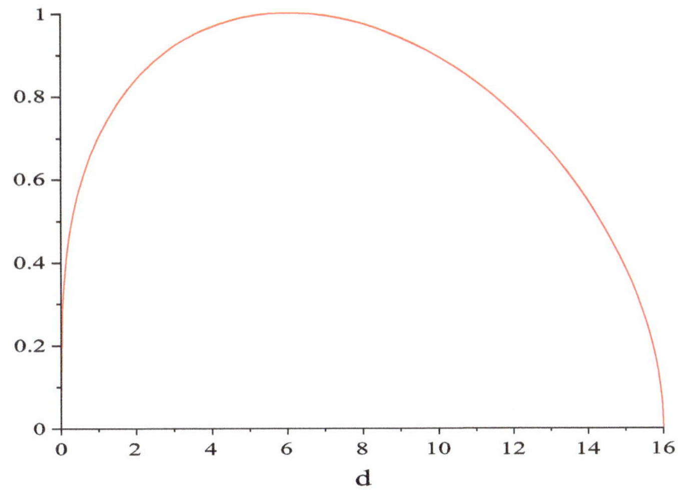

Fig. 7 Graph of $\mathbb{V}(d)$ defined in (5.9)

where $f \in H^n(\mathbb{R})$, holds only for $\theta = 1 - \frac{1}{2n}$. The sharp constant was found in [17]:

$$\mathrm{C}_{1,n}(\theta) = \frac{1}{\theta^\theta(1-\theta)^{1-\theta}} \frac{1}{\pi} \int_0^\infty \frac{dx}{1+x^{2n}} = \frac{1}{\theta^\theta(1-\theta)^{1-\theta}} \frac{1}{2n} \frac{1}{\sin\frac{\pi}{2n}}. \tag{5.12}$$

Thus, for first-order inequalities both in the discrete and continuous cases the constants are equal to 1, while for the second-order inequalities we see from (5.8) and (5.12) that

$$\mathrm{C}_{1,2}(3/4) = \left(\frac{4}{27}\right)^{1/4} < \frac{\sqrt{2}}{2} = \mathrm{K}_{1,2}(3/4).$$

The next theorem states that for higher order inequalities the constants in the discrete case are always strictly greater than those in the continuous case.

Theorem 5.2 *Let $n \geq 1$ and let $u \in l^2(\mathbb{Z})$. The inequality*

$$u(0)^2 \leq \mathrm{K}_{1,n}(\theta)\|u\|^{2\theta}\|\mathcal{D}^n u\|^{2(1-\theta)}, \tag{5.13}$$

where

$$\mathcal{D}^n := \begin{cases} \Delta^{n/2}, & \text{for } n \text{ even}, \\ \nabla\Delta^{(n-1)/2}, & \text{for } n \text{ odd}, \end{cases} \tag{5.14}$$

holds for $\theta \in [1 - 1/(2n), 1]$ and

$$\mathrm{K}_{1,n}(\theta) = \frac{1}{\theta^\theta(1-\theta)^{1-\theta}} \frac{1}{\pi} \sup_{\lambda>0} \lambda^\theta \int_0^\pi \frac{dx}{\lambda + 2^{2n}\sin^{2n}\frac{x}{2}}. \tag{5.15}$$

For all $\theta \in [1 - 1/(2n), 1]$ *supremum is the maximum. If* $n \geq 2$, *then for* $\theta = \theta_* :=$ $1 - 1/(2n)$ *the constants in the continuous and discrete inequalities satisfy*

$$\mathrm{C}_{1,n}(\theta_*) < \mathrm{K}_{1,n}(\theta_*). \tag{5.16}$$

Proof Following the scheme developed above we look for the solution of the equation

$$(-1)^n \Delta^n G_\lambda + \lambda G_\lambda = \delta,$$

and as in (5.3) find that

$$G_\lambda(0) = \frac{1}{\pi} \int_0^\pi \frac{dx}{\lambda + 2^{2n} \sin^{2n} \frac{x}{2}},$$

which proves (5.15) (whenever the supremum is finite). Using $\sin^2 \frac{x}{2} = \tan^2 \frac{x}{2}/(1 + \tan^2 \frac{x}{2})$ and changing the variable $\tan \frac{x}{2} = \sqrt{\mu} t$, where $\mu = \lambda^{1/n}$ we have

$$\lambda^{\theta_*} \int_0^\pi \frac{dx}{\lambda + 2^{2n} \sin^{2n} \frac{x}{2}} = \int_0^\infty \frac{2dt}{(1 + \mu t^2)(1 + \frac{2^{2n} t^{2n}}{(1+\mu t^2)^n})} =: S(\mu).$$

Clearly $S(\infty) = 0$, and we have to study $S(\mu)$ as $\mu \to 0$. The integral converges uniformly for $\mu \in [0, 1]$, since the denominator is greater then 1 for $t \in [0, 1]$ and is greater then $1 + c(n)t^2$ for $t \geq 1$ observing that $t^{2n}/(1+\mu t^2)^{n-1} \geq c_1(n)t^2$ uniformly for $\mu \in [0, 1]$. Therefore

$$\lim_{\mu \to 0} S(\mu) = S(0) = \int_0^\infty \frac{2dt}{1 + 2^{2n} t^{2n}} = \int_0^\infty \frac{dx}{1 + x^{2n}},$$

which proves, in the first place, that the right-hand side in (5.15) is finite if and only if $\theta \in [\theta_*, 1]$ and, secondly, that non-strict inequality (5.16) holds. Finally, for $n \geq 2$ we have strict inequality since

$$S'_\mu(0) = 2 \int_0^\infty \left[\frac{n 2^{2n} t^{2n+2}}{(1 + 2^{2n} t^{2n})^2} - \frac{t^2}{1 + 2^{2n} t^{2n}} \right] dt = \frac{1}{16n} \frac{\pi}{\sin \frac{3\pi}{2n}} > 0.$$

For $n = 1$ we have $\lambda = \mu$, $S'_\mu(0) < 0$ and

$$S(\mu) = \frac{\pi}{\sqrt{\mu + 4}}$$

is strictly decreasing not only at $\mu = 0$ but for all $\mu \geq 0$, the fact that we have already seen in (2.24). □

Remark 5.2 Inequality (5.13) holds for $\theta \in [1 - 1/(2n), 1]$, that is, when the weight of the stronger norm, which is the l_2-norm, increases. Accordingly, inequality (5.11) for periodic functions with mean value zero holds for θ in the *complementary* interval $\theta \in [0, 1 - 1/(2n)]$, when the weight of the stronger norm, which is the L_2-norm of the n-th derivative, increases:

$$\|f\|_{L_\infty}^2 \le \mathrm{C}_{1,n}^{\mathrm{per}}(\theta)\|f\|^{2\theta}\|f^{(n)}\|^{2(1-\theta)}, \quad \theta \in [0, 1 - \tfrac{1}{2n}], \tag{5.17}$$

where

$$f \in H^n_{\mathrm{per}}(\mathbb{S}^1), \qquad \int_0^{2\pi} f(x)dx = 0.$$

A general method for finding sharp constants in interpolation inequalities of L_∞-L_2-L_2-type was developed in [1,11,18], which was also used in the discrete case in the present paper. For example, for $n = 1$

$$\mathrm{C}_{1,1}^{\mathrm{per}}(\theta) = \frac{1}{\theta^\theta(1-\theta)^{1-\theta}} \sup_{\lambda\ge 0} \lambda^\theta G(\lambda), \quad \theta \in [0, 1/2], \tag{5.18}$$

where $G_\lambda(x,\xi) = \frac{1}{2\pi}\sum_{x\in\mathbb{Z}\setminus\{0\}} \frac{e^{ik(x-\xi)}}{k^2+\lambda}$ is the Green's function of the equation

$$-G_\lambda(x,\xi)'' + \lambda G_\lambda(x,\xi) = \delta(x-\xi), \quad x, \xi \in [0, 2\pi]^{\mathrm{per}},$$

and

$$G(\lambda) := G_\lambda(\xi,\xi) = \frac{1}{\pi}\sum_{k=1}^{\infty}\frac{1}{k^2+\lambda} = \frac{1}{2\pi}\frac{\pi\sqrt{\lambda}\coth(\pi\sqrt{\lambda}) - 1}{\lambda}.$$

For the limiting $\theta = \theta_* = 1/2$ the constant is the same as on $\mathbb{R}$: $\mathrm{C}_{1,1}^{\mathrm{per}}(\theta_*) = \mathrm{C}_{1,1}(\theta_*) = 1$. The graph of $\mathrm{C}_{1,1}^{\mathrm{per}}(\theta)$ on the interval $\theta \in [0, 1/2]$ is shown in Fig. 1 on the left. Observe that $\mathrm{C}_{1,1}(0) = \pi/6$.

6 Applications

6.1 Discrete and integral Carlson inequalities

We now discuss applications of the inequalities for the discrete operators, and our first group of results concerns Carlson inequalities. The original Carlson inequality [5] is as follows:

$$\left(\sum_{k=1}^{\infty} a_k\right)^2 \le \pi \left(\sum_{k=1}^{\infty} a_k^2\right)^{1/2} \left(\sum_{k=1}^{\infty} k^2 a_k^2\right)^{1/2}, \tag{6.1}$$

where the constant π is sharp and cannot be attained at a non identically zero sequence $\{a_k\}_{k=1}^{\infty}$. This inequality has attracted a lot of interest and has been a source of generalizations and improvements (see, for example, [10,12] and the references therein, and also [18] for the most recent strengthening of (6.1)). Inequality (6.1) has an integral analog (with the same sharp constant)

$$\left(\int_0^\infty g(t)dt\right)^2 \leq \pi \left(\int_0^\infty g(t)^2 dt\right)^{1/2} \left(\int_0^\infty t^2 g(t)^2 dt\right)^{1/2}. \tag{6.2}$$

As was first observed in [9], inequality (6.1) is equivalent to the inequality

$$\|f\|_\infty^2 \leq 1 \cdot \|f\| \|f'\|, \tag{6.3}$$

for periodic functions $f \in H^1_{\text{per}}(0, 2\pi)$, $\int_0^{2\pi} f(x)dx = 0$, by setting for a sequence $\{a_k\}_{k=1}^\infty$

$$f(x) = \sum_{k=-\infty}^{\infty} a_{|k|} e^{ikx}, \quad a_0 = 0.$$

Accordingly, inequality (6.3) for $f \in H^1(\mathbb{R})$ is equivalent (as was first probably observed in [15]) to (6.2) by setting $g = \mathcal{F}f$ and further restricting g (and f) to even functions. Furthermore, the unique (up to scaling) extremal function $f_*(x) = e^{-|x|}$ in (6.3) on the whole axis produces the extremal function $g_*(t) = 1/(1 + t^2)$ in (6.2).

In the similar way, discrete inequalities have equivalent integral analogs. Let $\mathscr{F}$ be the discrete Fourier transform $\mathscr{F} : \{a(n)\} \to \widehat{a}(x)$, where

$$\widehat{a}(x) = \sum_{n\in\mathbb{Z}^d} a(n)e^{inx}, \quad a(n) = (2\pi)^{-d} \int_{\mathbb{T}^d} \widehat{a}(x)e^{-inx}dx.$$

Then for $e_j = (0, \dots, 0, 1, 0, \dots, 0)$ with 1 on the jth place

$$\begin{aligned} \mathrm{D}_j a(n) = a(n + e_j) - a(n) &= (2\pi)^{-d} \int_{\mathbb{T}^d} \widehat{a}(x)(e^{-i(n+e_j)x} - e^{-inx})dx \\ &= (2\pi)^{-d} \int_{\mathbb{T}^d} \widehat{a}(x)e^{-ix/2}(-2i)\sin\tfrac{x_j}{2}\, e^{-inx}dx. \end{aligned}$$

Therefore

$$\|\mathrm{D}_j a\|^2 = (2\pi)^{-d} \int_{\mathbb{T}^d} |\widehat{a}(x)|^2 4 \sin^2 \tfrac{x_j}{2}\, dx. \tag{6.4}$$

and, finally,

$$\|a\|^2 = (2\pi)^{-d}\|\widehat{a}\|^2, \ \|\nabla a\|^2 = (2\pi)^{-d} \int_{\mathbb{T}^d} |\widehat{a}(x)|^2 4 \sum\nolimits_{j=1}^d \sin^2 \tfrac{x_j}{2}\, dx. \tag{6.5}$$

Thus, we have proved the following result.

Theorem 6.1 *Let $1/2 < \theta \le 1$. The inequality*

$$u(0)^2 \le \mathrm{K}_1(\theta)\|u\|^{2\theta}\|\mathrm{D}u\|^{2(1-\theta)}, \quad u \in l_2(\mathbb{Z})$$

established in Theorem 2.2 is equivalent to the inequality

$$\left(\int_0^{2\pi} g(x)dx\right)^2 \le 2\pi \mathrm{K}_1(\theta)\left(\int_0^{2\pi} g(x)^2dx\right)^{\theta}\left(\int_0^{2\pi} 4\sin^2\tfrac{x}{2}\, g(x)^2dx\right)^{1-\theta}, \tag{6.6}$$

for $g \in L_2(0, 2\pi)$. Here $\mathrm{K}_1(\theta) = \frac{1}{2}(2/\theta)^{\theta}(2\theta-1)^{\theta-1/2}$ [see (2.28)]. In the limiting case inequality (2.27) is equivalent to

$$\left(\int_0^{2\pi} g(x)dx\right)^2 \le \pi\sqrt{4 - \frac{\int_0^{2\pi} 4\sin^2\frac{x}{2}\, g(x)^2dx}{\int_0^{2\pi} g(x)^2dx}}$$
$$\times\left(\int_0^{2\pi} g(x)^2dx\right)^{1/2}\left(\int_0^{2\pi} 4\sin^2\tfrac{x}{2}\, g(x)^2dx\right)^{1/2} \tag{6.7}$$

Proof The proof follows from Theorem 2.2 and (6.4). We also point out that for $\theta \in (1/2, 1)$ inequality (6.6) saturates for

$$g_{\lambda_*}(x) = \frac{1}{\lambda_* + 4\sin^2\frac{x}{2}}, \quad \lambda_* = \lambda_*(\theta) = \frac{4\theta-2}{1-\theta}.$$

for $\theta = 1/2$ no extremals exist and maximizing sequence is obtained by letting $\lambda_* \to 0$; finally for $\theta = 1$, (6.6) saturates at constants.

For each d, $0 < d < 4$ and $\lambda(d) = \frac{2d}{2-d}$ inequality (6.7) saturates at

$$g_{\lambda(d)}(x) = \frac{1}{\lambda(d) + 4\sin^2\frac{x}{2}}, \quad \text{with} \quad \frac{\int_0^{2\pi} 4\sin^2\frac{x}{2}\, g_{\lambda(d)}(x)^2dx}{\int_0^{2\pi} g_{\lambda(d)}(x)^2dx} = d.$$

For $d = 2$, $g = \text{const}$. □

Remark 6.1 Corresponding to (5.8) is the integral inequality

$$\left(\int_0^{2\pi} g(x)dx\right)^2 \le \pi\sqrt{2}\left(\int_0^{2\pi} g(x)^2dx\right)^{3/4}\left(\int_0^{2\pi} 16\sin^4\tfrac{x}{2}\, g(x)^2dx\right)^{1/4}, \tag{6.8}$$

which turns into equality for

$$g_*(x) = \frac{1}{\frac{16}{3} + 16\sin^4\frac{x}{2}}.$$

Remark 6.2 The integral analog of the two dimensional discrete inequality is

$$\left(\int_{\mathbb{T}^2} g(x,y)dxdy\right)^2 \le (2\pi)^2 \mathrm{K}_2(\theta) \left(\int_{\mathbb{T}^2} g(x,y)^2 dxdy\right)^\theta \times \left(\int_{\mathbb{T}^2} 4(\sin^2\tfrac{x}{2}+\sin^2\tfrac{y}{2})\, g(x,y)^2 dxdy\right)^{1-\theta}, \tag{6.9}$$

where $\theta \in (0, 1]$, and $\mathrm{K}_2(\theta)$ is defined in (3.11).

Remark 6.3 In the d-dimensional case, $d \ge 3$, for the exponent $\theta = 0$ the Parseval's identities (6.5) provide an independent *elementary* proof of (4.8). In fact, setting $g_0(x) = 4\sum_{j=1}^d \sin^2 \frac{x_j}{2}$ we have

$$\begin{aligned}(2\pi)^{2d}|a(0)|^2 &= \left|\int_{\mathbb{T}^d} \widehat{a}(x)dx\right|^2 \\ &\le \left(\int_{\mathbb{T}^d} |\widehat{a}(x)| g_0(x)^{1/2} g_0(x)^{-1/2} dx\right)^2 \\ &\le \int_{\mathbb{T}^d} |\widehat{a}(x)|^2 g_0(x)dx \int_{\mathbb{T}^d} g_0(x)^{-1}dx = (2\pi)^{2d}\mathrm{K}_d(0)\|\nabla a\|^2, \end{aligned} \tag{6.10}$$

which proves (4.8).

This approach can be generalized to the l^p-case for the proof of the discrete Sobolev type inequality in the *non-limiting* case (1.12). Here in addition to the Parseval's identity we also use the Hausdorff–Young inequality (see, for instance, [2]):

$$\|\widehat{a}\|_{L_p(\mathbb{T}^d)} \le (2\pi)^{d/p}\|a\|_{l^{p'}(\mathbb{Z}^d)}, \tag{6.11}$$

where $p \ge 2$ and $p' = p/(p-1)$.

In fact, we have $\|\mathscr{F}\|_{l^2\to L_2} = (2\pi)^{d/2}$ and $\|\mathscr{F}\|_{l^1\to L_\infty} = 1$ and by the Riesz–Thorin interpolation theorem

$$\|\mathscr{F}\|_{l^{p'}\to L_p} \le (2\pi)^{d\theta/2} = (2\pi)^{d/p},$$

where $\frac{1}{p'} = \frac{\theta}{2} + \frac{1-\theta}{1}$, $\frac{1}{p} = \frac{\theta}{2} + \frac{1-\theta}{\infty}$. We also observe that (6.11) becomes an equality for $\widehat{a}(x) = 1$ and $a = \delta$.

Setting $q = 2p$ in (1.12), $v(n) := \overline{u(n)^p u(n)^{p-1}}$, and using the auxiliary inequality (6.12), (6.13) below, we obtain

$$\begin{aligned}\|u\|_{l^{2p}}^{2p} &= \sum_{n\in\mathbb{Z}^d} v(n)u(n) = \sum_{n\in\mathbb{Z}^d} (\mathrm{D}^*\mathrm{D}\Delta^{-1}v(n))u(n)\\ &= \sum_{n\in\mathbb{Z}^d} \mathrm{D}\Delta^{-1}v(n)\mathrm{D}u(n) \le \left(\sum_{n\in\mathbb{Z}^d} |\mathrm{D}\Delta^{-1}v(n)|^2\right)^{1/2} \|\mathrm{D}u\|\\ &\le \left(\frac{1}{4}(2\pi)^{d/p'} I_{p',d}\right)^{1/2} \|v\|_{l^{(2p)'}} \|\mathrm{D}u\| = \left(\frac{1}{4}(2\pi)^{d/p'} I_{p',d}\right)^{1/2} \|u\|_{l^{2p}}^{2p-1} \|\mathrm{D}u\|.\end{aligned}$$

It remains to prove (6.12). By Hölder's inequality and (6.11) we have

$$\begin{aligned}\sum_{n\in\mathbb{Z}^d} |\mathrm{D}\Delta^{-1}v(n)|^2 &= \frac{(2\pi)^d}{4}\int_{\mathbb{T}^d} \frac{|\widehat{v}(x)|^2 dx}{\sum_{j=1}^d \sin^2\frac{x_j}{2}}\\ &\le \frac{(2\pi)^d}{4} I_{p',d}\left(\int_{\mathbb{T}^d} |\widehat{v}(x)|^{2p}dx\right)^{1/p}\\ &\le \frac{(2\pi)^d}{4} I_{p',d}(2\pi)^{d/p}\|v\|_{l^{(2p)'}}^2 = \frac{1}{4}(2\pi)^{d(p+1)/p} I_{p',d}\|v\|_{l^{(2p)'}}^2 \qquad (6.12)\end{aligned}$$

where

$$I_{p',d} = \left(\int_{\mathbb{T}^d} \frac{dx}{(\sum_{j=1}^d \sin^2\frac{x_j}{2})^{p'}}\right)^{1/p'} < \infty \quad \text{for } p' < d/2 \Leftrightarrow p > d/(d-2). \qquad (6.13)$$

Thus we obtain the following result.

Theorem 6.2 *Let $d \ge 3$ and $2p > 2d/(d-2)$. Then*

$$\|u\|_{l^{2p}(\mathbb{Z})^d}^2 = \left(\sum_{n\in Z^d} |u(n)|^{2p}\right)^{1/p} \le \frac{1}{4}(2\pi)^{d(p+1)/p} I_{p',d}\|\mathrm{D}u\|^2,$$

where $I_{p',d}$ is defined in (6.13).

Remark 6.4 We do not claim that the constant here is sharp. Moreover, it blows up for $2p = 2d/(d-2)$, while it can be shown that the inequality still holds. However, the constant is sharp in the opposite limit $p = \infty$, see (4.8).

6.2 Spectral inequalities for discrete operators

Interpolation inequalities characterizing imbeddings of Sobolev spaces into the space of bounded continuous functions have important applications in spectral theory. The original fruitful idea in [7] has been generalized in [6] to give best-known estimates for the Lieb–Thirring constants in estimates for the negative trace of Schrödinger operators.

In this section we apply our sharp interpolation inequalities with the method of [7] for estimates of the negative trace of the discrete operators [16].

We write the inequalities obtained above in the unform way

$$\sup_{k\in\mathbb{Z}^d} u(k)^2 \leq \mathrm{K}(\theta)\|u\|^{2\theta}\|\mathcal{D}^n u\|^{2(1-\theta)}, \tag{6.14}$$

where $\mathcal{D}^n$ is as in (5.14) and θ belongs to a certain subinterval of [0, 1] uniquely defined in the corresponding theorem:

$$\theta \in \begin{cases} [1-1/(2n), 1], & d=1, n\geq 1; \\ (0, 1], & d=2, n=1; \\ [0, 1], & d\geq 3, n=1. \end{cases} \tag{6.15}$$

Theorem 6.3 *Let* $\{u^{(j)}\}_{j=1}^N \in l^2(\mathbb{Z}^d)$ *be a family of N sequences that are orthonormal with respect to the natural scalar product in* $l^2(\mathbb{Z}^d)$. *We set*

$$\rho(k) := \sum_{j=1}^N u^{(j)}(k)^2, \quad k\in\mathbb{Z}^d. \tag{6.16}$$

Then for θ *as in* (6.15) *and* $\theta < 1$

$$\|\rho\|_{l^{\frac{2-\theta}{1-\theta}}}^{\frac{2-\theta}{1-\theta}} = \sum_{k\in\mathbb{Z}^d} \rho(k)^{\frac{2-\theta}{1-\theta}} \leq \mathrm{K}(\theta)^{\frac{1}{1-\theta}} \sum_{j=1}^N \|\mathcal{D}^n u^{(j)}\|^2. \tag{6.17}$$

Proof For arbitrary $\xi_1, \ldots, \xi_N \in \mathbb{R}$ we construct a sequence $f\in l^2(\mathbb{Z}^d)$

$$f(k) := \sum_{j=1}^N \xi_j u^{(j)}(k), \quad k\in\mathbb{Z}^d.$$

Applying (6.14) and using orthonormality we obtain for a fixed k

$$f(k)^2 \leq \mathrm{K}(\theta)\left(\sum_{j=1}^N \xi_j^2\right)^{\theta}\left(\sum_{i,j=1}^N \xi_i\xi_j\big(\mathcal{D}^n u^{(i)}, \mathcal{D}^n u^{(j)}\big)\right)^{1-\theta}.$$

We now set $\xi_j := u^{(j)}(k)$:

$$\rho(k)^2 \leq \mathrm{K}(\theta)\rho(k)^{\theta}\left(\sum_{i,j=1}^N u^{(i)}(k)u^{(j)}(k)\big(\mathcal{D}^n u^{(i)}, \mathcal{D}^n u^{(j)}\big)\right)^{1-\theta},$$

or

$$\rho(k)^{\frac{2-\theta}{1-\theta}} \leq \mathrm{K}(\theta)^{\frac{1}{1-\theta}} \sum_{i,j=1}^{N} u^{(i)}(k)u^{(j)}(k)\big(\mathcal{D}^n u^{(i)}, \mathcal{D}^n u^{(j)}\big).$$

Summing over $k \in \mathbb{Z}^d$ and using orthonormality we obtain (6.17). □

Corollary 6.1 *Setting $N = 1$ in Theorem 6.3 we obtain a family of interpolation inequalities for $u \in l^2(\mathbb{Z}^d)$*

$$\|u\|_{l^{\frac{2(2-\theta)}{1-\theta}}(\mathbb{Z}^d)} \leq \mathrm{K}(\theta)^{\frac{1}{2(2-\theta)}} \|u\|^{\frac{1}{2-\theta}} \|\mathcal{D}^n u\|^{\frac{1-\theta}{2-\theta}}. \tag{6.18}$$

In particular, to mention a few examples with limiting θ

$$\|u\|_{l^6(\mathbb{Z})} \leq 1 \cdot \|u\|^{2/3} \|\mathrm{D}u\|^{1/3}, \quad \theta = 1/2,$$
$$\|u\|_{l^{10}(\mathbb{Z})} \leq 2^{-1/5} \|u\|^{4/5} \|\Delta u\|^{1/5}, \quad \theta = 3/4,$$

in dimension $d \geq 3$

$$\|u\|_{l^4(\mathbb{Z}^d)} \leq \mathrm{K}_d(0)^{1/4} \|u\|^{1/2} \|\nabla u\|^{1/2}, \quad \theta = 0.$$

Remark 6.5 The last inequality holding in dimension three and higher curiously resembles the celebrated Ladyzhenskaya inequality that is vital for the uniqueness of the weak solutions of the *two-dimensional* Navier–Stokes system:

$$\|f\|_{L_4(\Omega)} \leq c_{\mathrm{L}} \|f\|^{1/2} \|\nabla f\|^{1/2}, \quad f \in H_0^1(\Omega),\ \Omega \subseteq \mathbb{R}^2.$$

We now exploit the equivalence between the inequalities for orthonormal families and spectral estimates for the negative trace of the Schrödinger operators [14].

We consider the discrete Schrödinger operator

$$H := (-1)^n \Delta^n - V, \tag{6.19}$$

acting on $u \in l^2(\mathbb{Z}^d)$ as follows

$$Hu(k) = (-1)^n \Delta^n u(k) - V(k)u(k).$$

Theorem 6.4 *Let $V(k) \geq 0$ and let $V(k) \to 0$ as $|k| \to \infty$, then the negative spectrum of H is discrete and satisfies the estimate*

$$\sum |\lambda_j| \leq \mathrm{K}(\theta) \frac{(1-\theta)^{1-\theta}}{(2-\theta)^{2-\theta}} \sum_{k \in \mathbb{Z}^d} V(k)^{2-\theta}. \tag{6.20}$$

Proof Suppose that there exists N negative eigenvalues $-\lambda_j < 0$, $j = 1, \ldots, N$ with corresponding N orthonormal eigenfunctions $u^{(j)}$:

$$(-1)^n \Delta^n u^{(j)}(k) - V(k)u^{(j)}(k) = -\lambda_j u^{(j)}(k).$$

Taking the scalar product with $u^{(j)}$, summing the the results with respect to j, and using (6.16), Hölder inequality and (6.17), we obtain

$$\begin{aligned}
\sum_{j=1}^{N} \lambda_j &= (V, \rho) - \sum_{j=1}^{N} \|\mathcal{D}^n u^{(j)}\|^2 \\
&\le \|V\|_{l_{2-\theta}} \|\rho\|_{l_{\frac{2-\theta}{1-\theta}}} - \mathrm{K}(\theta)^{-\frac{1}{1-\theta}} \|\rho\|_{l_{\frac{2-\theta}{1-\theta}}}^{\frac{2-\theta}{1-\theta}} \\
&\le \max_{y>0} \left(\|V\|_{l_{2-\theta}} y - \mathrm{K}(\theta)^{-\frac{1}{1-\theta}} y^{\frac{2-\theta}{1-\theta}} \right) \\
&= \mathrm{K}(\theta) \frac{(1-\theta)^{1-\theta}}{(2-\theta)^{2-\theta}} \sum_{k \in \mathbb{Z}^d} V(k)^{2-\theta}.
\end{aligned}$$

□

6.3 Examples

$d = 1, n = 1, \theta = 1/2$. Then $\mathrm{K} = 1$ and the negative trace of the operator

$$-\Delta - V \quad \text{in } l^2(\mathbb{Z})$$

satisfies

$$\sum |\lambda_j| \le \frac{2}{3\sqrt{3}} \sum_{\alpha=-\infty}^{\infty} V^{3/2}(\alpha).$$

$d = 1, n = 2, \theta = 3/4$. Then $\mathrm{K} = \sqrt{2}/2$ and the negative trace of the operator

$$\Delta^2 - V \quad \text{in } l^2(\mathbb{Z})$$

satisfies

$$\sum |\lambda_j| \le \frac{2\sqrt{2}}{5^{5/4}} \sum_{\alpha=-\infty}^{\infty} V^{5/4}(\alpha).$$

$d \ge 3, n = 1, \theta = 0$. Then $\mathrm{K} = \mathrm{K}_d(0)$ and the negative trace of the operator

$$-\Delta - V \quad \text{in } l^2(\mathbb{Z}^d)$$

satisfies

$$\sum |\lambda_j| \leq \frac{\mathrm{K}_d(0)}{4} \sum_{\alpha \in \mathbb{Z}^d} V^2(\alpha).$$

In particular, in three dimensions

$$\frac{\mathrm{K}_3(0)}{4} = 0.0631\ldots.$$

Acknowledgments Funding for this research was provided by the grant of the Russian Federation Government to support scientific research under the supervision of leading scientist at Siberian Federal University, no. 14.Y26.31.0006, and from the Russian Foundation for Basic Research (grant no. 12-01-00203).

References

1. Bartuccelli, M.V., Deane, J., Zelik, S.V.: Asymptotic expansions and extremals for the critical Sobolev and Gagliardo–Nirenberg inequalities on a torus. Proc. R. Soc. Edinb. **143A**, 445–482 (2013)
2. Bergh, J., Löfström, J.: Interpolation Spaces. An Introduction. Grundlehren der Mathematischen Wissenschaften, No. 223. Springer, Berlin (1976)
3. Borwein, J.M., Glasser, M.L., McPhedran, R.C., Wan, J.G., Zucker, I.J.: Lattice Sums Then and Now. Encyclopedia of Mathematics and its Applications, vol. 150. Cambridge University Press, Cambridge (2013)
4. Brezis, H., Gallouet, T.: Nonlinear Schrödinger evolution equations. Nonlinear Anal. **4**, 677–681 (1980)
5. Carlson, F., Une inégalité. Ark. Mat. Astr. Fysik **25B**(1) (1934)
6. Dolbeault, J., Laptev, A., Loss, M.: Lieb–Thirring inequalities with improved constants. J. Eur. Math. Soc. **10**(4), 1121–1126 (2008)
7. Eden, A., Foias, C.: A simple proof of the generalized Lieb–Thirring inequalities in one space dimension. J. Math. Anal. Appl. **162**, 250–254 (1991)
8. Gradshteyn, I.S., Ryzhik, I.M.: Table of Integrals, Series and Products. Academic Press, London (2007)
9. Hardy, G.H.: A note on two inequalities. J. Lond. Math. Soc. **11**, 167–170 (1936)
10. Hardy, G.H., Littlewood, J.E., Pólya, G.: Inequalities. Cambridge University Press, Cambridge (1934) [Addendum by Levin, V.I., Stechkin, S.B.: To the Russian translation, GIIL, Moscow (1948); English transl. in Levin, V.I., Stechkin, S.B.: Inequalities. Am. Math. Soc. Transl. **14**, 1–22 (1960)]
11. Ilyin, A.A.: Best constants in multiplicative inequalities for sup-norms. J. Lond. Math. Soc. **2**(58), 84–96 (1998)
12. Larsson, L., Maligranda, L., Pečarić, J., Persson, L.-E.: Multiplicative inequalities of Carlson type and interpolation. World Scientific, Singapore (2006)
13. Laptev, A., Weidl, T.: Sharp Lieb–Thirring inequalities in high dimensions. Acta Math. **184**, 87–111 (2000)
14. Lieb, E., Thirring, W.: Inequalities for the moments of the eigenvalues of the Schrödinger Hamiltonian and their relation to Sobolev inequalities. In: Studies in Mathematical Physics. Essays in Honor of Valentine Bargmann, pp. 269–303. Princeton University Press, Princeton (1976)
15. Sz.-Nagy, B.: Über integralungleichungen zwischen einer funktion und ihrer ableitung. Acta Univ. Szeged Sect. Sci. Math. **10**, 64–74 (1941)
16. Sahovic, A.: New constants in discrete Lieb–Thirring inequalities for Jacobi matrices. Problems in mathematical analysis. No. 45. J. Math. Sci. (N. Y.) **166**(3), 319–327 (2010)
17. Taikov, L.V.: Kolmogorov-type inequalities and the best formulas for numerical differentiation. Mat. Zametki **4**, 233–238 (1968) [English transl. Math. Notes **4**, 631–634 (1968)]
18. Zelik, S.V., Ilyin, A.A.: Green's function asymptotics and sharp interpolation inequalities. Uspekhi Mat. Nauk **69**(2), 23–76 (2014) [English transl. in Russ. Math. Surv. **69**, 2 (2014)]

9

Road systems and betweenness

Paul Bankston

Abstract A *road system* is a collection of subsets of a set—the *roads*—such that every singleton subset is a road in the system and every doubleton subset is contained in a road. The induced ternary (betweenness) relation is defined by saying that a point c *lies between* points a and b if c is an element of every road that contains both a and b. Traditionally, betweenness relations have arisen from a plethora of other structures on a given set, reflecting intuitions that range from the order-theoretic to the geometric and topological. In this paper we initiate a study of road systems as a simple mechanism by means of which a large majority of the classical interpretations of betweenness are induced in a uniform way.

Keywords Betweenness · axiomatizations · Road systems · Intervals · R-relations · Partial orders · Real vector spaces · Metric spaces · Connected topological spaces

Mathematics Subject Classification (2010) Primary 08A02; Secondary 03B10 · 06A06 · 52A05 · 54D05 · 54E35 · 54F15

1 Introduction

The ancient notion of a point lying between two given points on a geometric line or a totally ordered set has strong intuitive appeal, and has been generalized in a number of

---

Communicated by S.K. Jain.

The author would like to thank the anonymous referee for valuable suggestions regarding the exposition of this paper.

P. Bankston (✉)
Department of Mathematics, Statistics and Computer Science,
Marquette University, Milwaukee, WI 53201-1881, USA
e-mail: paulb@mscs.mu.edu

interesting directions. In all of these, betweenness is taken to be a ternary relation that satisfies certain conditions, usually expressible via sentences of the first-order language L_t, built up using a single ternary predicate symbol, plus equality (see, e.g., [7] for background). The earliest example of this known to us occurs in the monograph [18] of M. Pasch in 1882, and there is an extensive body of work from various authors since then (see, e.g., [9] for an excellent bibliography of betweenness-related work before 1970), but no definitive axiom system for betweenness has emerged. The reason for this is not particularly surprising; there are many well-studied—often higher-order—structures that naturally give rise to notions of betweenness, but which also behave in their own indiosyncratic ways. These other structures reflect intuitions that range from the order-theoretic to the geometric and topological, and are what we informally refer to here as *pre-betweenness structures*.

In this article we take the intuitive view of a point c lying between points a and b if "c blocks all roads connecting a and b". This naïve approach suggests that what we are looking for in a natural pre-betweenness structure is a formalized system of "roads" that "connect" (or, "allow travel from") one point to another.

The essential idea is that a road connecting points a and b of a set X is a special subset of X which contains both points. Many roads may connect a and b; if it so happens that a point c is contained in each of them, then we say that *c lies between a and b*. The notion of *road system*—i.e., that of a set of points together with a collection of roads connecting them—provides a way of viewing most, if not all, of the pre-betweennes structures encountered in the literature as instances of the same general phenomenon.

2 Betweenness from road systems

More formally, a *road system* is a pair $\langle X, \mathcal{R}\rangle$, where X is a nonempty set and $\mathcal{R}$ is a collection of nonempty subsets of X—the *roads*—such that: (1) for each $a \in X$, the singleton set $\{a\}$ is a road (the most efficient way to get from here to here is to stay put); and (2) each two points $a, b \in X$ belong to at least one road (you can always get there from here). We denote by $\mathcal{R}(a, b)$ the set of roads that contain both a and b; colloquially, roads in $\mathcal{R}(a, b)$ *connect* the two points.

Each road system $\langle X, \mathcal{R}\rangle$ naturally induces a ternary (betweenness) relation $[\ ,\ ,\ ]_{\mathcal{R}} \subseteq X^3$ as follows: $[a, c, b]_{\mathcal{R}}$ holds precisely when every road containing a and b also contains c. Another way of saying this is to define the *$\mathcal{R}$-interval* $[a, b]_{\mathcal{R}}$ to be $\bigcap \mathcal{R}(a, b)$, and to take $[a, c, b]_{\mathcal{R}}$ to mean that $c \in [a, b]_{\mathcal{R}}$. We call the elements a, b *bracket points* for the interval $[a, b]_{\mathcal{R}}$, noting that it may well be the case for an interval to have more than one set of bracket points (see Subsect. 3.2 below).

If $\langle X, [\ ,\ ,\ ]\rangle$ is any ternary relational structure, we use interval notation just as above: for fixed $a, b \in X$, $[a, b] := \{x \in X : [a, x, b] \text{ holds in } X\}$. A ternary relation on X is defined to be an *R-relation* if it is induced by a road system on X.

Some obvious L_t-formulable facts about R-relations are (the universal closures of) the following, where $\rightarrow$ and $\wedge$ denote implication and logical conjunction, respectively:

(R1) Symmetry: $[a, c, b] \to [b, c, a]$ (so $[a, b] = [b, a]$);
(R2) Reflexivity: $[a, b, b]$ (so—with R1—$[a, b] \supseteq \{a, b\}$);
(R3) Minimality: $[a, b, a] \to b = a$ (so—with R2—$[a, a] = \{a\}$); and
(R4) Transitivity: $([a, c, b] \wedge [a, x, c]) \to [a, x, b]$ (so if $c \in [a, b]$, then $[a, c] \subseteq [a, b]$).

By R1 and R2, intervals are guaranteed to contain their bracket points. An interval that comprises at most two points is called *trivial*; trivial intervals containing exactly two points are called *gaps*.

Note that, in the presence of R1, R4 implies that $[a, c] \cup [c, b] \subseteq [a, b]$ whenever $c \in [a, b]$. We will see later that the reverse inclusion does not generally hold for road systems, but that it does hold for a significant class of them. A ternary relation on a set is called *basic* if it satisfies R1, R2, and R3, and is τ-*basic* if it satisfies R4 as well.

We remark that there are many statements which could justifiably be called "transitivity"; see the paper [19] of Pitcher and Smiley for a list that is *exhaustive* in a precise sense. (In particular, our R4 is their axiom t_2.)

If we fix $a \in X$ and define the binary relation $\leq_a$ by saying that $x \leq_a y$ just in case $[a, x, y]$ holds, then R2 and R4 say, respectively, that the relation $\leq_a$ is reflexive and transitive in the usual sense, and is hence a pre-order. Also the conjunction of R1, R2, and R3 says that a is the unique $\leq_a$-minimal element of X. We will return later to consider when these pre-orders arising from τ-basic relations are also antisymmetric, and hence partial orders.

Another obvious first-order fact about R-relations is the following strengthened transitivity axiom:

(R5) Strong transitivity: $([a, c, b] \wedge [a, d, b] \wedge [c, x, d]) \to [a, x, b]$ (so if $c, d \in [a, b]$, then $[c, d] \subseteq [a, b]$).

Strong transitivity appears in [19] as axiom T_5. It clearly implies transitivity in the presence if R1 and R2: just replace each occurrence of d with a.

We now show that the class of road systems is L_t-*axiomatizable*, meaning that there is a set of L_t-sentences characterizing when a ternary relation on a set is an R-relation.

Theorem 2.0.1 *A ternary relation on a set is an R-relation if and only if it satisfies conditions R1–R5; i.e., it is basic and strongly transitive.*

Proof The less trivial direction is to assume the conditions R1—R5 hold for $\langle X, [\ ,\ ,\ ]\rangle$; and it turns out that the naïve choice of taking $\mathcal{R}$ to be $\{[a, b] : a, b \in X\}$ is a good one. From R1 and R2 we know each $[a, b] \in \mathcal{R}$ contains both a and b, so every doubleton set from X is contained in a member of $\mathcal{R}$. If $a \in X$, then R3 gives us $\{a\} = [a, a] \in \mathcal{R}$. Thus $\mathcal{R}$ is a road system, and it remains to show that $[a, b]$ always equals the intersection of $\mathcal{R}(a, b)$. Clearly $[a, b] \supseteq \bigcap \mathcal{R}(a, b)$ since $[a, b] \in \mathcal{R}(a, b)$. For the reverse inclusion, suppose $R \in \mathcal{R}(a, b)$ is chosen. Then, by definition, there are $c, d \in X$ with $R = [c, d]$. Since $a, b \in [c, d]$, we infer by R5 that $[a, b] \subseteq [c, d]$. Hence $[a, b] \subseteq \bigcap \mathcal{R}(a, b)$, and the desired equality is established. □

Remarks 2.0.2 (i) In [19] there is a simple example of a τ- basic relation on a five-element set that is not strongly transitive. (Just declare $[a, c, b]$, $[a, d, b]$, and

$[c, e, d]$ as the only generating triples in the relation.) Hence being transitive is not enough to ensure that a basic ternary relation is an R-relation (see also Subsects. 3.2 and 3.6 below).

(ii) Bearing in mind that R1—R4 involve four variables and that R5 involves five, it would be interesting to see whether R-relations may be L_t- axiomatized using fewer than five variables. This question is related to the topic of [19], but is not answered there. (See also Remark 4.0.6 (ii) below.)

(iii) The road system offered in the proof of Theorem 2.0.1 is merely the most convenient to make the proof work; generally there are many road systems giving rise to the same R-relation. It is clearly always the case that if $\mathcal{R} \subseteq \mathcal{S}$ then $[\ ,\ ,\ ]_{\mathcal{R}} \supseteq [\ ,\ ,\ ]_{\mathcal{S}}$, but it is possible for there to be road systems $\mathcal{R}$ and $\mathcal{S}$ on X such that $[\ ,\ ,\ ]_{\mathcal{R}} = [\ ,\ ,\ ]_{\mathcal{S}}$, with $\mathcal{R} \cap \mathcal{S}$ comprising the singletons of X. (See Theorem 4.0.7 and Example 5.0.9 below.)

3 Some classical pre-betweenness structures

In this section we survey some classical interpretations of betweenness arising from order-theoretic and geometric structures. We show that all of these ternary relations are basic, and that the vast majority of them are R-relations.

3.1 General partial orders

If $\langle X, \leq \rangle$ is a partially ordered set, then one may define, as does Birkhoff [5], the betweenness interpretation $[\ ,\ ,\ ]_P$ on X by the condition that $[a, c, b]_P$ holds for $\langle a, b, c \rangle \in X^3$ just in case either $a \leq c \leq b$ or $b \leq c \leq a$ is true. (Note: In the sequel, when we define betweenness relations in a manner other than that given by road systems, we use subscripts as a mnemonic. Thus the subscripted letter "P" just means "partially ordered set in the Birkhoff interpretation".)

As it stands, this interpretation agrees with that given in the Huntington-Kline paper [12] when the partial ordering is total; but it does not define a basic ternary relation on X otherwise, as R2 clearly fails when there are incomparable elements. However, if we modify the Birkhoff interpretation by defining $[a, c, b]_O$ to hold just in case $([a, c, b]_P \vee (c = a) \vee (c = b))$ does (where $\vee$ is logical disjunction), then we clearly have an R-relation whose collection of order intervals of the form $[a, b]_O := [a, b]_P \cup \{a, b\}$ provides an inducing road system. So under the modified interpretation, what used to be empty intervals now become gaps.

One of the main results of [12] is that one may characterize—using a finite list of L_t-axioms that each involve no more than four variables—when a ternary relation is the betweenness relation naturally induced by a total ordering.

Birkhoff [5] gives as exercises several L_t-sentences that follow from properties of general partial orders, and Altwegg [2] later provides an infinite L_t-axiomatization of the class of partially ordered sets under the Birkhoff interpretation.

Düntsch and Urquhart [8] do likewise for betweenness interpretations induced in the same way by reflexive antisymmetric binary relations, no transitivity implied.

They show the Altwegg axiomatization result to be a corollary of theirs; they also show that in any such axiomatization, it is necessary that infinitely many variables be used.

While it is true that $[\ ,\ ,\ ]_O$ is first-order definable from $[\ ,\ ,\ ]_P$, the opposite is not the case: both a total ordering and an antichain ordering—i.e., no two elements are comparable—on a two-element set give rise to the same interpretation for $[\ ,\ ,\ ]_O$, but not for $[\ ,\ ,\ ]_P$. In spite of this, it seems quite likely that the axiomatizability conclusions in [8] may be extended to hold for the modified interpretation.

3.2 Doubly directed partial orders

If $\langle X, \leq \rangle$ is a partially ordered set and $d \leq e$, then the order interval $[d, e]_O$ defined above is just $\{x \in X : d \leq x \leq e\}$. We call the ordering *doubly directed* if, for each $a, b \in X$, there are $d, e \in X$ with $d \leq e$ and $a, b \in [d, e]_O$; i.e., the ordering is directed both downwardly and upwardly. So, in doubly directed partial orders, each two-element set is contained in at least one order interval, and we define $[a, c, b]_D$ to mean that c is contained within each order interval also containing a and b. Thus, by taking $\mathcal{R}$ to be the class of order intervals $[c, d]_O$ with $c \leq d$, we obtain a road system inducing $[\ ,\ ,\ ]_D$.

Note that this interpretation of betweenness *enhances* the one given by Birkhoff [5], in the sense that there is agreement when the bracket points are $\leq$-comparable. As far as we know, there is no L_t-axiomatization of doubly directed partial orderings under this interpretation.

When the doubly directed partial ordering is a lattice, the interval $[a, b]_D$ is just $[a \sqcap b, a \sqcup b]_O$ (where $\sqcap$ and $\sqcup$ are the lattice operations of meet and join, respectively). Here is an illustration of how intervals may not determine their sets of bracket points, as $[a, b]_D$ always equals $[a \sqcap b, a \sqcup b]_D$.

However, there is another notion of betweenness for lattices, itself well known [19], and the two notions agree exactly when the lattice is distributive. Note that, in general lattices, the inequalities $(a \sqcap b) \sqcup (c \sqcap b) \leq (a \sqcup c) \sqcap b$ and $(a \sqcap c) \sqcup b \leq (a \sqcup b) \sqcap (c \sqcup b)$ hold. Hence, if we define $[a, c, b]_G$ to mean that $(a \sqcap c) \sqcup (b \sqcap c) = c = (a \sqcup c) \sqcap (b \sqcup c)$, we may easily see that $[a, c, b]_G$ implies $[a, c, b]_D$. The definition of $[\ ,\ ,\ ]_G$ is due to Glivenko [10,11], who was studying metric lattices. He showed that this purely lattice-theoretic condition coincides with the metric definition of betweenness (see Subsect. 3.6 below). It is easy to show that $[\ ,\ ,\ ]_G$ is a basic ternary relation, but—from Theorem 9.3 in [19]—strong transitivity holds just in case the lattice is distributive (i.e., when $[\ ,\ ,\ ]_G = [\ ,\ ,\ ]_D$). Even R4 is not guaranteed: By Theorem 9.1 in [19], transitivity holds for $[\ ,\ ,\ ]_G$ if and only if the lattice is modular (i.e., satisfies the condition $c \leq b \rightarrow (c \sqcup (a \sqcap b) = (c \sqcup a) \sqcap b)$). In view of Theorem 2.0.1, this immediately gives us:

Corollary 3.2.1 *The basic ternary relation on a lattice arising from the Glivenko interpretation of betweenness is an R-relation precisely when the lattice is distributive.*

3.3 Trees

Another betweenness interpretation enhancing the order-theoretic one used in [12] and [5] arises when the partial ordering $\langle X, \leq\rangle$ is a *tree*; i.e., the ordering is downwardly directed, but no two order-incomparable elements have a common upper bound. In trees the predecessors of any element form a totally ordered subset, or *chain*. If $a, b, c \in X$, define $[a, c, b]_T$ to hold just in case $c \in [d, a]_O \cup [d, b]_O$ for every common lower bound d of a, b. By taking $\mathcal{R}(a, b)$ to consist of sets $V(a, b, d) := [d, a]_O \cup [d, b]_O$ for $d \leq a, b$, we clearly have a road system inducing $[\ ,\ ,\ ]_T$. Betweenness relations arising from trees are extensively studied in the Adeleke-Neumann paper [1].

3.4 Branch sets of trees

This topic is also explored in [1]. Here we define our betweenness relation not on the tree ordering $\langle Y, \leq\rangle$ itself, but on the set $X = \mathfrak{B}(Y)$ of its *branches*; i.e., maximal chains. If $\alpha, \beta, \gamma \in X$, define $[\alpha, \gamma, \beta]_{\mathfrak{B}}$ to hold just in case $\alpha \cap \gamma \supseteq \alpha \cap \beta$ (equivalently, just in case $\gamma \cap \beta \supseteq \alpha \cap \beta$). In this way the interval $[\alpha, \beta]_{\mathfrak{B}}$ consists of all branches that "coalesce" with α and β above where α and β "coalesce" with each other. If we view branches also as leaf nodes of Y (so $X \subseteq Y$), then $[\alpha, \gamma, \beta]_{\mathfrak{B}}$ holds exactly when $[\alpha, c, \beta]_T$ holds for some common lower bound c of α, γ (i.e., some $c \in \alpha \cap \gamma$). By taking our roads to consist of sets $V(a) := \{\alpha \in X : a \in \alpha\}$, $a \in Y$, we obtain a road system that induces $[\ ,\ ,\ ]_{\mathfrak{B}}$.

3.5 Real vector spaces

If X is a vector space over the real field $\mathbb{R}$ and $a, b \in X$ are two vectors, the *linear betweenness relation* is given by $[a, c, b]_L$ just in case $c = ta + (1 - t)b$ for some $0 \leq t \leq 1$; i.e., c is a convex linear combination of a and b.

Proposition 3.5.1 *A linear betweenness relation is an R-relation.*

Proof A linear betweenness relation is clearly basic; all we need to show is that R5 holds, and then use Theorem 2.0.1. (This will show that letting $\mathcal{R}$ consist of the intervals $[a, b]_L$ will yield an inducing road system.) Indeed, suppose a and b are given and $c, d \in [a, b]_L$. Then we have $c = sa + (1 - s)b$ and $d = ta + (1 - t)b$ for some $0 \leq s, t \leq 1$. And if $x \in [c, d]_L$, say $x = uc + (1 - u)d$, $0 \leq u \leq 1$, we then have $x = va + (1 - v)b$, where $v = us + (1 - u)t$. Since $0 \leq s, t, u \leq 1$, so is v, and hence $x \in [a, b]_L$. □

3.6 Metric spaces

If $\langle X, \rho\rangle$ is a metric space—so $\rho : X^2 \to [0, \infty)$; and $\rho(x, y) = \rho(y, x)$, $\rho(x, y) = 0$ iff $x = y$, and $\rho(x, z) \leq \rho(x, y) + \rho(y, z)$ universally hold (see, e.g., [15])—then the *strict metric betweenness relation* is defined by saying $[a, c, b]_{SM}$ holds just in case $a \neq c \neq b$ and $\rho(a, b) = \rho(a, c) + \rho(c, b)$. This notion of betweenness was first studied by Menger [14]; we modify the definition—as is done in [19]—by including R2 instead of its negation. That is, we define the *metric betweenness relation* by saying $[a, c, b]_M$ holds just in case $\rho(a, b) = \rho(a, c) + \rho(c, b)$. Hence we have the straightforward interdefinability

$$[a, c, b]_M \Leftrightarrow ([a, c, b]_{SM} \vee (c = a) \vee (c = b))$$
$$[a, c, b]_{SM} \Leftrightarrow ([a, c, b]_M \wedge (c \neq a) \wedge (c \neq b)).$$

As with linear betweenness, one might reasonably expect that an appropriate road system for metric betweenness comes about by letting $\mathcal{R}$ consist of the metric intervals $[a, b]_M$. Surprisingly, this approach does not work in general: in [14] it is shown that while strict metric betweenness is transitive—and hence metric betweenness—is τ-basic—strong transitivity does not necessarily hold. (See the "railroad space", [14], p. 80). This means that $[\ ,\ ,\ ]_M$ is not always an R-relation. And while $\mathcal{R}$ is a perfectly good road system, it is possible for $[a, b]_M$ to properly contain $\bigcap \mathcal{R}(a, b)$.

Smiley [20] considers betweenness relations that arise from multiple structures on the same underlying set. For example, if a real vector space has a norm $\|\ \|$, then it has the metric $\rho(a, b) := \|a - b\|$, and hence the two betweenness notions $[\ ,\ ,\ ]_L$ and $[\ ,\ ,\ ]_M$. Betweenness in the linear sense generally entails betweenness in the metric, but the converse is true exactly when the normed space is *rotund*; i.e., when $\|a\| + \|b\| = \|a + b\|$ for nonzero a, b happens exactly when one of the two is a non-negative scalar multiple of the other. Hence, using Proposition 3.5.1, we obtain a weak analogue of Corollary 3.2.1.

Corollary 3.6.1 *In a rotund normed vector space, the ternary relation arising from the Menger interpretation of betweenness is an R-relation.*

4 Betweenness from additive road systems

A road system is *additive* if the union of two overlapping roads is a road. Additive road systems may be regarded as arising from topological intuitions rather than geometric; indeed any connected topological space may be seen as a road system, with its connected subsets for roads. We will have more to say about topological interpretations of betweenness in Sect. 6 below.

A ternary relation on X is defined to be an *AR-relation* if it is induced by an additive road system on X. An important first-order betweenness feature of AR-relations is the following:

(R6) Disjunctivity: $[a, x, b] \rightarrow ([a, x, c] \vee [c, x, b])$ (so for any $a, b, c \in X$, $[a, b] \subseteq [a, c] \cup [c, b]$).

To see this, suppose $\mathcal{R}$ is an additive road system and that $x \notin [a, c]_{\mathcal{R}} \cup [c, b]_{\mathcal{R}}$. Then there are $R \in \mathcal{R}(a, c)$ and $S \in \mathcal{R}(c, b)$ such that x is in neither R nor S. Since $c \in R \cap S$, we have $R \cup S \in \mathcal{R}(a, b)$. Since $x \notin R \cup S$, we know $x \notin [a, b]_{\mathcal{R}}$.

The ternary relation $\langle X, [\ ,\ ,\ ]\rangle$ is *weakly disjunctive* if $[a, b] \subseteq [a, c] \cup [c, b]$ for any $c \in [a, b]$. Since the reverse inclusion is automatic for all τ-basic relations, we have the equality $[a, b] = [a, c] \cup [c, b]$, for $c \in [a, b]$, under the added assumption of weak disjunctivity.

Proposition 4.0.1 *Every weakly disjunctive τ-basic relation is an R-relation.*

Proof In view of Theorem 2.0.1, what we need to show is that weakly disjunctive τ-relations $\langle X, [\ ,\ ,\ ]\rangle$ are strongly transitive. So suppose $c, d \in [a, b]$, with $x \in [c, d]$. By weak disjunctivity, either $c \in [a, d]$ or $c \in [d, b]$. If $c \in [a, d]$, then $x \in [a, d]$, by R4 (modulo R1). A second application of R4 gives $[a, d] \subseteq [a, b]$, and so $x \in [a, b]$. If $c \in [d, b]$, we make the same argument to infer $x \in [a, b]$. □

Here are some examples of R-relations which are not weakly disjunctive.

Examples 4.0.2 (i) *Lattice road systems:* For lattice $\langle X, \sqcup, \sqcap\rangle$, we take the road system $\mathcal{R} = \{[a \sqcap b, a \sqcup b]_O : a, b \in X\}$, as in Subsect. 3.2. The induced R-relation $[\ ,\ ,\ ]_D$ is obviously weakly disjunctive if the lattice is a chain, but is not otherwise: Indeed, if $a, b \in X$ are order-incomparable, then we have $a \sqcup b \in [a, b]_D$, but $a \sqcap b \in [a, b]_D \setminus ([a, a \sqcup b]_D \cup [a \sqcup b, b]_D)$.

(ii) *Metric road systems:* If $\langle X, \rho\rangle$ is a metric space, we saw in Subsect. 3.6 that $\mathcal{R} = \{[a, b]_M : a, b \in X\}$ is not necessarily a road system because R5 may fail. However, even if it succeeds, weak disjunctivity may still fail: Consider the unit circle $X = \{a \in \mathbb{R}^2 : \|a\| = 1\}$. For $a, b \in X$, define $\rho(a, b) := 2\arcsin(\frac{1}{2}\|a - b\|)$. This gives the "shortest arc" distance from a to b, and it is easy to see that R5 holds for this metric space. However, if a and b are antipodal points, then the ρ-induced interval $[a, b]_M$ is all of X. So if c is any third point, then $[a, c]_M \cup [c, b]_M$ is a proper subset.

The R-relations arising from trees are disjunctive.

Examples 4.0.3 (i) *Tree road systems:* If $\langle X, \leq\rangle$ is a tree ordering, we saw in Subsect. 3.3 that the associated road system $\mathcal{R}$ consists of the sets $V(a, b, d) := [d, a]_O \cup [d, b]_O$, $d \leq a, b$, and the intervals are of the form $[a, b]_T = \bigcap\{V(a, b, d) : d \leq a, b\}$. For any $a \in X$, let $\overline{a}$ denote the order ideal $\{x \in X : x \leq a\}$. Given $a, b, c \in X$, the three order ideals $\overline{a} \cap \overline{b}$, $\overline{a} \cap \overline{c}$, and $\overline{b} \cap \overline{c}$ form a chain of sets under inclusion, so there are six possible ways to arrange them in order. However, if $\overline{a} \cap \overline{c} \subseteq \overline{b} \cap \overline{c} \subseteq \overline{a} \cap \overline{b}$, say, and d is a common lower bound of b and c, then d is also a lower bound of a. Hence we have $\overline{a} \cap \overline{c} = \overline{b} \cap \overline{c}$, and the six original arrangements collapse to three.

So if a, b, c are given and we wish to show $[a, b]_T \subseteq [a, c]_T \cup [c, b]_T$, then there are three cases to consider, dictated by which of the ideals $\overline{a} \cap \overline{b}$, $\overline{a} \cap \overline{c}$, $\overline{b} \cap \overline{c}$ includes the others. Now suppose $x \notin [a, c]_T \cup [c, b]_T$. Then there are $d \in \overline{a} \cap \overline{c}$ and $e \in \overline{b} \cap \overline{c}$ such that x is in neither $[d, a]_O \cup [d, c]_O$ nor $[e, b]_O \cup [e, c]_O$. Since d and e are in the chain $\overline{c}$, they are comparable. If $\overline{a} \cap \overline{b}$ contains the other two (equal) ideals, then both d and e are common lower bounds of a and b. If $d \leq e$ then we have $x \notin [e, a]_O \cup [e, b]_O = V(a, b, e)$; if $e \leq d$, then $x \notin V(a, b, d)$. In either instance, we have $x \notin [a, b]_T$.

Each of the remaining two cases is handled the same as the other, so suppose $\overline{a} \cap \overline{c} \supseteq \overline{b} \cap \overline{c} = \overline{a} \cap \overline{b}$. We know e is a common lower bound of a and b, so if

$d \leq e$ we may argue as above to infer $x \notin [a, b]_T$. If $e \leq d$, then we have $[e, a]_O = [e, d]_O \cup [d, a]_O$ and $[e, c]_O = [e, d]_O \cup [d, c]_O$. From these two equalities and what we assume above about x, we conclude $x \notin [e, a]_O \cup [e, b]_O = V(a, b, e)$; so $x \notin [a, b]_T$.

(ii) *Branch set road systems:* Let $\langle Y, \leq \rangle$ be a tree ordering, with $X = \mathfrak{B}(Y)$. Then the associated road system $\mathcal{R}$ consists of the sets $V(a) := \{\alpha \in X : a \in \alpha\}$, $a \in Y$. If $a, b \in Y$ and $\alpha \in V(a) \cap V(b)$, then both a and b belong to the chain α, and hence either $a \leq b$ or $b \leq a$. This tells us that $V(a) \cup V(b)$ is either $V(a)$ or $V(b)$, and therefore that $\mathcal{R}$ is additive.

In light of Theorem 2.0.1, it is natural to ask whether every disjunctive (or maybe weakly disjunctive) R-relation is an AR-relation. The following example shows that weak disjunctivity is not enough.

Example 4.0.4 Suppose X is a real vector space, with associated road system consisting of intervals $[a, b]_L$. We first show weak disjunctivity holds: indeed fix $c \in [a, b]_L$, say $c = sa + (1 - s)b$, for $0 \leq s \leq 1$; and let $x \in [a, b]$, say $x = ta + (1 - t)b$, for $0 \leq t \leq 1$. If either s or t is 0 or 1, or if $s = t$, there is nothing to prove; so we first assume $0 < t < s < 1$. Then $0 < \frac{t}{s} < 1$ and $x = \frac{t}{s}c + (1 - \frac{t}{s})b$, so $x \in [c, b]_L$. In the event $s < t$, we have $x \in [a, c]_L$; hence $[a, b]_L \subseteq [a, c]_L \cup [c, b]_L$ whenever $c \in [a, b]_L$.

Now assume the vector space dimension of X is ≥ 2. Then there exist three points $a, b, c \in X$ such that $[a, b]_L \cap ([a, c]_L \cup [b, c]_L) = \{a, b\}$. Since $[a, b]_L$ contains points other than a and b, this shows the failure of disjunctivity. So no additive road system can induce $\langle X, [\ ,\ ,\ ]_L \rangle$.

But full disjunctivity does turn out to be enough.

Theorem 4.0.5 *An R-relation on a set is disjunctive if and only if each of its inducing road systems is contained in an inducing road system that is additive. In particular, a ternary relation on a set is an AR-relation if and only if it satisfies axioms R1—R6; i.e., it is a disjunctive R-relation.*

Proof We have already seen that additive road systems induce AR-relations, so we prove the converse.

Given an AR-relation $\langle X, [\ ,\ ,\ ] \rangle$, let $\mathcal{R}$ be a road system inducing $[\ ,\ ,\ ]$. $\mathcal{R}$ may easily fail to be additive (see Remark 4.0.6 (i) below). Recognizing this, let $\mathcal{R}^*$ consist of all $R \subseteq X$ such that there is a nonempty finite family $\{R_i : i \in I\}$ of roads from $\mathcal{R}$ such that $R = \bigcup_{i \in I} R_i$, and whenever $I = J \cup K$ with $J \neq \emptyset \neq K$, we have $(\bigcup_{i \in J} R_i) \cap (\bigcup_{i \in K} R_i) \neq \emptyset$. The family $\{R_i : i \in I\}$ representing $R \in \mathcal{R}^*$ is said to satisfy the *intersection property*.

Clearly $\mathcal{R}^* \supseteq \mathcal{R}$; and since every superfamily of a road system is also a road system, it remains to show: (1) $\mathcal{R}^*$ is additive; and (2) $\mathcal{R}^*$ induces $[\ ,\ ,\ ] = [\ ,\ ,\ ]_{\mathcal{R}}$.

Ad (1): Suppose $R, S \in \mathcal{R}^*$ have nonempty intersection; we need to show $R \cup S \in \mathcal{R}^*$. By how $\mathcal{R}^*$ is defined, we may write $R = \bigcup_{1 \leq i \leq m} R_i$, $S = \bigcup_{m+1 \leq i \leq n} R_i$, where $1 \leq m < n$, each R_i is a road in $\mathcal{R}$, $1 \leq i \leq n$, and both families $\{R_1, \ldots, R_m\}$ and $\{R_{m+1}, \ldots, R_n\}$ satisfy the intersection property. Fix p, q such that $1 \leq p \leq m <$

$q \le n$ and $R_p \cap R_q \ne \emptyset$, and suppose $\{1, \dots, n\} = J \cup K$, with $J \ne \emptyset \ne K$. Our goal is to show $(\bigcup_{i \in I} R_i) \cap (\bigcup_{i \in J} R_i) \ne \emptyset$. If $r \in J \cap K$, we clearly have $(\bigcup_{i \in J} R_i) \cap (\bigcup_{i \in K} R_i) \supseteq R_r \ne \emptyset$, so we may assume J and K form a partition of $\{1, \dots, n\}$. If p and q lie in different blocks of the partition $\{J, K\}$, then $(\bigcup_{i \in J} R_i) \cap (\bigcup_{i \in K} R_i) \supseteq R_p \cap R_q \ne \emptyset$. If p and q lie in the same block, say, $p, q \in J$, let $J_m = J \cap \{1, \dots, m\}$ and $K_m = K \cap \{1, \dots, m\}$. Then $J_m \ne \emptyset$ because $p \in J_m$. If $K_m \ne \emptyset$ also, then $\{J_m, K_m\}$ forms a partition of $\{1, \dots, m\}$ and $(\bigcup_{i \in J} R_i) \cap (\bigcup_{i \in K} R_i) \supseteq (\bigcup_{i \in J_m} R_i) \cap (\bigcup_{i \in K_m} R_i) \ne \emptyset$ because $\{R_1, \dots, R_m\}$ satisfies the intersection property. If K_m is empty, then—because $q \in J$—we know $\{J \cap \{m+1, \dots, n\}\}, K\}$ forms a partition of $\{m+1, \dots, n\}$. Now use the fact that $\{R_{m+1}, \dots, R_n\}$ satisfies the intersection property to infer $(\bigcup_{i \in J} R_i) \cap (\bigcup_{i \in K} R_i) \ne \emptyset$. Thus $R \cup S \in \mathcal{R}^*$.

Ad (2): Let $a, b \in X$. Since $[a, b] = [a, b]_{\mathcal{R}} = \bigcap \mathcal{R}(a, b)$ and $\mathcal{R} \subseteq \mathcal{R}^*$, we immediately have $[a, b] \supseteq [a, b]_{\mathcal{R}^*} = \bigcap \mathcal{R}^*(a, b)$. So it remains to show that $[a, b] \subseteq [a, b]_{\mathcal{R}^*}$; and for this it suffices to prove the

Claim. If $R \in \mathcal{R}^*(a, b)$, then $[a, b] \subseteq R$.

For each $R \in \mathcal{R}^*$, let $\|R\|$ be the least number $n \ge 1$ such that R may be written as the union $R = \bigcup_{i \in I} R_i$ of roads from $\mathcal{R}$, where $\{R_i : i \in I\}$ satisfies the intersection property.

In the interests of obtaining a contradiction, suppose $R \in \mathcal{R}^*$ is a "least counterexample" to our Claim; namely that there exist $a, b \in R$ with $[a, b] \not\subseteq R$, but whenever $R' \in \mathcal{R}^*$ is such that $\|R'\| < \|R\|$, it follows that $[x, y] \subseteq R'$ for all $x, y \in R'$.

Since $[a, b] \not\subseteq R$, we know $R \notin \mathcal{R}$; so we may write R as a union $\bigcup_{i \in I} R_i$ of members of $\mathcal{R}$, where $|I| = \|R\| \ge 2$ and $\{R_i : i \in I\}$ satisfies the intersection property. Pick $j, k \in I$, with $a \in R_j$ and $b \in R_k$. If there is some $x \in R_j \cap R_k$, then disjunctivity ensures that $[a, b] \subseteq [a, x] \cup [x, b] \subseteq R_j \cup R_k \subseteq R$, and we have a contradiction right away.

Thus it must be the case that R_j and R_k are disjoint, and therefore that $\{R_j, R_k\}$ does not satisfy the intersection property. This means that I properly contains $\{j, k\}$.

Let $S = \bigcup_{i \in I \setminus \{j,k\}} R_i$. It is quite possible for $\{R_i : i \in I \setminus \{j, k\}\}$ to fail to satisfy the intersection property. In that case there is a partition $\{J', K'\}$ of $I \setminus \{j, k\}$ such that $(\bigcup_{i \in J'} R_i) \cap (\bigcup_{i \in K'} R_i) = \emptyset$. Continuing inductively in this way—since our index sets are finite—we obtain a partition $\{I_1, \dots, I_m\}$ if $I \setminus \{j, k\}$ such that $\{R_i : i \in I_l\}$ has the intersection property, $1 \le l \le m$, and the unions $T_l = \bigcup_{i \in I_l} R_i$ are pairwise disjoint.

If some T_l misses both R_j and R_k, then $J = I_l$ and $K = I \setminus I_l$ are nonempty, and give a partition of I such that $(\bigcup_{i \in J} R_i) \cap (\bigcup_{i \in K} R_i) = \emptyset$. This contradicts the assumption that $\{R_i : i \in I\}$ satisfies the intersection property.

So each T_l must intersect at least one of R_j, R_k. Let $J' = \bigcup\{I_l : T_l \cap R_j \ne \emptyset\}$, $K' = \bigcup\{I_l : T_l \cap R_k \ne \emptyset\}$. Then clearly $J' \cup K' = I \setminus \{j, k\}$. Hence, if no T_l intersects both R_j and R_k, we have $J' \cap K' = \emptyset$; and $J = J' \cup \{j\}$ and $K = K' \cup \{k\}$ give a partition of I such that $(\bigcup_{i \in J} R_i) \cap (\bigcup_{i \in K} R_i) = \emptyset$. Again a contradiction.

We have now shown that there is some T_l intersecting both R_j and R_k; say $x \in T_l \cap R_j$ and $y \in T_l \cap R_k$. But $T_l \in \mathcal{R}^*$ and $\|T_l\| < \|R\|$, and our induction hypothesis tells us that $[x, y] \subseteq T_l$. So using disjunctivity twice, we have $[a, b] \subseteq [a, x] \cup [x, b] \subseteq [a, x] \cup [x, y] \cup [y, b] \subseteq R_j \cup T_l \cup R_k \subseteq R$. This last contradiction completes the proof.

The second sentence of the theorem follows immediately from the argument above, together with Theorem 2.0.1. □

Remarks 4.0.6 (i) In Example 4.0.3 (i), the road system $\mathcal{R}$ consisting of sets of the form $V(a, b, d)$, $d \leq a, b$, induces an AR-relation while not in general being additive. In this case we may describe the sets in $\mathcal{R}^*$ as being of the form $V(a_1, \ldots, a_n, d)$, where d is a common lower bound of $\{a_1, \ldots, a_n\}$ and $V(a_1, \ldots, a_n, d) := \bigcup_{1 \leq i \leq n} [d, a_i]_O$.
As we saw in Example 4.0.3 (ii), the roads $V(a)$, $a \in Y$, do provide an additive road system for $\mathfrak{B}(Y)$.

(ii) In Remark 2.0.2 (ii) we asked whether being an R-relation may be L_t-axiomatized using at most four variables. In view of Proposition 4.0.1 and Theorems 2.0.1, 4.0.5, being an AR-relation is so axiomatizable.

If $\mathcal{R}$ and $\mathcal{S}$ are road systems on a set, write $\mathcal{R} \leq_B \mathcal{S}$ to mean that $[\ ,\ ,\ ]_{\mathcal{R}} \supseteq [\ ,\ ,\ ]_{\mathcal{S}}$. $\mathcal{R}$ and $\mathcal{S}$ are *B-equivalent* if $[\ ,\ ,\ ]_{\mathcal{R}} = [\ ,\ ,\ ]_{\mathcal{S}}$.

The construction $\mathcal{R} \mapsto \mathcal{R}^*$ described in the proof of Theorem 4.0.5 works for any road system $\mathcal{R}$. $\mathcal{R}^*$ is always an additive road system; it contains $\mathcal{R}$, and hence we always have $\mathcal{R} \leq_B \mathcal{R}^*$. Theorem 4.0.5 then tells us that $\mathcal{R}$ and $\mathcal{R}^*$ are B-equivalent precisely when $[\ ,\ ,\ ]_{\mathcal{R}}$ is disjunctive.

The following result justifies the designation of $\mathcal{R}^*$ as the *additive closure* of $\mathcal{R}$.

Theorem 4.0.7 *For any road system $\mathcal{R}$, $\mathcal{R}^*$ is an additive road system that contains $\mathcal{R}$. If $\mathcal{S}$ is an additive road system and $\mathcal{R} \subseteq \mathcal{S}$ (resp., $\mathcal{R} \leq_B \mathcal{S}$), then $\mathcal{R}^* \subseteq \mathcal{S}$ (resp., $\mathcal{R}^* \leq_B \mathcal{S}$). Moreover, if $\mathcal{R}$ and $\mathcal{S}$ are road systems with $\mathcal{R} \leq_B \mathcal{S}$, then $\mathcal{R}^* \leq_B \mathcal{S}^*$.*

Proof Suppose road system $\mathcal{R}$ is contained in road system $\mathcal{S}$, where $\mathcal{S}$ is additive. To show $\mathcal{R}^* \subseteq \mathcal{S}$, we use induction on the number $\|R\|$ for $R \in \mathcal{R}^*$ (as defined in the proof of Theorem 4.0.5). If $R \in \mathcal{R}^*$ and $\|R\| = 1$, then $R \in \mathcal{R}$. Hence $R \in \mathcal{S}$ by assumption. Now suppose $\|R\| = n > 1$ and that $R' \in \mathcal{S}$ for every $R' \in \mathcal{R}^*$ with $\|R'\| < n$. We write $R = R_1 \cup \cdots \cup R_n$, where each R_i is in $\mathcal{R}$, $n = \|R\|$, and $\{R_1, \ldots, R_n\}$ satisfies the intersection property. Set $S = R_2 \cup \cdots \cup R_n$. Then, as in the proof of Theorem 4.0.5, $\{2, \ldots, n\}$ has a partition $\{I_1, \ldots, I_m\}$, where $\{R_i : i \in I_l\}$ has the intersection property for each $1 \leq l \leq m$, and the unions $T_l = \bigcup_{i \in I_l} R_i$ are pairwise disjoint.

Now for each $1 \leq l \leq m$, we have $T_l \in \mathcal{R}^*$, $\|T_l\| < n$, and hence $T_l \in \mathcal{S}$, by our induction hypothesis.

Also R_1 must intersect each T_l. Indeed, if it happens that $R_1 \cap T_l = \emptyset$ for some $1 \leq l \leq m$, then $J = I_l$ and $K = I \backslash I_l$ are both nonempty (note that $1 \in K$). Thus $\{J, K\}$ provides a partition of I such that $(\bigcup_{i \in J} R_i) \cap (\bigcup_{i \in K} R_i) = \emptyset$, contradicting the assumption that $\{R_i : i \in I\}$ satisfies the intersection property.

Since R_1 and $T_1, \ldots, T_m$ all belong to the additive road system $\mathcal{S}$ and R_1 intersects each T_l, $1 \leq l \leq m$, it follows that $R = R_1 \cup T_1 \cup \cdots \cup T_m \in \mathcal{S}$.

Next suppose $\mathcal{R} \leq_B \mathcal{S}$, where $\mathcal{S}$ is additive. We know that for any $a, b \in X$ and any $R \in \mathcal{R}(a, b)$, we have $[a, b]_{\mathcal{S}} \subseteq R$, and we wish to show that this is also true for any $R \in \mathcal{R}^*(a, b)$. To do this, we mimic the induction part of the proof of Theorem 4.0.5, using the fact that $[\ ,\ ,\ ]_{\mathcal{S}}$ is disjunctive. The exact same argument—with only trivial changes—works.

Finally suppose $\mathcal{R}$ and $\mathcal{S}$ are road systems, such that $\mathcal{R} \leq_B \mathcal{S}$. Then $\mathcal{R} \leq_B \mathcal{S}^*$ because $\mathcal{S} \subseteq \mathcal{S}^*$. Since $\mathcal{S}^*$ is additive, we infer $\mathcal{R}^* \leq_B \mathcal{S}^*$ by the preceding argument. □

Example 4.0.8 Suppose $\langle X, \mathcal{R}\rangle$ is a road system, with $a, b \in X$. If disjunctivity fails to the extent that for all $x \in [a, b]_{\mathcal{R}} \backslash \{a, b\}$ there is some $c \in X$ with $x \notin [a, c]_{\mathcal{R}} \cup [c, b]_{\mathcal{R}}$, then $[a, b]_{\mathcal{R}^*} = \{a, b\}$; i.e., it is trivial.

Harking back to Example 4.0.4: If X is a real vector space of dimension ≥ 2 and $\mathcal{R}$ is the road system of intervals $[a, b]_L$, then disjunctivity fails in this sense for all pairs; so $\mathcal{R}^*$-intervals are all trivial.

Harking back to Example 4.0.2 (i): If X is a lattice and $\mathcal{R}$ is the road system of order intervals (so $[a, b]_D = [a \sqcap b, a \sqcup b]_O$), then disjunctivity may easily fail in this sense: suppose, for example, that $a \leq b$ and $[a, b]_D = [a, b]_O$ is such that for each $x \in [a, b]_D \backslash \{a, b\}$ there is some $c \in [a, b]_D$ order-incomparable with x. Then $[a, b]_{\mathcal{R}^*}$ is trivial.

5 Betweenness from separative road systems

A road system $\langle X, \mathcal{R}\rangle$ is *separative* if for any $a, b, c \in X$ with $b \neq c$, there is some $R \in \mathcal{R}$ such that either $a, b \in R$ but $c \notin R$ or $a, c \in R$ but $b \notin R$. A ternary relation on X is an *SR-relation* (resp., *SAR-relation*) if it is induced by a separative (resp., separative additive) road system on X.

Recall from above the pre-orders $\leq_a$ on τ-basic relations; reflexive because of R2 and transitive because of R4. They become antisymmetric—and hence partial orders—in the presence of the following condition:

(R7) Antisymmetry: $([a, b, c] \wedge [a, c, b]) \rightarrow b = c$ (so for any $a, b, c \in X$, if $c \in [a, b]$ and $b \in [a, c]$, then b=c).

The following is now obvious.

Proposition 5.0.1 *A road system is separative if and only if its induced R-relation is antisymmetric.*

And, as an immediate consequence of Proposition 5.0.1 and Theorem 4.0.5, we obtain

Corollary 5.0.2 *An AR-relation on a set is antisymmetric if and only if each of its inducing road systems is contained in an inducing road system that is additive and separative.*

Remarks 5.0.3 (i) Antisymmetry—our terminology—appears in many betweenness studies as a fundamental axiom [see, e.g., [1,5,8,12,19]. In [19] it is condition β (closure)]. It is an easy exercise to verify that antisymmetry holds for the betweenness interpretations arising from linear and metric structures; as for the order-theoretic ones, the verdict is mixed: $[\ ,\ ,\ ]_G$ is antisymmetric, by Lemma 8.2 in [19]; consequently, $[\ ,\ ,\ ]_D$ is antisymmetric for distributive lattices because it coincides with $[\ ,\ ,\ ]_G$ then. As for trees, it is easy to show $[\ ,\ ,\ ]_T$ is antisymmetric, but that $[\ ,\ ,\ ]_{\mathfrak{B}}$ is not. (If α, β, and γ all have the same pairwise intersection, then each is between the other two.) In [3,4] we investigate betweenness in the context of continuum theory, where antisymmetry often fails as well. So from our point of view this condition is not axiomatic; rather it is as honored in the breach as in the observance.

(ii) From Remark 4.0.6 (i), there is a nonadditive road systems $\langle X, \mathcal{R}\rangle$ which induces an AR-relation. Theorem 4.0.5 then tells us that $\mathcal{R}^*$, an additive road system on X, also induces $[\ ,\ ,\ ]_{\mathcal{R}}$. In particular, we may infer that additivity in a road system—in contrast with separativity (Proposition 5.0.1)—cannot be captured just using information—first-order or otherwise—about its induced AR-relation.

Proposition 5.0.4 *If $\langle X, [\ ,\ ,\ ]\rangle$ is a τ-basic relation that is antisymmetruc and weakly disjunctive, then for each $a \in X$ the ordering $\leq_a$ is a tree ordering with least element a (the root of the tree).*

Proof Fix $a \in X$. By antisymmetry, plus conditions R1—R4, $\leq_a$ is a partial ordering with least element a. It remains to show it is a tree ordering. Downward directedness is immediate, so suppose $c \leq_a b$ and $d \leq_a b$ both hold. We need to show that either $c \leq_a d$ or $d \leq_a c$. But we have $c, d \in [a, b]$. Thus, by weak disjunctivity, either $d \in [a, c]$, in which case $d \leq_a c$, or $d \in [c, b]$. In the first case, we are done; in the second, we wish to show that $c \in [a, d]$. But if $c \notin [a, d]$, then—by weak disjunctivity again—$c \in [d, b]$. But then, by antisymmetry, we have $c = d$, a contradiction. □

For any ternary relation $\langle X, [\ ,\ ,\ ]\rangle$ and $a, b \in X$, define the binary relation $\leq_{ab}$ to be the restriction of $\leq_a$ to the interval $[a, b]$.

Proposition 5.0.5 *In τ-basic relations that are antisymmetric and weakly disjunctive, the orderings $\leq_{ab}$ are total orderings. Moreover, $\leq_{ba}$ is the relation-inverse of $\leq_{ab}$.*

Proof That each $\leq_{ab}$ is a total ordering is obvious from the proof of Proposition 5.0.4. For the rest, suppose $c, d \in [a, b]$. We wish to show that $c \leq_{ab} d$ if and only if $d \leq_{ba} c$. But the first condition is that $c \in [a, d]$ and the second is that $d \in [b, c] = [c, b]$. If the second condition fails, then weak disjunctivity forces $d \in [a, c]$, and we infer again—by antisymmetry—that $c = d$. □

Antisymmetry has several interesting variants.

Theorem 5.0.6 *For a τ-basic relation that is weakly disjunctive, the following conditions are equivalent:*

(i) *(R7) Antisymmetry.*
(ii) *(R7.1) Slenderness: the property that if $c \in [a, b]$, then $[a, c] \cap [c, b] = \{c\}$.*
(iii) *(R7.2) Reciprocity: the property that if $c, d \in [a, b]$ and $c \in [a, d]$, then $d \in [c, b]$.*
(iv) *(R7.3) Uniqueness of centroids: the property that the set $[abc] := [a, b] \cap [b, c] \cap [a, c]$—the centroids of $\{a, b, c\}$—has at most one element.*
(v) *(R7.4) Uniqueness of bracket point sets: the property that if $[a, b] = [c, d]$, then $\{a, b\} = \{c, d\}$.*

Proof Let $\langle X, [\ ,\ ,\ ]\rangle$ be a weakly disjunctive ternary relation satisfying R1—R4, and assume (i) holds. Reciprocity is merely a restatement that the total ordering $\leq_{ba}$ is the relation-inverse of $\leq_{ab}$, so (iii) holds by virtue of the proof of Proposition 5.0.5.

Still assuming antisymmetry, suppose $c \in [a, b]$ and $d \in [a, c] \cap [c, b]$. Then $d \in [a, b]$ as well; hence, by reciprocity, we have $c \in [d, b]$. But then $d = c$, by antisymmetry. Hence (ii) holds.

Suppose (ii) holds, and assume $c \in [a, b]$ and $b \in [a, c]$. Since $[a, c] = [a, b]$, we have—by slenderness—that $\{b\} = [a, b] \cap [b, c] = [a, c] \cap [c, b] = \{c\}$. Thus (i) holds.

Next, suppose (iii) holds for $\langle X, [\ ,\ ,\ ]\rangle$, and that $b \in [a, c]$ and $c \in [a, b]$ are both true. Pick $d = b$. Then $c, d \in [a, b]$ and $d \in [a, c]$; so—by reciprocity—we have $c \in [d, b] = [b, b] = \{b\}$. Hence (i) holds.

So we now have the equivalence of (i), (ii), and (iii); next we show the equivalence of (iv) with any one of the first three.

Start by showing (iv) implies slenderness. Indeed, suppose $c \in [a, b]$, with $d \in [a, c] \cap [c, b]$. Since both $[a, c]$ and $[c, b]$ are contained in $[a, b]$, we have that $c, d \in [abc]$. by (iv) we infer that $b = c$, so slenderness follows from uniqueness of centroids.

Assuming (i), (ii), and (iii), suppose d_1 and d_2 are centroids of $\{a, b, c\}$. Since $d_1, d_2 \in [a, b]$, Proposition 5.0.5 tells us that either $d_1 \in [a, d_2]$ or $d_2 \in [d_1, b]$. Without loss of generality, let us suppose $d_1 \in [a, d_2]$. By reciprocity, we now have $d_2 \in [d_1, b]$. We also know $d_1 \in [b, c]$. Since $d_2 \in [b, d_1]$, we infer $d_1 \in [d_2, c]$, again by reciprocity. Lastly, we know that $d_2 \in [a, c]$, so—reciprocity again—we have $d_2 \in [a, d_1]$. But now $d_1 = d_2$ by direct application of antisymmetry.

Finally we show the equivalence of (i),...,(iii) and (v). Clearly (v) implies antisymmetry; just take $d = a$. Then, given $[a, c] = [a, b]$, $c = b$ immediately follows.

Now suppose $[a, b] = [c, d]$ and (i),...,(iii) hold. Then we have $c, d \in [a, b]$; so, by weak disjunctivity, either $d \in [a, c]$ or $d \in [c, b]$. In the first case, $[c, d] \subseteq [a, c] \subseteq [a, b]$, and so $[a, c] = [a, b]$. Thus $b = c$ by (i), and so $[d, b] = [a, b]$. A second appeal to (i) gives $a = d$; so $\{a, b\} = \{c, d\}$ in this case.

If it happens that $d \in [c, b]$, then reciprocity implies that $c \in [a, d]$. Argue as above to obtain $b = d$ and $a = c$. Hence $\{a, b\} = \{c, d\}$ in this case too. □

Remark 5.0.7 R7.4 is not an automatic consequence of R7, as witnessed by the lattice interpretation $[\ ,\ ,\ ]_D$ of betweenness [see Subsect. 3.2 and Example 4.0.2 (i)]. This shows the necessity of weak disjunctivity in the base assumption of Theorem 5.0.6.

As we saw in Proposition 5.0.4, if $\langle X, [\ ,\ ,\ ]\rangle$ is a τ-basic relation that is antisymmetric and weakly disjunctive, and $a \in X$ is a fixed root point, then the binary relation $\leq_a$ defines a tree ordering on X. Using the notation of Subsect. 3.3, we then have an SAR-relation $[\ ,\ ,\ ]_T$ induced by this tree structure; and a natural question to ask is how the tree notion of betweenness relates to the original one. First we note that, by Example 4.0.3 (i), the original τ-basic relation—automatically an R-relation, by Proposition 4.0.1—must be disjunctive (not just weakly so) if there is to be any hope that the two notions coincide.

Corollary 5.0.8 *Let $\langle X, [\ ,\ ,\ ]\rangle$ be a τ-basic relation that is antisymmetric and disjunctive, with root point $a \in X$. Then each interval $[b, c]$ is contained in the* *corresponding tree interval* $[b, c]_T$ *induced by* $\leq_a$. *Furthermore, if the centroid set* $[abc]$ *is nonempty, then* $[b, c] = [b, c]_T$.

Proof First we note that if $d \leq_a b$, then $[d, b]$ equals the order interval $[d, b]_O = \{x \in X : d \leq_a x \leq_a b\}$. Indeed, suppose $[a, d, b]$ and that $x \in [d, b]$; i.e., $[d, x, b]$ holds. Then we get $[a, x, b]$ by R4 (modulo R1), so $x \leq_a b$. Since $d, c \in [a, b]$ and $[d, x, b]$, we have $[a, d, x]$, by antisymmetry (in the form of reciprocity R7.2, see Theorem 5.0.6). Hence $d \leq_a x$, and thus $x \in [d, b]_O$. Conversely, if $x \in [d, b]_O$, we know that $[a, d, x]$ and $[a, x, b]$. Since $[a, d, b]$ also, we have $[d, x, b]$, again by reciprocity. So $x \in [d, b]$.

Now suppose $b, c \in X$ and $d \in X$ is arbitrary. Then disjunctivity tells us $[b, c] \subseteq [d, b] \cup [d, c]$. And if also $d \leq_a b, c$, then the right-hand side equals $[d, b]_O \cup [d, c]_O = V(b, c, d)$. This shows $[b, c] \subseteq [b, c]_T$.

Finally suppose there is a centroid $d \in [abc]$. Since $d \in [b, c]$, weak disjunctivity tells us $[b, c] = [d, b] \cup [d, c]$. And because $d \in [a, b] \cap [a, c]$, the right-hand side becomes $V(b, c, d)$. This shows that $[b, c] \supseteq [b, c]_T$. □

We do not know whether the centroid existence hypothesis is necessary for equality to hold in Corollary 5.0.8, but the following example shows it cannot be dropped.

Example 5.0.9 Let X be a *simple closed curve*; i.e., a homeomorphic copy of the unit circle from Example 4.0.2 (ii) (topologized as a subspace of the euclidean plane), and let $\mathcal{C}$ be the road system of connected subsets of X. Then (see Theorem 6.12 below) $\mathcal{C}$ is separative and additive; hence $[\ ,\ ,\ ]_{\mathcal{C}}$ satisfies the hypothesis of Corollary 5.0.8. But all the intervals in this case are clearly trivial; so if $a \in X$ is any root point, the tree order $\leq_a$ is described by saying $x \leq_a y$ precisely when either $x = y$ or $x = a$. If $b, c \in X \backslash \{a\}$, then $[b, c]_{\mathcal{C}} = \{b, c\}$, while $[b, c]_T = \{a, b, c\}$

Also worthy of note is that the road system you get from the trivial betweenness relation according to the proof of Theorem 4.0.4 is the collection $\mathcal{R}$ of nonempty subsets of X with at most two points. $\mathcal{R}^*$, then, is the collection of all finite nonempty subsets of X. $\mathcal{R}^*$ and $\mathcal{C}$ are as disjoint as possible, though, since thay have only the singleton subsets of X in common.

Remarks 5.0.10 (i) All first-order properties so far considered in this paper are universal; i.e., of the form $\forall \overline{x} \varphi$, where φ is quantifier-free. Centroid existence, on the other hand, is our first example of a more complex statement, as its format is universal-existential. Higher complexity L_t-formulas arise in considerations of topological betweenness notions in Hausdorff continua—this being taken up in [3,4].

(ii) When centroids exist in antisymmetric, weakly disjuctive τ-basic relations, they are unique, by Theorem 5.0.6. This brings up the notion of the centroid as a ternary operation, akin to the median operations studied in [13] and later papers, in connection with graphs and other structures.

In light of Theorem 4.0.7, it is tempting to ask whether every road system has a "separative closure" in a sense analogous to how it has an additive closure. We are grateful to Aisling McCluskey for bringing up this issue and to Jorge Bruno [6] for providing a negative answer.

Example 5.0.11 Let X be a set with at least three elements, and fix distinct $b, c \in X$. Let $\mathcal{R}$ consist of all singletons of X, along with all nonempty subsets of $X \backslash \{b, c\}$, along with all sets of the form $A \cup \{b, c\}$ for $A \subseteq X \backslash \{b, c\}$. Clearly $\mathcal{R}$ is a road system. And if we take $a \in X \backslash \{b, c\}$, we see that any road containing a must contain both b and c if it contains either. Hence $\mathcal{R}$ is not separative. However, both $\mathcal{R} \cup \{\{a, b\} : a \in X \backslash \{b, c\}\}$ and $\mathcal{R} \cup \{\{a, c\} : a \in X \backslash \{b, c\}\}$ are separative road systems, neither is contained in the other, and neither properly contains a separative road system extension of $\mathcal{R}$.

6 Three topological pre-betweenness structures

In this section we consider three different interpretations of betweenness in a connected topological space. (See, e.g., [15] for topological background.)

If Y is any topological space, recall that a subset of Y that is both open and closed is called *clopen*. Y is *connected* just in case the only nonempty clopen subset of Y is Y itself. If Y is not connected, a *disconnection* of Y is a pair $\langle A, B \rangle$ of disjoint nonempty clopen subsets of Y, with $Y = A \cup B$. A *component* of Y is a connected subset of Y that is not properly contained in any other connected subset of Y. Each point of Y is contained in a unique component, namely the union of all connected subsets containing the point. The union of two overlapping connected sets is connected; hence the components of a space form a partition of the space. Since the closure of any connected subset is also connected, each component of Y is a closed subset. A point c of a connected space X is a *cut point* if $X \backslash \{c\}$ is disconnected; i.e., has more than one component.

6.1 The C-interpretation

Let X be a connected topological space, and define the ternary relation $[a, c, b]_C$ to hold for $a, b, c \in X$ just in case either $c \in \{a, b\}$ or a and b lie in different components of $X \backslash \{c\}$. This is the *C-interpretation* of betweenness in a connected topological space; a point of X is a cut point if and only if it lies properly between two other points in the C-interpretation.

For the next result, we state an old chestnut concerning connected spaces. It occurs as Theorems 3.3 and 3.4 in [17]; a weak version of it can also be found in [16].

Lemma 6.11 *If X is a connected topological space, A is a connected subset of X, and B is either a component of $X \backslash A$ or a clopen set therein, then $B \cup A$ is connected.*

Theorem 6.12 *For any connected space X, the C-interpretation of betweenness is an SAR-relation.*

Proof Let $\mathcal{C}$ consist of the connected subsets of X. Since singletons are connected and X itself is in $\mathcal{C}$, we know $\mathcal{C}$ is a road system. As mentioned above, the union of two overlapping connected sets is connected; so $\mathcal{C}$ is additive.

Next we show $[\ ,\ ,\ ]_C = [\ ,\ ,\ ]_{\mathcal{C}}$. Fixing $a, b \in X$, we show $[a, b]_C = \bigcap \mathcal{C}(a, b)$. Indeed, suppose $c \in [a, b]_C$. If $c \in \{a, b\}$, then $c \in \bigcap \mathcal{C}(a, b)$; otherwise a and b lie in different components of $X \backslash \{c\}$. But then no connected subset of $X \backslash \{c\}$ can be a member of $\mathcal{C}(a, b)$; i.e., we have $c \in \bigcap \mathcal{C}(a, b)$ in this case as well. If $c \notin [a, b]_C$, then a and b lie in the same component A of $X \backslash \{c\}$. Thus A witnesses the fact that $c \notin \bigcap \mathcal{C}(a, b)$.

Finally we show $\mathcal{C}$ is separative by showing $[\ ,\ ,\ ]_C$ to be antisymmetric (see Proposition 5.0.1). So assume $c \in [a, b]_C$, with $c \neq b$. It suffices to show $b \notin [a, c]_C$. If $c \in \{a, b\}$ then $c = a$ and $[a, c]_C = \{c\}$. Thus $b \notin [a, c]_C$. If $c \notin \{a, b\}$, then there are distinct components A and B of $X \backslash \{c\}$ with $a \in A$ and $b \in B$. By Lemma 6.11, $A \cup \{c\}$ is a connected set containing both a and c, but not b. Hence here too, $b \notin [a, c]_C$. □

6.2 The Q-interpretation

In a topological space Y, the *quasicomponent* of a point $a \in Y$ is the intersection of all clopen subsets containing a. Points a and b lie in different quasicomponents precisely when there is a disconnection $\langle A, B \rangle$ of Y with $a \in A$ and $b \in B$. The quasicomponents of a space also form a partition of the space into closed subsets; every quasicomponent is a union of components. If the space is locally connected—i.e., components of open sets are open—the components and the quasicomponents coincide.

Again we start with a connected topological space X, but we now define $[a, c, b]_Q$ to hold for $a, b, c \in X$ just in case either $c \in \{a, b\}$ or a and b lie in different quasicomponents of $X \backslash \{c\}$. This is the *Q-interpretation* of betweenness in a connected topological space, and is generally more restrictive than the C-interpretation. The two interpretations do agree in a locally connected T_1 space; regardless of agreement, though, a point $c \in X$ is a cut point of X if and only if c lies properly between two other points in either the C- or the Q-interpretation.

The Q-interpretation—our terminology—was first studied from the point of view of betweenness by Ward [21], whose focus was to outline an abstract approach to Whyburn's theory of *cyclic elements* in continuum theory (see [22]). (In the parlance of [22], if c lies properly between a and b in the Q-interpretation, *c separates a from b*.) The paper contains no proofs, but instead promises a more comprehensive treatment in a later work (which sadly has never appeared).

Ward defined a *cut point structure* to be a ternary relational structure $\langle X, [\ ,\ ,\ ] \rangle$ satisfying conditions R1, R2, R3, R6, and R7. He claimed that the Q-interpretation of betweenness on a connected topological space defines a cut point structure, and went on to assert that a number of other properties—including R4 (transitivity)—hold in any cut point structure.

The goal of this subsection is an analogue of Theorem 6.12 for the Q-interpretation. The difficulty is that we know of no obvious choice for a suitable road system; only that such a system, in general, could not be contained within the family $\mathcal{C}$ of connected subsets. This is where Corollary 5.0.2 proves most useful.

Theorem 6.21 *For any connected space X, the Q-interpretation of betweenness is an SAR-relation.*

Proof The existence of an appropriate road system is guaranteed by Corollary 5.0.2, once we establish that the Q-interpretation satisfies conditions R1—R7. R1 (symmetry), R2 (reflexivity), and R3 (minimality) obviously hold for $\langle X, [\ ,\ ,\ ]_Q \rangle$. Since transitivity always follows from strong transitivity, it is R5, R6 and R7 that we need to concentrate on. It will be convenient to do this in reverse order.

Ad R7 (antisymmetry): Assume $c \in [a, b]_Q$, with $c \neq b$. Proceeding as in the proof of Theorem 6.12, we may jump to the assumption that $c \notin \{a, b\}$. Then there is a disconnection $\langle A, B \rangle$ of $X \backslash \{c\}$ with $a \in A$ and $b \in B$. Again we may apply Lemma 6.11. Since A is clopen in $X \backslash \{c\}$, we infer that $A \cup \{c\}$ is a connected set containing both a and c, but not b. This immediately shows that $b \notin [a, c]_C$, and hence that $b \notin [a, c]_Q$.

Ad R6 (disjunctivity): Suppose $[a, x, b]_Q$ holds, with $c \in X$ arbitrary. We wish to show that either $[a, x, c]_Q$ or $[c, x, b]_Q$ holds. If $x \in \{a, b\}$, we are done; so assume there is a disconnection $\langle A, B \rangle$ of $X \backslash \{x\}$ with $a \in A$ and $b \in B$. If $[a, x, c]_Q$ fails, then it must be the case that $c \in A$. Hence the disconnection $\langle A, B \rangle$ witnesses that $[c, x, b]_Q$ indeed holds.

Ad R5 (strong transitivity): Suppose $\langle X, [\ ,\ ,\ ] \rangle$ is any ternary relational structure satisfying R1, R6, and R7. We show R5 necessarily follows.

First we show R4 (transitivity). Suppose $[a, c, b]$ and $[a, x, c]$ are both true; we wish to show that $[a, x, b]$ is true also. By R6 and $[a, x, c]$ we have either $[a, x, b]$ or $[b, x, c]$. In the first case we are done; in the second, we use $[a, c, b]$ and R6 to infer that either $[a, c, x]$ or $[b, c, x]$. But $[a, x, c]$ and $[b, x, c]$ are both true. Hence, by antisymmetry, we have $x = c$. Thus $[a, x, b]$ holds in this case too.

Now we may argue as in Proposition 4.0.1, since only R1, R4, and weak disjunctivity are used. This gives us R5. □

As a corollary to Theorems 6.21 and 2.0.1, we see that in a connected topological space the collection $\mathcal{Q}$ of Q-intervals gives us a road system inducing $[\ ,\ ,\ ]_Q$. $\mathcal{Q}$ is automatically separative, but is not necessarily additive. Since $[\ ,\ ,\ ]_Q$ is disjunctive, though, the additive closure $\mathcal{Q}^*$ is additive and separative, and is B-equivalent to $\mathcal{Q}$ (see Theorems 4.0.5 and 4.0.7). What may seem a bit odd is that no members of $\mathcal{Q}^*$, other than singletons, are guaranteed to be connected; hence $\mathcal{Q}^*$ and $\mathcal{C}$ may have very little to do with one another. We may get $\mathcal{C}$ back into the game, however, with the following.

Corollary 6.22 *Let X be a connected topological space. Then $[\ ,\ ,\ ]_Q$ is an SAR-relation on X. And if $\mathcal{R}$ is any road system inducing $[\ ,\ ,\ ]_Q$, then $\mathcal{R} \cup \mathcal{C}$ is also a road system inducing $[\ ,\ ,\ ]_Q$. Consequently, $(\mathcal{R} \cup \mathcal{C})^*$ is a separative, additive, $[\ ,\ ,\ ]_Q$-inducing road system, which also contains $\mathcal{C}$.*

Proof In light of Theorems 4.0.5 and 4.0.7, all we need to show is that $\mathcal{R} \cup \mathcal{C} \leq_B \mathcal{R}$. So pick $a, b \in X$, and let $S \in (\mathcal{R} \cup \mathcal{C})(a, b)$. If $S \in \mathcal{R}$, then $[a, b]_Q \subseteq S$ because $\mathcal{R}$ induces $[\ ,\ ,\ ]_Q$. If $S \in \mathcal{C}$, then $[a, b]_Q \subseteq [a, b]_C \subseteq S$ because $\mathcal{C}$ induces $[\ ,\ ,\ ]_C$. □

6.3 The K-interpretation

For the purposes of this paper, a *continuum* is a connected space that is also compact; a *subcontinuum* of a space is a subset which is a continuum in its subspace topology.

(Since we do not assume any separation conditions, subcontinua need not be closed subsets.) A space is *continuumwise connected* if each two of its points are contained in a subcontinuum. (Continuumwise connectedness implies connectedness; the converse is true if the space is compact Hausdorff.) The union of all subcontinua containing a given point is the *continuum component* of the point; each continuum component is contained within a unique component. The union of two overlapping subcontinua is a subcontinuum, so no two continuum components can overlap. A space is continuumwise connected just in case it has exactly one continuum component. Continuum components are connected, but are not necessarily open, closed, or compact.

We now start with a continuumwise connected topological space X, and define $[a, c, b]_K$ to hold for $a, b, c \in X$ just in case either $c \in \{a, b\}$ or a and b lie in different continuum components of $X \backslash \{c\}$. This is the *K-interpretation* of betweenness in a continuumwise connected topological space, and is generally less restrictive than the C-interpretation. A point whose complement has more than one continuum component is known as a *weak cut point*; so a point is a weak cut point if and only if it lies properly between two other points in the K-interpretation of betweenness.

Theorem 6.31 *For any continuumwise connected space X, the K-interpretation of betweenness is an AR-relation.*

Proof Let $\mathcal{K}$ consist of the subcontinua of X. Then it is routine to show that $\mathcal{K}$ is an additive road system. To show $[a, b]_K = \bigcap \mathcal{K}(a, b)$ for $a, b \in X$, mimic the proof of Theorem 6.12. □

Significant in its absence for the K-interpretation of betweenness is antisymmetry; the road system $\mathcal{K}$ is not generally separative, even when the space is a metrizable continuum.

Example 6.32 X is the $\sin(1/x)$*-curve*, defined to be the union $A \cup S \subseteq \mathbb{R}^2$, where $A = \{0\} \times [-1, 1]$ and $S = \{\langle x, \sin(1/x)\rangle : 0 < x \leq 1\}$. X is an example of a metrizable continuum whose K-interpretation of betweenness is not antisymmetric: if $a \in S$ and $b, c \in A$, then $c \in [a, b]_K$ and $b \in [a, c]_K$; hence $\leq_a$ (see Sect. 5) fails to be a partial ordering.

On the other hand, if $a \in A$ and a is a cut point of the arc A, then $\leq_a$ is a partial ordering which is not a tree ordering: the two noncut points of A are $\leq_a$-incomparable, but any element of S is a common $\leq_a$-upper bound for them.

7 Concluding remarks

The notion of *road system* is a new and very simple way to interpret betweenness uniformly, over a wide spectrum of mathematical environments. We have shown that the vast majority of betweenness interpretations in the literature may each be viewed as being naturally induced by a suitable road system; and have begun a serious investigation of road systems themselves, especially in establishing Theorems 4.0.5 and 4.0.7.

In [3,4] we take up a more extended study of the topological interpretations of betweenness in the context of Hausdorff continua. In [3] the focus is on when there are no gaps, and in [4] we look at the role of antisymmetry.

References

1. Adeleke, S.A., Neumann, P.M.: Relations related to betweenness: their structure and automorphisms, Mem. Amer. Math. Soc. **131** (163), viii + pp. 125 (1998)
2. Altwegg, M.: Zur Axiomatik der teilweise geordneten Mengen. Commentarii Mathematici Helvitici **24**, 149–155 (1950)
3. Bankston, P.: When Hausdorff continua have no gaps, Top. Proc. (to appear)
4. Bankston, P.: The antisymmetry betweenness axiom and Hausdorff continua (in preparation)
5. Birkhoff, G.: Lattice Theory, 3rd edn. Colloquium publications, Amer. Math. Soc, Providence, R.I (1995). (Originally published in 1948.)
6. Bruno, J.: (private communication)
7. Chang, C.C., Keisler, H.J.: Model theory, 3rd edn. North Holland, Amsterdam (1990)
8. Düntsch, I., Urquhart, A.: Betweenness and comparability obtained from binary relation. Lecture notes in Comput. Sci. **4136**, 148–161 (2006)
9. Fishburn, P.C.: Betweenness, orders and interval graphs. J. Pure Appl. Alg. **1**, 159–178 (1971)
10. Glivenko, V.: Géometrie des systèmes de choses normées. Amer. J. Math. **58**, 799–828 (1936)
11. Glivenko, V.: Contributions à l'étude des systèmes de choses normées. Amer. J. Math. **59**, 941–956 (1937)
12. Huntington, E.V., Kline, J.R.: Sets of independent postulates for betweenness. Trans. Amer. Math. Soc. **18**, 301–325 (1917)
13. Isbell, J.R.: Median algebra. Trans. Amer. Math. Soc. **260**, 319–362 (1980)
14. Menger, K.: Untersuchungen über allgemeine Metrik. Mathematische Annalen **100**, 75–163 (1928)
15. Munkres, J.R.: Topology. Prentice Hall, New Jersey (2000)
16. Nadler, Jr, S.B.: Continuum theory, an introduction. Marcel Dekker, New York (1992)
17. Newman, M.H.A.: Elements of the topology of plane sets of points. Cambridge University Press, Cambridge (1939)
18. Pasch, M.: Vorlesungen über neuere Geometrie, 1st edn. Teubner, Leipzig (1882)
19. Pitcher, E., Smiley, M.: Transitivities of betweenness. Trans. Amer. Math. Soc. **52**, 95–114 (1942)
20. Smiley, M.: A comparison of algebraic, metric, and lattice betweenness. Bull. Amer. Math. Soc. **49**, 246–252 (1943)
21. Ward, Jr, L.E.: Axioms for cutpoints. In: Topology, general, analysis, modern, conference proceedings, University of California, Riverside, Academic Press, New York, 1980, pp. 327–336 (1981)
22. Whyburn, G.T.: Analytic topology, American Mathematical Society Colloquium Publications, Providence, **28** (1971)

Permissions

All chapters in this book were first published in BMS, by Springer; hereby published with permission under the Creative Commons Attribution License or equivalent. Every chapter published in this book has been scrutinized by our experts. Their significance has been extensively debated. The topics covered herein carry significant findings which will fuel the growth of the discipline. They may even be implemented as practical applications or may be referred to as a beginning point for another development.

The contributors of this book come from diverse backgrounds, making this book a truly international effort. This book will bring forth new frontiers with its revolutionizing research information and detailed analysis of the nascent developments around the world.

We would like to thank all the contributing authors for lending their expertise to make the book truly unique. They have played a crucial role in the development of this book. Without their invaluable contributions this book wouldn't have been possible. They have made vital efforts to compile up to date information on the varied aspects of this subject to make this book a valuable addition to the collection of many professionals and students.

This book was conceptualized with the vision of imparting up-to-date information and advanced data in this field. To ensure the same, a matchless editorial board was set up. Every individual on the board went through rigorous rounds of assessment to prove their worth. After which they invested a large part of their time researching and compiling the most relevant data for our readers.

The editorial board has been involved in producing this book since its inception. They have spent rigorous hours researching and exploring the diverse topics which have resulted in the successful publishing of this book. They have passed on their knowledge of decades through this book. To expedite this challenging task, the publisher supported the team at every step. A small team of assistant editors was also appointed to further simplify the editing procedure and attain best results for the readers.

Apart from the editorial board, the designing team has also invested a significant amount of their time in understanding the subject and creating the most relevant covers. They scrutinized every image to scout for the most suitable representation of the subject and create an appropriate cover for the book.

The publishing team has been an ardent support to the editorial, designing and production team. Their endless efforts to recruit the best for this project, has resulted in the accomplishment of this book. They are a veteran in the field of academics and their pool of knowledge is as vast as their experience in printing. Their expertise and guidance has proved useful at every step. Their uncompromising quality standards have made this book an exceptional effort. Their encouragement from time to time has been an inspiration for everyone.

The publisher and the editorial board hope that this book will prove to be a valuable piece of knowledge for researchers, students, practitioners and scholars across the globe.

List of Contributors

Guy Barles
Laboratoire de Mathématiques et Physique Théorique (UMR CNRS 7350), Fédération Denis Poisson, Université de Tours (FR CNRS 2964), Place de Grandmont, 37200 Tours, France

Hitoshi Ishii
Faculty of Education and Integrated Arts and Sciences, Waseda University, Nishi-Waseda, Shinjuku, Tokyo 169-8050, Japan
Faculty of Science, King Abdulaziz University, P.O. Box 80203, Jeddah 21589, Saudi Arabia

Hiroyoshi Mitake
Department of Applied Mathematics, Faculty of Science, Fukuoka University, Fukuoka 814-0180, Japan

Ping Kang
Department of Mathematics, Tianjin Polytechnic University, Tianjin 300387, People's Republic of China

Claus Michael Ringel
Department of Mathematics, Shanghai Jiao Tong University, Shanghai 200240, People's Republic of China
King Abdulaziz University, P O Box 80200, Jeddah, Saudi Arabia

Ö. Gökmen Yıldız, Sıddıka Ö. Karakuş and H. Hilmi Hacısalihoğlu
Department of Mathematics, Faculty of Sciences and Arts, Bilecik ¸Seyh Edebali University, 11210 Bilecik, Turkey

Feida Jiang and Neil S. Trudinger
Centre for Mathematics and Its Applications, The Australian National University, Canberra, ACT 0200, Australia

Michael Ruzhansky
Department of Mathematics, Imperial College London, 180 Queen's Gate, London SW7 2AZ, UK

Mitsuru Sugimoto
Graduate School of Mathematics, Nagoya University, Furocho Chikusa-ku, Nagoya 464-8602, Japan

A. Ghorbani and Z. Nazemian
Department of Mathematical Sciences, Isfahan University of Technology, P.O.Box: 84156-83111, Isfahan, Iran

S. K. Jain
Department of Mathematical Sciences, Isfahan University of Technology, P.O.Box: 84156-83111, Isfahan, Iran
Department of Mathematics, Ohio University, Athens, OH 45701, USA
King Abdulaziz University, Jeddah, Saudi Arabia

Alexei Ilyin
Keldysh Institute of Applied Mathematics, Moscow, Russia
Institute for Information Transmission Problems, Moscow, Russia

Ari Laptev
Imperial College London, London, UK
Institute Mittag-Leffler, Djursholm, Sweden

Sergey Zelik
Department of Mathematics, University of Surrey, Surrey, UK

Paul Bankston
Department of Mathematics, Statistics and Computer Science, Marquette University, Milwaukee, WI 53201-1881, USA